Ferroic Materials-Based Technologies

Scrivener Publishing
100 Cummings Center, Suite 541J
Beverly, MA 01915-6106

Publishers at Scrivener
Martin Scrivener (martin@scrivenerpublishing.com)
Phillip Carmical (pcarmical@scrivenerpublishing.com)

Ferroic Materials-Based Technologies

Edited by
Inamuddin
Tariq Altalhi
and
Mohammad Abu Jafar Mazumder

Scrivener Publishing

WILEY

This edition first published 2024 by John Wiley & Sons, Inc., 111 River Street, Hoboken, NJ 07030, USA and Scrivener Publishing LLC, 100 Cummings Center, Suite 541J, Beverly, MA 01915, USA

For more information about Scrivener publications please visit www.scrivenerpublishing.com.

Wiley Global Headquarters
111 River Street, Hoboken, NJ 07030, USA

For details of our global editorial offices, customer services, and more information about Wiley products visit us at www.wiley.com.

Library of Congress Cataloging-in-Publication Data

ISBN 978-1-394-23815-6

Cover image: Pixabay.Com
Cover design by Russell Richardson

Set in size of 11pt and Minion Pro by Manila Typesetting Company, Makati, Philippines

Printed in the USA

10 9 8 7 6 5 4 3 2 1

Contents

Preface

Ferroic materials have sparked widespread attention because they represent a broad spectrum of elementary physics and are employed in a plethora of fields, including flexible memory, enormous energy harvesting/storage potential, spintronic functionalities, spin caloritronics, and a large range of other multi-functional devices. Moreover, in multiferroic materials, two or more primary ferroic orderings—such as ferroelectrics; Dia-, Para-, Ferro-, Ferri-, and Antiferromagnetic materials; ferroelastic materials; and ferrotoroidicity as the fourth form of ferroic order—are all present simultaneously in the similar phase. Magnetic materials exhibit significant electro-mechanical coupling, pyro-electricity, and electro-optical characteristics, as well as spontaneous but switchable magnetic behavior, whereas ferroelectric materials exhibit spontaneous but switchable electric polarization.

With the application of new ferroic materials, a strong room-temperature ferroelectricity with high saturation polarization may be established in ferroelectric materials, and magnetism with significant magnetization can be accomplished in magnetic materials. Furthermore, magnetoelectric interaction between ferroelectric and magnetic orderings is high in multiferroic materials, which could enable a wide range of innovative devices. Magnetic, ferroelectric, and multiferroic 2D materials with ultrathin characteristics above ambient temperature are often expected to enable future miniaturization of electronics beyond Moore's law for energy-efficient nanodevices. This book addresses the prospective, relevant, and original research developments in the ferroelectric, magnetic, and multiferroic fields. Readers of this book will can gain a better understanding of ferroic materials, their related technologies, and their various applications. In fields both specific and related to ferroic materials, this book serves as a great reference tool for students, physicists, academics, researchers, and professionals.

Chapter 1 gives a complete overview of ferroic materials, their history, and the latest progress. It discusses four main types, ferromagnetic, ferroelastic, ferroelectric, and multiferroic materials, while explaining their properties and various applications like data storage, sensors, and more. The chapter also covers the current research pertaining to these materials.

Chapter 2 encapsulates the history and current presence of ferroic materials. This chapter gives details of ferroic materials, e.g., primary and secondary types of ferroic materials. Primary ferroic materials include ferromagnetic, ferroelectric, and ferroelastic, while secondary ferroic materials include multiferroics, ferromagnetoelectric, ferromagnetoelastic, and ferroelastoelectric materials, and all are addressed. The origin of ferroelectricity, ferroelasticity, ferromagnetism, scaling, and recent advances in ferroic materials is also outlined.

Chapter 3 covers an in-depth evaluation of the state of these materials currently and their promising future possibilities. This chapter offers a greater understanding of the possible improvements and applications of ferroic and multiferroic materials in a variety of industries, making it an invaluable resource for researchers and scientists.

Chapter 4 presents a brief history, working principle, theory, and various measurement techniques of electrocaloric effect (ECE) in ferroelectrics. Furthermore, it focuses on different strategies that are adopted to enhance ECE in various lead-based and lead-free ferroelectrics, relaxor ferroelectrics, and hafnia-zirconia-based ferroelectrics.

Chapter 5 discusses the preparation, improvement, and characterizations of ferroelectric and ferroelastoelectric materials. Several synthesis methods for ferroelectric and ferroelastoelectric materials are explained in detail. Improvements and applications of ferroelectric and ferroelastoelectric materials are also discussed.

Chapter 6 examines the elastocaloric impact (ECE) in the discipline of solid-state cooling. Ferroelectric materials are employed for solid-state cooling purposes due to their distinctive thermodynamic features and phase transition uses. Additionally, ECE is further divided into uniaxial and biaxial based on its behavior in the microstructure of materials.

Chapter 7 discusses the flexoeffect alterations in the ferroic nanosystem and their attributes that result from the inhomogeneity of order parameters. This effect exists spontaneously. The equation framework of FCT (flexoelectric coupling tensor) for various point groups using a direct matrix approach, including the quantity of non-zero independent components, is also covered.

Chapter 8 covers recent advancements in ferroic thin films, multi-layers, and hetero-structures, including their historical background, properties, and characterization methods. The significance of atomically controlled thin films and interfaces is emphasized, along with their applications in microelectronics, data storage, sensing, and energy conversion. The chapter also reviews the ongoing research into the enhancement of their characteristics and the expansion of their usage in the future.

Chapter 9 deals with multiferroic materials, their origin, history, and present development, along with the origin of ferromagnetic, antiferromagnetic, and ferroelectric phases. Classification of multiferroics in single-phase and composite multiferroic materials based on their coupling is also addressed. Single-phase multiferroics classification in type I and type II multiferroics and their prominent examples are discussed, as are possible applications, such as magnetic field sensors, memory devices, and laser applications of multiferroics.

Chapter 10 delves into the basic concepts, formation mechanism, phase structure, feasibility, and classifications of ferroic materials. Furthermore, it elaborates on comparing primary and secondary ferroics along with multi-ferroics formation and their relation with both classes of ferroic materials. Applications of these ferroic materials in various fields, including nanotechnology, are also elaborated upon.

Chapter 11 discusses the importance of domain boundary engineering in multiferroic materials and how it affects the properties of these materials. Domain boundary engineering can be used to manipulate the domain structure and affect the interaction between various ferroic orders. The advantages and challenges of domain boundary engineering are also discussed.

In chapter 12, magneto-electric, dielectric, and optical features of ferro/antiferroelectric materials are explained in detail. Polarization curves for 0.96BaTi–0.04BaFO were studied after the samples had been treated at different temperatures. Additionally, dielectric characteristics of La-modified AFE PbZrO3 thin films, BNTC22, PLZST-AFE material and BiMgFeCeO6 are reviewed.

Chapter 13 details the rapid growth in smart electronic and optical devices based on metal-halide perovskites (MHPs), due to their auspicious and innovative semiconducting nature to show ferroelectric behavior. The theoretical and experimental evidence of the ferroic nature of MHPs, the dimension dimension-dependent ferroelectricity within MHPs, and their applications in devices are all covered thoroughly.

We are deeply grateful to everyone who helped with this book and greatly appreciate the dedicated support and valuable assistance rendered by Martin Scrivener and the Scrivener Publishing team during its publication.

The Editors
March 2024

1

Ferroic Materials: From Past to Present

Sandeep Yadav[1], Pallavi Jain[1*] and Prashant Singh[2†]

[1]*Department of Chemistry, SRM Institute of Science and Technology, Delhi-NCR Campus, Modinagar, India*
[2]*Department of Chemistry, Atma Ram Sanatan Dharma College, University of Delhi, New Delhi, India*

Abstract
The chapter provides an overview of the development and applications of ferroic materials. The chapter begins with a brief history of ferroic materials and their discovery and a detailed discussion of the four major types of ferroic materials: ferromagnetic, ferroelastic, ferroelectric, and multiferroic materials. The section on ferromagnetic materials covers their magnetic properties and applications in data storage, sensing, and medical imaging. The section on ferroelastic materials describes their mechanical properties and applications in actuators, sensors, and energy-harvesting devices. The section on ferroelectric materials explains their electrical properties and applications in capacitors, transducers, and memory devices. Finally, the section on multiferroic materials discusses their unique combination of multiple ferroic properties and their potential applications in next-generation devices such as spintronics, magneto-electric sensors, and nonvolatile memory. The chapter concludes by mentioning the present state of research and probable future directions for developing and applying ferroic materials. This chapter comprehensively introduces ferroic materials, their historical development, and their diverse applications across multiple fields. It will interest researchers, students, and materials science, physics, and engineering professionals.

Keywords: Ferromagnetic, ferroelectric, ferroelastic, multiferroic, piezoelectric sensors, magnetic sensors

**Corresponding author*: palli24@gmail.com
†*Corresponding author*: psingh@arsd.du.ac.in

Inamuddin, Tariq Altalhi and Mohammad Abu Jafar Mazumder (eds.) Ferroic Materials-Based Technologies, (1–18) © 2024 Scrivener Publishing LLC

1.1 Introduction

Ferroic materials exhibit ferroic ordering, a long-range ordered state in which the material shows a spontaneous polarization, magnetization, strain, or other physical properties. Intense research has been done on these materials over the past few decades due to their potential for various technological applications.

One of the most critical applications of ferroic materials is in the field of data storage. Ferromagnetic materials, for example, are used extensively in hard disk drives, where they store digital information in magnetic domains. Ferroelectric materials, on the other hand, are used in nonvolatile memory devices. One example of a nonvolatile memory device is ferroelectric random access memory (FeRAM). It has faster access times and lower power consumption than conventional memory devices. In data storage, ferroelastic materials have some potential applications because they can undergo reversible strain, and this property can be utilized to develop data storage devices with high-density data [1, 2].

Besides data storage, ferroic materials have found applications in various other fields. For example, ferromagnetic materials are used in motors, transformers, and generators due to their ability to generate a magnetic field.

The study of ferroic materials has a long and rich history, dating back to the discovery of ferromagnetism in lodestones by the ancient Greeks. However, it was in the 1930s that the concept of ferroic materials as a class of materials with similar properties was established. Since then, the field has proliferated with the discovery of new materials and phenomena and the development of new characterization techniques [3].

Recently, ferroic materials came into the limelight due to their potential use in technologies such as nanoelectronics and spintronics. Developing new synthesis and fabrication techniques has caused the formation of new materials with novel properties and functionalities, such as multiferroics, which exhibit more than one type of ferroic ordering. The study of ferroic materials is a crucial research field with a wide range of practical applications across different industries and disciplines. The history of ferroic materials research dates back centuries, and recent developments in synthesis and characterization techniques have led to the discovery of new materials with novel properties and functionalities. The remaining sections of this chapter will provide a detailed overview of the different types of ferroic materials, their properties, and their applications.

1.2 Types of Ferroic Materials

Ferroic materials encompassed various types of ordered materials. Ferromagnetic materials, such as iron and nickel, possessed spontaneous magnetization even without an external magnetic field. Ferroelectric materials, like barium titanate and lead zirconate titanate, exhibited spontaneous electric polarization that could be reversed with an electric field. Ferroelastic materials, including shape memory alloys like Nitinol, showcased reversible deformation and strain. Moreover, multiferroic materials exhibited multiple ferroic properties simultaneously, allowing for coupling between different types of orderings. These materials offered promising potential for applications in diverse fields, including magnetoelectric sensors and memory devices [4]. Figure 1.1 shows the types of ferroic materials.

1.2.1 Ferromagnetic Materials

Ferromagnetic materials are a type of ferroic materials that exhibit spontaneous magnetization. This means that these materials have some magnetic behavior even when no external magnetic field exists. Ferromagnetic materials have been extensively studied over the past century due to their importance in technological applications.

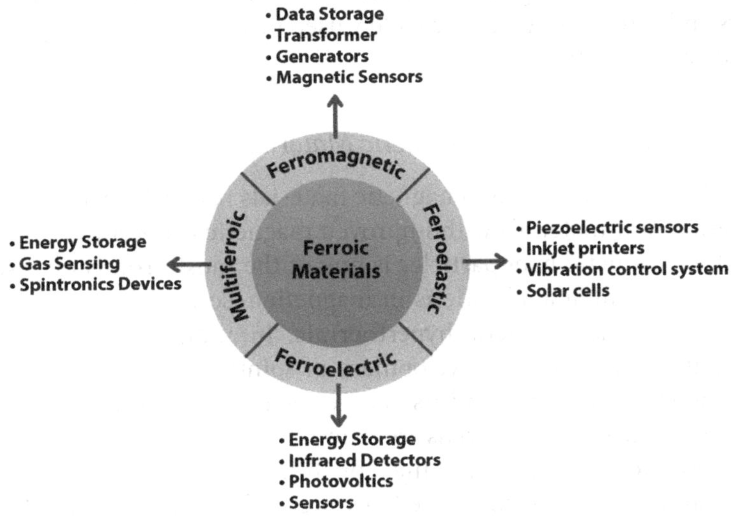

Figure 1.1 Types of ferroic materials.

1.2.1.1 Past of Ferromagnetic Materials

Ferromagnetism dates back to ancient times when the Greeks discovered that certain minerals could attract iron. However, it was not until the 19th century that the phenomenon was studied in detail. In 1820, Hans Christian Oersted saw a magnetic needle gets deflected when placed near a current-carrying wire, indicating a relationship between electricity and magnetism. This led to the development of electromagnetic theory and the discovery of the connection between electricity and magnetism.

In 1825, William Sturgeon developed the first practical electromagnet, which used a coil of wire to create a magnetic field. This was followed by the development of permanent magnets made of ferromagnetic materials such as iron, cobalt, and nickel. In 1915, the German physicist Wilhelm Lenz proposed the concept of ferromagnetism, which explained the phenomenon of spontaneous magnetization in certain materials [5].

Ferromagnetic materials have found numerous technological applications. One of the most important applications is in the field of data storage. Hard disk drives, for example, use ferromagnetic materials to store digital information in the form of magnetic domains. The development of high-density magnetic storage devices has been made possible by discovering new ferromagnetic materials and fabrication techniques.

Ferromagnetic materials are also used in various electrical and electronic devices, such as motors, transformers, and generators, due to their ability to generate a magnetic field. In addition, they are used in magnetic sensors and actuators, which are widely used in automotive, aerospace, and biomedical applications.

1.2.1.2 Present of Ferromagnetic Materials

Recent developments in ferromagnetic materials research have focused on discovering new materials with improved magnetic properties. The development of magnetic nanoparticles is one of the most promising research areas, with potential applications in magnetic resonance imaging (MRI), drug delivery, and magnetic hyperthermia. In addition, there has been significant progress in the development of spintronics. This field aims to utilize the electronic spin and its charge for information processing and storage. Ferromagnetic materials are crucial in spintronics, as they are used as magnetic electrodes and spin injectors [6].

Chiu and colleagues conducted research where they explored ferromagnetic shape memory alloys (FSMA) in actuating robots. Their findings indicated that Ni-Mn-Ga alloys were well-suited for this purpose, as they exhibited a high work:volume ratio with rapid response rates. The alloys had a 5-modulated (5M) martensite phase, which could provoke martensite variant reorientation (MVR) when an external magnetic or stress field is applied. Additionally, the study revealed that 7–13 vol.% of the Ni-Mn-Ga alloy was necessary to initiate MVR [7].

Wang *et al.* found that the heterogeneous nanocrystalline Co/Ni sample exhibited a high permeability and good impedance similarity between permeability and permittivity, resulting in good absorption in the Ku-band (12–18 GHz). The CoNi microsphere sample displayed good absorption in the microwave region, also known as C-band (4–8 GHz) and X-band (8–11.5 GHz). On the other hand, CoNi microspheres with diameters greater than 2 μm formed chain-like structures that demonstrated inadequate microwave absorption capability [8].

A Fe_3O_4-based aerogel was produced by Lo *et al.* via the sol-gel method and supercritical freezing. Loaded Ag enhanced its photocatalytic Fenton-oxidation and reduction activities, leading to the efficient removal of benzoic acid and the production of 4-aminophenol under visible light. The Ag-loaded porous structure of the aerogel provided a high surface area, many active sites, and many active radicals. The research provides valuable information on creating a highly effective aerogel catalyst that can be used for environmental clean-up purposes [9].

Choi *et al.* added carbon to FePt ferromagnetic grains to improve magnetic isolation and enable smaller grain sizes. This is important for developing high-density magnetic recording media, such as heat-assisted magnetic recording (HAMR) technology. This technology is used in hard disk drives and other data storage devices to increase capacity. By identifying the optimal carbon concentration for L10-ordered FePt films, the researchers have provided a valuable guide for designing media with high areal densities of 4 Tb/in^2 and beyond, which could lead to even higher capacity and more advanced data storage technologies in the future [10]. Fe/N codoping was used by Li *et al.* to alter the magnetic and electrical properties of graphdiyne (GDY) materials. The Fe-N-GDY achieved high conductivity as a ferromagnetic semiconductor with a Curie temperature of about –68°C. Fe-N-GDY could be used as a conductive ink in printers and flexible field-effect transistors with a high mobility of 215 $cm^2 V^{-1} s^{-1}$. These results demonstrate an easy way of modulating the properties of GDY and promoting its potential in electronic devices [11]. A few more recent studies are mentioned in Table 1.1.

In conclusion, ferromagnetic materials are a class of ferroic materials that exhibit spontaneous magnetization. They have been extensively studied for their applications in modern-day technologies, such as

Table 1.1 Some recent studies of various ferromagnetic materials.

S. no.	Application	Ferromagnetic material	Ref.
1	Soft robotics	Ni-Mn-Ga	[7]
2	Microwave absorption materials	Co/Ni nanocrystalline	[8]
3	Environmental remediation	Fe_3O_4 aerogel	[9]
4	Magnetic data storage	FePt-C	[10]
5	Spin caloritronics	$NiFe_2O_4$	[12]
6	Flexible electronics	Fe-N-GDY	[11]
8	Energy storage	Fe_3O_4@C	[13]
9	Cancer therapy	FeNi-CNT	[14]
10	Magnetic sensors	VSe_2-graphite/MoS_2	[15]
11	Electromagnetic interference shielding	$Co_{0.5}Zn_{0.5}Fe_2O_4$/PANI-PTSA	[16]
14	Magnetic nano-devices	$NiFe_2O_4$ nanocrystals	[17]

motors, sensors, and data storage. The history of ferromagnetic materials research dates back to the 19th century, and recent developments have focused on discovering new materials with improved magnetic properties and the development of spintronics. The remaining sections of this chapter will provide a detailed overview of the properties, applications, and recent developments in ferroelectric, ferroelastic, and multiferroic materials.

1.2.2 Ferroelectric Materials

Ferroelectric materials exhibit immediate electric polarization, which can be changed by applying an external electric field. This behavior arises due to the displacement of ions within the crystal lattice, leading to a net dipole moment. The ferroelectric behavior can occur at a specific temperature or below a critical size, known as the Curie temperature and Curie size. Figure 1.2 shows the working of ferroelectric materials

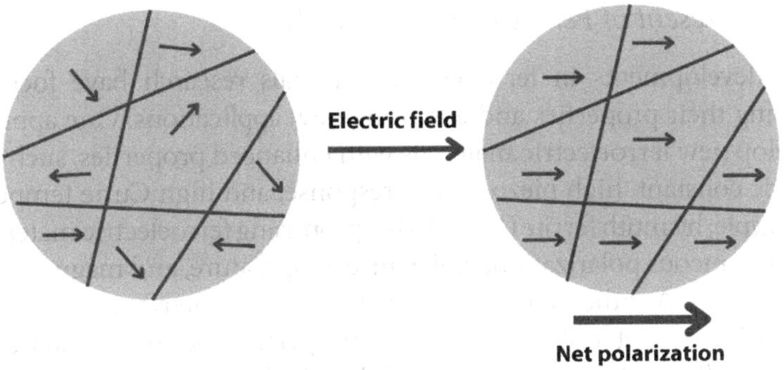

Figure 1.2 Effect of electric field on ferroelectric materials.

1.2.2.1 Past of Ferroelectric Materials

The study of ferroelectric materials began in the early 20th century with the discovery of piezoelectricity. The first ferroelectric material discovered was Rochelle salt, which exhibits ferroelectric and piezoelectric behavior. In 1920, the ferroelectric properties of Rochelle salt (potassium sodium tartrate) were successfully elucidated by the Curie brothers, marking a significant breakthrough. Subsequent investigations carried out during the 1940s by Valasek and Stevenson revealed the reversible polarization behavior that characterizes ferroelectric materials, with barium titanate being a prominent example. A momentous milestone arose in the 1950s with the discovery of lead zirconate titanate (PZT) by Jaffe, Cook, and Jaffe. PZT quickly garnered acclaim as a versatile ferroelectric material due to its exceptional attributes, including its piezoelectric, pyroelectric, and electrostrictive properties [18, 19].

Since then, extensive research has been carried out to explore the fundamental properties of ferroelectric materials and their potential applications.

Ferroelectric materials have numerous applications in various fields, including electronics, memory devices, energy harvesting, and sensors. For example, lead zirconate titanate (PZT) is a commonly used ferroelectric material for constructing electronic devices, including piezoelectric actuators and sensors. Another example is lithium niobate ($LiNbO_3$), commonly used in optical modulators, switches, and sensors. Barium titanate ($BaTiO_3$) is another crucial ferroelectric material used in capacitors, electromechanical transducers, and nonlinear optics.

1.2.2.2 Present of Ferroelectric Materials

Recent developments in ferroelectric materials research have focused on improving their properties and exploring new applications. One approach is to develop new ferroelectric materials with enhanced properties, such as high dielectric constant, high piezoelectric response, and high Curie temperature. For example, bismuth ferrite ($BiFeO_3$) is a promising ferroelectric material with high spontaneous polarization, high Curie temperature, and magneto-electric solid coupling. Another approach is to engineer the interface between different ferroelectric materials to create novel properties and functionalities.

A metal/ferroelectric capacitor [(FE)-HfO_2/IGZO/metal] was constructed by Mo *et al.* without atomic interdiffusion for three-dimensional high-density memory. The composite exhibited ferroelectricity with many endurance cycles and retention of up to 10 years. The lack of the wake-up effect improves the design and manufacturing process of the circuit. The asymmetric effect was caused by the varying band modulations in different states of IGZO [20].

In another example, researchers found a new method to develop a three-dimensional memory structure by self-rolling up single-crystalline ferroelectric oxides, which enhanced the information density more than 40 times. The "$Pb(Zr,Ti)O_3$/stressor" membranes also had a strong self-rolling up force, which increased the area ratio and allowed for high-density information storage of 102 Tbit. The study provided a new method for developing concise memory devices having high-density and three-dimensional memories from oxide materials [21].

In another study, the researchers achieved enormous pyroelectric performance in ceramics by changing the phase boundary made of ferroelectric-relaxor. They engineered the phase boundaries by introducing $BaTiO_3$, which broke the long-range antiferroelectric order. Due to the planned phase boundaries, ceramics exhibited a significant pyroelectric coefficient and figures of merit. The results promised potential applications for ultra-sensitive infrared detectors and some technological insights into pyroelectric ceramic design [22].

Huang *et al.* used a coaxial electrospinning technique to manage the distribution of polarization gradient in a TiO_2 nanofiber and polymer hybrid nanocomposite system, significantly boosting the discharged energy storage density (Ue) from 6.5 J cm^{-3} to 12.7 J cm^{-3} [23].

Habib *et al.* achieved excellent room-temperature piezoelectric performance in $0.67Bi_{1.03}FeO_3$-$0.33Ba_{1-x}La_xTiO_3$ ceramics with a very high temperature of 482°C. The researchers conducted a study where they slightly modified the piezoelectric strain within a temperature range of 25°C to

125°C, which resulted in hysteresis of low strain and a significant average electrostrictive coefficient. This development in non-lead ceramics provided an opportunity to advance temperature-insensitive piezoelectric properties, thus making it viable for high-temperature commercial applications [24]. Table 1.2 lists some ferroelectric materials with their applications.

Ferroelectric materials also have been explored for use in energy storage and conversion. For example, using ferroelectric materials in capacitors can lead to high energy-density storage devices with fast charging and discharging times. Ferroelectric materials have also been studied for photovoltaic devices, where their polarization can enhance the transformation efficiency of solar to electrical energy.

In summary, ferroelectric materials have significantly influenced various technological advancements and scientific discoveries. Studying these materials has led to a better understanding of their fundamental properties and potential for various applications. Ongoing research in this field continues to explore new materials, improve existing properties, and develop new applications for ferroelectric materials.

Table 1.2 Some recent studies of various ferroelectric materials.

S. no.	Application	Ferroelectric material	Ref.
1	Nonvolatile memory	HfO_2/IGZO	[20]
2	Ultrahigh-density data storage	$PbZr_{0.3}Ti_{0.7}O_3$	[21]
3	Infrared detectors	Ferroelectric $BaTiO_3$	[22]
4	Energy storage	Ferroelectric poly(vinylidene fluoride-trifluoroethylene)	[23]
5	Piezoelectric actuators	$BiFeO_3$–$BaTiO_3$ ceramics	[24]
6	Ferroelectric memory	Ferroelectric thin films of $BaTiO_3$	[25]
7	Sensor and actuator technology	Ferroelectric single crystals of $LiNbO_3$	[26]
9	Ferroelectric tunnel junctions	Ferroelectric $Hf_{0.5}Zr_{0.5}O_2$	[27]
10	Energy storage	$Bi_{0.5}Na_{0.5}TiO_3$–$Ba(Zr_{0.2}Ti_{0.8})O_3$ film	[28]
12	Photovoltaics	$Pb_{0.93}La_{0.07}(Zr_{0.6}Ti_{0.4})_{0.9825}O_3$ (PLZT) thin films	[29]

1.2.3 Ferroelastic Materials

Ferroelastic materials are another class of ferroic materials that exhibit reversible deformation because of the electric or magnetic field. The term "ferroelastic" comes from the Greek word "elasticos," meaning capable of being stretched or compressed without breaking. In ferroelastic materials, the deformation is caused by the movement of domain walls, which can change the shape of the material.

The definition of ferroelastic materials has been refined over time, but it is generally accepted that these materials exhibit reversible deformation due to the applied electric or magnetic field. Ferroelastic materials are typically classified into soft and hard ferroelastic materials. Soft ferroelastic materials exhibit large deformations at low applied fields, while hard ferroelastic materials require higher areas to induce deformation. Figure 1.3 shows how ferroelastic materials function in stress.

1.2.3.1 Past of Ferroelastic Materials

The history of ferroelastic materials research dates back to the early 20th century when researchers first discovered the phenomenon of reversible deformation in certain materials.

During the early 20th century, the term "ferroelasticity" was introduced by Auguste Brusconi to describe the reversible deformation observed in specific materials. Extensive research conducted in the 1950s and 1960s, led by James G. Simmons and colleagues, focused on investigating ferroelasticity in crystals and identifying materials that exhibited spontaneous strain and reversible phase transitions when subjected to external influences. Significant studies were carried out on potassium dihydrogen

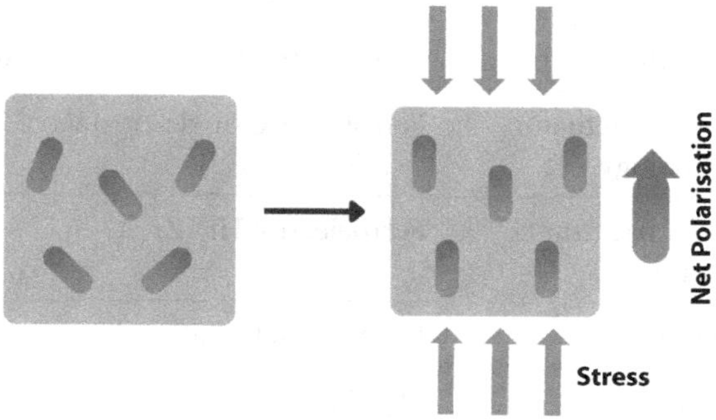

Figure 1.3 Effect of mechanical stress on ferroelastic materials.

phosphate (KDP) and ammonium dihydrogen phosphate (ADP) to explore their ferroelastic properties, revealing the intricate relationship between crystal structure and ferroelastic behavior. Subsequent research delved into the exploration of complex perovskite oxides, shape memory alloys, and organic compounds, leading to the discovery of remarkable ferroelastic materials such as lead zirconate titanate (PZT) and metal-organic frameworks (MOFs) [30]. Over the years, the understanding of ferroelastic materials has grown, and scientists have identified several materials that exhibit ferroelastic behavior. Some earliest examples of ferroelastic materials include lead zirconate (PZT) and barium titanate ($BaTiO_3$).

Ferroelastic materials have various applications, including medicine, electronics, and energy. One of the most essential applications of ferroelastic materials is in medical devices. For example, ferroelastic materials can create implants that can change shape in response to an applied field, making them ideal for use in devices such as stents and catheters.

In the electronics industry, ferroelastic materials are used to create piezoelectric sensors and actuators. The devices mentioned above could convert energy from mechanical form to electrical and vice versa, proving useful in several applications such as automotive sensors, inkjet printers, and vibration control systems.

1.2.3.2 Present of Ferroelastic Materials

Recent developments in ferroelastic materials research have focused on improving the properties of existing materials and discovering new ferroelastic materials. For example, researchers have developed ferroelastic composites that exhibit enhanced properties over traditional ferroelastic materials. They have also discovered new materials, such as ferroelastic graphene, that exhibit unique ferroelastic properties.

The researchers aimed to enhance the performance of metal halide perovskite (MHP) solar cells by inducing the creation of ferroelectric and ferroelastic domains in $CsPbBr_3$ nanowires.

In the study, an AFM or atomic force microscope tip was utilized to trigger the development of ferroelastic portions at room temperature, and the fields were observed by using nano-focused X-ray diffraction. The researchers noticed orthorhombic 004 reflections along the nanowire axis in areas observed with a low force of AFM tip, while areas under higher forces displayed 220 reflections. The lattice tilts caused by the applied stress defined the ferroelastic areas, which terminated at {112}-type domain walls. This achievement has opened up significant potential for new designs of devices and basic experimental studies [31].

Xie *et al.* investigated the mechanical properties linked to ferroelastic behaviors of W/Cr co-doped BIT ceramics, a type of bismuth layer-structured ferroelectrics, to provide design considerations for practical applications of the ceramics. The researchers synthesized BTWC ceramics using optimal chemical composition and analyzed the structures of the ferroelastic domains, their switching characteristics, and some harmful properties of the sintered ceramics, such as mechanical failure. It was observed that the ferroelastic switching caused by mechanical stress enhanced the fracture toughness of BTWC ceramics. Also, the ceramics sintered at elevated temperatures were expected to demonstrate enhanced ferroelastic properties, even at a lower failure stress level [32].

Zednik *et al.* investigated the properties of ferroelastic areas in thin films of $Pb(Zr,Ti)O_3$ (PZT). The traditional method of probing using electric fields was limited because of the mechanical clamping effect of the substrate. To explore this issue, the researchers applied biaxial stress to 70-nm-thick $Pb(Zr_{0.25}Ti_{0.75})O_3$ films, inducing 90° domain wall motion and resulting in crystallographic reorientation. The team used X-ray diffraction and piezo force microscopy to observe the phenomenon. The researchers found that the mobile 90° domain walls were responsible for the orientation switching in the grains without wall pinning, demonstrating that bulk-like ferroelastic behavior was extended to nanocrystalline films without substrate clamping [33]. Table 1.3 contains a few examples of ferroelastic materials with their respective applications.

In conclusion, ferroelastic materials are a type of ferroic materials that exhibit reversible deformation in the presence of an external electric or magnetic field. They have various applications in various fields, including

Table 1.3 Some recent studies of various ferroelastic materials.

S. no.	Application	Ferroelastic material	Ref.
1	Solar cells	$CsPbBr_3$ perovskite	[31]
	Ferroelectric devices	$Bi_4Ti_{2.95}W_{0.05}O_{12.05}$+0.2 wt% Cr_2O_3	[32]
2	Displacive ferroelectrics	$Pb(Zr,Ti)O_3$ films	[33]
3	Energy harvester	8/65/35 PLZT	[34]
4	Ferroic switching	$BaTiO_3$	[35]
5	Flexible electronics	$BiFeO_3$	[36]

medicine, electronics, and energy. The history of ferroelastic materials research dates back to the early 20th century, and recent developments in this field have focused on improving the properties of existing materials and discovering new materials.

1.2.4 Multiferroic Materials

Multiferroic materials possess both magnetic and electric properties that are coupled together, making them attractive for various technological applications. A multiferroic material can exhibit ferromagnetism, ferroelectricity, and ferroelasticity, all in one material, providing enhanced functionalities compared to their counterparts.

1.2.4.1 Past of Multiferroic Materials

The study of multiferroic materials dates back to the 1960s when researchers discovered that some materials exhibit magnetic and electric properties. Since then, multiferroic materials have gained significant attention in the scientific community because of their possible applications in information storage, sensing, and energy conversion.

The definition of multiferroic materials can vary, but generally, a material is considered multiferroic if it exhibits more than one ferroic property. Ferroic properties refer to spontaneous magnetization, polarization, or strain that can be changed in the presence of an external magnetic, electric, or stress field, respectively. A multiferroic material can have a magnetization coupled with its polarization or strain, leading to interesting properties that can be exploited for technological applications.

The history of multiferroic materials research can be traced back to the discovery of bismuth ferrite ($BiFeO_3$) in the 1960s. $BiFeO_3$ is the first material discovered to exhibit both ferroelectricity and antiferromagnetism, making it a promising candidate for multiferroic applications. Since then, researchers have discovered other materials that exhibit multiferroic properties, such as $TbMnO_3$ and $CoCr_2O_4$.

Multiferroic materials have shown great promise in various technological applications. For example, multiferroic materials can be used in magnetic random-access memory (MRAM) devices, which use magnetic fields to write data and electric fields to read data, providing high-speed data storage with low power consumption. Multiferroic materials can also be used in sensors and actuators, where their magnetic and electric properties can convert energy from one form to another.

Table 1.4 Some recent studies of various multiferroic materials.

S. no.	Application	Multiferroic material	Ref.
1	Gas sensing	$BiFeO_3$	[37]
2	Energy storage	$CoFe_2O_4/BaTiO_3$	[38]
4	Magnetic field sensing	$BiFeO_3$ and $BaTiO_3$	[39]
5	Modulation of anomalous Hall effect	$BiFeO_3/SrRuO_3$	[40]
7	Spintronic devices	$La_{0.7}Sr_{0.3}MnO_3/BiFeO_3$	[41]

1.2.4.2 Present of Multiferroic Materials

Recent developments in multiferroic materials research have focused on discovering new materials with improved multiferroic properties and understanding the fundamental physics that governs their behavior. Researchers have also been exploring ways to manipulate the coupling between multiferroic materials' magnetic and electric properties to enhance their functionality. For example, by applying an electric field to a multiferroic material, researchers can change its magnetic properties, leading to new possibilities for device designs. Table 1.4 contains a few multiferroic materials with their respective applications.

In conclusion, multiferroic materials possess unique magnetic and electric properties that make them attractive for various technological applications. The study of multiferroic materials dates back to the 1960s with the discovery of bismuth ferrite. Since then, researchers have discovered new materials and gained a deeper understanding of their fundamental properties. As research in this field continues, new and exciting applications for multiferroic materials will likely be discovered, further advancing technology and scientific understanding.

1.3 Conclusion

In summary, this chapter provided an overview of ferroic materials and their significance in various fields of science and technology. Ferroic materials show properties like magnetization, polarization, and strain, making them useful for many applications. Ferromagnetic materials have been extensively researched due to their technological importance in magnetic storage, sensing, and energy conversion. Similarly, ferroelectric materials have applications in actuators, capacitors, and memory devices. Ferroelastic materials are used in sensors, transducers, and switches. Multiferroic materials exhibit multiple ferroic properties and have promising potential for developing next-generation memory and sensing devices. The history of ferroic materials research dates back to the early 19th century, and since then, there have been significant advancements in their understanding and applications. Recent developments in ferroic materials research have led to the discovery of new materials and the optimization of their properties, leading to the development of novel devices. Future research directions in ferroic materials include discovering new materials with enhanced properties, developing novel synthesis and fabrication techniques, and exploring new applications. In particular, integrating ferroic materials with other emerging technologies such as nanotechnology, 2D materials, and biotechnology is an exciting area of research with significant potential for advancing scientific understanding and technology.

In conclusion, the study of ferroic materials has contributed significantly to advancing science and technology, and its importance is expected to grow further. Continued research in this field will enable the development of new materials and technologies with improved performance and functionalities.

References

1. Wu, H., Li, L., Liang, L.Z., Liang, S., Zhu, Y.Y., Zhu, X.H., Recent progress on the structural characterizations of domain structures in ferroic and multiferroic perovskite oxides: A review. *J. Eur. Ceram. Soc.*, 35, 411, 2015.
2. Ajayan, J., Mohankumar, P., Nirmal, D., Joseph, L.M.I.L., Bhattacharya, S., Sreejith, S., Kollem, S., Rebelli, S., Tayal, S., Mounika, B., Ferroelectric Field Effect Transistors (FeFETs): Advancements, challenges and exciting prospects for next generation Non-Volatile Memory (NVM) applications. *Mater. Today Commun.*, 35, 105591, 2023.
3. Stojanovic, B.D., *Magnetic, ferroelectric, and multiferroic metal oxides*, Elsevier, Amsterdam, 2018.
4. Cano, A., Meier, D., Trassin, M., *Multiferroics: Fundamentals and applications*, De Gruyter, Berlin, 2021.
5. Coey, J.M.D., History of magnetism and basic concepts, in: *Handbook of Magnetism and Magnetic Materials*, J.M.D. Coey, S.S.P.P. (Eds.), pp. 3–51, Springer International Publishing, Berlin, 2021.
6. Rafiq, S., Awan, S.U., Zheng, R.K., Wen, Z., Rani, M., Akinwande, D., Rizwan, S., Novel room-temperature ferromagnetism in Gd-doped 2-dimensional

Ti3C2Tx MXene semiconductor for spintronics. *J. Magn. Magn. Mater.*, 497, 165954, 2020.
7. Chiu, W.T., Goto, A., Tahara, M., Inamura, T., Hosoda, H., Effects of volume fraction between single crystal Ni-Mn-Ga ferromagnetic shape memory alloy and silicone rubber on the martensite variant reorientation. *J. Alloys Compd.*, 926, 166862, 2022.
8. Wang, Z., Yang, W., Lv, Q., Liu, S., Fang, Z., Ferromagnetic and excellent microwave absorbing properties of CoNi microspheres and heterogeneous Co/Ni nanocrystallines. *RSC Adv.*, 9, 13365, 2019.
9. Lo, L., Yang, Z.J., Hung, Y.C., Tseng, P.Y., Nomura, M., Lin, Y.F., Hu, C., Boosting photoassisted activity for catalytic oxidation of benzoic acid and reduction of 4-nitrophenol with Ag-supported Fe_3O_4 aerogel. *Chem. Eng. J.*, 405, 126641, 2021.
10. Choi, M., Hong, Y.K., Won, H., Mankey, G.J., Yeo, C.D., Shah, N.M.R., Lee, W., Jung, M.H., Thiele, J.U., Upper limit of carbon concentration in ferromagnetic L1-ordered FePt-C for Tb/in data storage density heat-assisted magnetic recording media. *IEEE Trans. Magn.*, 57, 1, 2021.
11. Li, R., Li, X., Zhang, M., Li, Y., Yang, Z., Huang, C., A universal Fe/N incorporated graphdiyne for printing flexible ferromagnetic semiconducting electronics. *J. Phys. Chem. Lett.*, 12, 204, 2021.
12. Rastogi, A., Li, Z., Singh, A.V., Regmi, S., Peters, T., Bougiatioti, P., Carsten Né Meier, D., Mohammadi, J.B., Khodadadi, B., Mewes, T., Mishra, R., Gazquez, J., Borisevich, A.Y., Galazka, Z., Uecker, R., Reiss, G., Kuschel, T., Gupta, A., Enhancement in thermally generated spin voltage at the interfaces between Pd and $NiFe_2O_4$ films grown on lattice-matched substrates. *Phys. Rev. Appl.*, 14, 014014, 2020.
13. Huang, Y., Li, Z., Zhu, T., Gao, X., Lv, X., Ling, M., Wan, Z., Xia, Y., Ferromagnetic 1D-Fe_3O_4@C microrods boost polysulfide anchoring for lithium-sulfur batteries. *ACS Appl. Energy Mater.*, 4, 3921, 2021.
14. Cui, X., Self-thermoregulating ferromagnetic FeNi filled carbon nanotubes for magnetic hyperthermia cancer therapy. *IOP Conf. Ser. Earth Environ. Sci.*, 310, 42016, 2019.
15. Jimenez, V.O., Kalappattil, V., Eggers, T., Bonilla, M., Kolekar, S., Huy, P.T., Batzill, M., Phan, M.H., A magnetic sensor using a 2D van der Waals ferromagnetic material. *Sci. Rep.*, 10, 4789, 2020.
16. Sulaiman, J.M.A., Ismail, M.M., Rafeeq, S.N., Mandal, A., Enhancement of electromagnetic interference shielding based on Co0.5Zn0.5Fe2O4/PANI-PTSA nanocomposites. *Appl. Phys. A Mater. Sci. Process.*, 126, 236, 2020.
17. Zhou, H., An, Z., Yuan, C., Luo, X., Light-modulated ferromagnetism of strained $NiFe_2O_4$ nanocrystals. *Ceram. Int.*, 45, 13319, 2019.
18. Haertling, G.H., Ferroelectric ceramics: History and technology. *J. Am. Ceram. Soc.*, 82, 797, 1999.
19. Sakka, S., History of ferroelectric materials prepared by sol-gel method. *J. Sol-Gel Sci. Technol.*, 101, 140, 2022.

20. Mo, F., Saraya, T., Hiramoto, T., Kobayashi, M., Reliability characteristics of metal/ferroelectric-HfO2/IGZO/metal capacitor for non-volatile memory application. *Appl. Phys. Express*, 13, 74005, 2020.
21. Guo, Y., Peng, B., Qiu, R., Dong, G., Yao, Y., Zhao, Y., Zhou, Z., Liu, M., Self-rolling-up enabled ultrahigh-density information storage in freestanding single-crystalline ferroic oxide films (Adv. Funct. Mater. 20/2023). *Adv. Funct. Mater.*, 33, 2213668, 2023.
22. Li, L., Wang, S., Li, X., Sha, Y., Liu, Z., Liu, L., Zhang, J., Zhang, S.T., Wang, Y., Large pyroelectricity via engineered ferroelectric-relaxor phase boundary. *J. Am. Ceram. Soc.*, 105, 5230, 2022.
23. Huang, C., Zhang, L., Liu, S., Wang, Y., Wang, N., Deng, Y., Double enhanced energy storage density via polarization gradient design in ferroelectric poly(vinylidene fluoride)-based nanocomposites. *Chem. Eng. J.*, 411, 128585, 2021.
24. Habib, M., Lee, M.H., Akram, F., Kim, M.H., Kim, W.J., Song, T.K., Temperature-insensitive piezoelectric properties of lead-free $BiFeO_3$–$BaTiO_3$ ceramics with high Curie temperature. *J. Alloys Compd.*, 851, 156788, 2021.
25. Choi, K.J., Biegalski, M., Li, Y.L., Sharan, A., Schubert, J., Uecker, R., Reiche, P., Chen, Y.B., Pan, X.Q., Gopalan, V., Che, L.Q., Schlom, D.C., Eom, C.B., Enhancement of ferroelectricity in strained $BaTiO_3$ thin films. *Sci. (80-.)*, 306, 1005, 2004.
26. Roshchupkin, D., Emelin, E., Plotitcyna, O., Rashid, F., Irzhak, D., Karandashev, V., Orlova, T., Targonskaya, N., Sakharov, S., Mololkin, A., Redkin, B., Fritze, H., Suhak, Y., Kovalev, D., Vadilonga, S., Ortega, L., Leitenberger, W., Single crystals of ferroelectric lithium niobate tantalate LiNb1 xTaxO3solid solutions for hightemperature sensor and actuator applications. *Acta Crystallogr. Sect. B Struct. Sci. Cryst. Eng. Mater.*, 76, 1071, 2020.
27. Long, X., Tan, H., Sánchez, F., Fina, I., Fontcuberta, J., Ferroelectric electroresistance after a breakdown in epitaxial Hf0.5Zr0.5O2 tunnel junctions. *ACS Appl. Electron. Mater.*, 5, 740, 2023.
28. Qian, J., Li, G., Zhu, K., Ge, G., Shi, C., Liu, Y., Yan, F., Li, Y., Shen, B., Zhai, J., Cheng, Z., High energy storage performances and large electrocaloric response in Bi0.5na0.5tio3-Ba(Zr0.2ti0.8)O3 thin films. *SSRN Electron. J.*, 14, 54012-54020, 2022.
29. Chen, G., Zou, K., Yu, Y., Zhang, Y., Zhang, Q., Lu, Y., He, Y., Effects of the film thickness and poling electric field on photovoltaic performances of $(Pb,La)(Zr,Ti)O_3$ ferroelectric thin film-based devices. *Ceram. Int.*, 46, 4148, 2020.
30. Salje, E.K.H., Ferroelastic materials. *Annu. Rev. Mater. Res.*, 42, 265, 2012.
31. Marçal, L.A.B., Benter, S., Irish, A., Dzhigaev, D., Oksenberg, E., Rothman, A., Sanders, E., Hammarberg, S., Zhang, Z., Sala, S., Björling, A., Unger, E., Mikkelsen, A., Joselevich, E., Timm, R., Wallentin, J., Inducing ferroelastic

domains in single-crystal CsPbBr3 perovskite nanowires using atomic force microscopy. *Phys. Rev. Mater.*, 5, L063001, 2021.

32. Xie, S., Tan, Z., Jiang, L., Nie, R., Xu, Q., Chen, Y., Zhu, J., Wang, Q., Ferroelastic properties and compressive stress-strain response of bismuth titanate based ferroelectrics. *Ceram. Int.*, 46, 1183, 2020.
33. Zednik, R.J., Varatharajan, A., Oliver, M., Valanoor, N., McIntyre, P.C., Mobile ferroelastic domain walls in nanocrystalline PZT films: The direct piezoelectric effect. *Adv. Funct. Mater.*, 21, 3104, 2011.
34. Kang, W. and Huber, J.E., Energy harvesting based on compressive stress-induced ferroelectric/ferroelastic switching in polycrystalline ferroelectric materials. *Cell Rep. Phys. Sci.*, 3, 100707, 2022.
35. Xu, Y., Shao, G., Pang, J., Zhou, Y., Ding, X., Sun, J., Lookman, T., Salje, E.K.H., Xue, D., Avalanches during ferroelectric and ferroelastic switching in barium titanate ceramics. *Phys. Rev. Mater.*, 6, 124413, 2022.
36. Dong, G., Li, S., Li, T., Wu, H., Nan, T., Wang, X., Liu, H., Cheng, Y., Zhou, Y., Qu, W., Zhao, Y., Peng, B., Wang, Z., Hu, Z., Luo, Z., Ren, W., Pennycook, S.J., Li, J., Sun, J., Ye, Z., Jiang, Z., Zhou, Z., Ding, X., Min, T., Liu, M., Flexible ferroelectrics: Periodic wrinkle-patterned single-crystalline ferroelectric oxide membranes with enhanced piezoelectricity (Adv. Mater. 50/2020). *Adv. Mater.*, 32, 2070377, 2020.
37. Ji, H., Zhang, L., Zhang, R., Gas sensitive performance and mechanism of multiferroic $BiFeO_3$ under thermal-magnetic synergetic excitation. *Inorg. Chem. Commun.*, 150, 110491, 2023.
38. Ahmed, S., Atif, M., Rehman, A.U., Bashir, S., Iqbal, N., Khalid, W., Ali, Z., Nadeem, M., Enhancement in the magnetoelectric and energy storage properties of core-shell-like $CoFe_2O_4$–$BaTiO_3$ multiferroic nanocomposite. *J. Alloys Compd.*, 883, 160875, 2021.
39. Sharma, S., Paliwal, A., Tomar, M., Gupta, V., Multiferroic BFO/BTO multi-layer structures based magnetic field sensor. *Phys. B Condens. Matter*, 571, 1, 2019.
40. Ren, Z.Y., Shao, F., Liu, P.F., Wang, M.X., Chen, J.K., Meng, K.K., Xu, X.G., Miao, J., Jiang, Y., Nonvolatile ferroelectric field control of the anomalous Hall effect in $BiFe\ O_3/SrRu\ O_3$ bilayer. *Phys. Rev. Appl.*, 13, 024044, 2020.
41. Jin, C., Zhu, Y., Han, W., Liu, Q., Hu, S., Ji, Y., Xu, Z., Hu, S., Ye, M., Chen, L., Exchange bias in flexible freestanding $La_{0.7}Sr_{0.3}MnO_3/BiFeO_3$ membranes. *Appl. Phys. Lett.*, 117, 252902, 2020.

2

An Overview of Ferroic Materials

M. Rizwan[1*], K. Nawaz[1], F. Arooj[1], F. Noor[1] and A. Ayub[2]

[1]*School of Physical Sciences, University of the Punjab, Lahore, Pakistan*
[2]*Department of Physics, University of Gujrat, Gujrat, Pakistan*

Abstract
In the last decades, there has been an extensive amount of research on ferroic materials, which show spontaneous polarization, magnetization, or strain. The discovery of Rochelle salt as the first ferroelectric material in the early 20th century marked the beginning of this field. Since then, the discovery of many new ferroic materials has led to the development of a range of technological applications, including piezoelectric actuators, sensors, and memory devices. This chapter provides an overview of the past and present of ferroic materials research, highlighting key discoveries and technological applications, as well as current research trends and challenges. Properties of ferroic materials, including ferromagnetic, ferroelectric, and ferroelastic materials, are also reviewed in this chapter, which can be tuned by varying their composition, structure, and processing conditions. Moreover, the scaling of these properties can be controlled by manipulating the size, shape, and orientation of the materials. Recent advancements in ferroic materials include the development of new synthesis methods, the discovery of new materials with enhanced properties, and the exploration of novel applications.

Keywords: Ferro-magnetoelectric materials, ferro-elastic, primary ferroics, ferro-magnetoelastic materials, ferroelectric polymers, ferro-elasticity

2.1 Introduction

Materials have their own symmetry, and occasionally, a change in symmetry happens on its own due to changes in the environment. For instance, the water turns into ice when it cools below a particular temperature (let us

**Corresponding author*: rizwan.sps@pu.edu.pk

Inamuddin, Tariq Altalhi and Mohammad Abu Jafar Mazumder (eds.) Ferroic Materials-Based Technologies, (19–34) © 2024 Scrivener Publishing LLC

say the freezing point). Liquid water has the same macroscopic properties in all directions thanks to its extremely great orientational or rotational symmetry. The directional symmetry is drastically reduced when water transforms into a crystalline substance (ice), and practically all macroscopic features of every crystal or ice are different in every direction. One way to define this freezing phenomenon is as a spontaneous breakdown or devaluation of directional symmetry. It happens frequently in nature for directional and other types of symmetry to spontaneously break down, even the moment the universe was created [1].

The term "spontaneous" refers to the fact that this ordering occurs without the application of an external field. This behavior arises from the interactions between individual atoms or ions in the material, which result in the alignment of their electric or magnetic moments. These moments can be described as vectors that indicate the direction of the magnetic or electric field at each point in the material. Ferroic materials, which are defined by their ferroic ordering, provide a useful way of connecting control of the mechanical, magnetic, and electrical subsystems in energy conversion. Due to the ferroic coupling effect's promise as a fresh, adaptable route for energy conversion, it has received attention. This article evaluates the use of ferroic materials with a focus on the underlying mechanisms that impose the energy conversion process. Research into the energy-conversion capabilities of ferroic materials lays the way for pervasive energy harvesting via magneto-mechano-electric-thermal coupling mechanisms [2].

The study of symmetry variations in crystals as a function of different variables (temperature, pressure, etc.) is a vast topic. Any crystal that experiences a change in symmetry is said to be transitioning from one phase to another. The term "phase transition" thus refers to a change in symmetry. "Ferroic materials" refer to crystals that experience at least one phase transition that modifies their directional symmetry.

The study of ferromagnetic, ferroelectric, ferroelastic, ferro-gyro tropic, and other materials is collectively referred to as "ferroic materials" [1].

2.2 Types of Ferroic Materials

Ferroic materials are a category of materials that exhibit spontaneous electric or magnetic ordering, which can be shifted by applying an external magnetic or electric field [3]. Ferroic behavior can be classified into two main types: primary ferroics and secondary ferroics. These materials have a broad range of applications in technology and engineering, and their study has been an important area of research in materials science and physics.

2.2.1 Primary Ferroics

So far, as the scientific literature is concerned, the primary types of ferroic materials are generally considered to be ferromagnetic, ferroelectric, and ferroelastic materials. These materials exhibit spontaneous magnetization, polarization, and deformation, respectively.

2.2.1.1 Ferromagnetic Materials

The definition of "ferromagnetic" has long been accepted to refer to iron's magnetic properties [4]. Ferromagnetic materials are commonly referred to as "magnetic materials". Although all the materials are magnetic to some extent, most of them are paramagnetic or diamagnetic in nature, having a very weak magnetic property. These materials can exist as conductors, insulators, or semiconductors depending on the operating conditions of the device [5]. Electric motors, generators, magnetic storage devices like hard drives, and a wide range of other technical applications all make use of ferromagnetic materials.

2.2.1.2 Ferroelectric Materials

A ferroelectric material exhibits one or more phases over an achievable pressure and temperature range [4]. Ferroelectric materials are dielectric materials with permanent polarization, even after the removal of the applied electric field. Additionally, by adding an electric field, the orientation of the dipole moment can be changed. Of the 32 crystal classes, 20 have direct piezoelectricity and 21 are non-centrosymmetric. Ten classes of piezoelectric crystals have impulsive electric polarization, which varies with the temperature change and is referred to as pyroelectricity. Some of the ferroelectric materials used in pyroelectricity have their polarization reversed by applying an electric field. Ferroelectric substances, therefore, possess both pyroelectricity and piezoelectricity. Figure 2.1 shows a Venn diagram of their relationships [6].

Ferroelectric materials have many uses, including the creation of actuators, which are devices that transform electrical energy into mechanical motion, and non-volatile memory systems like ferroelectric random-access memory (FeRAM), which are non-volatile memory systems.

2.2.1.3 Ferroelastic Materials

Ferroelastic materials are known to exhibit highly non-linear stress–strain behavior, commonly referred to as hysteresis. This class of materials is the

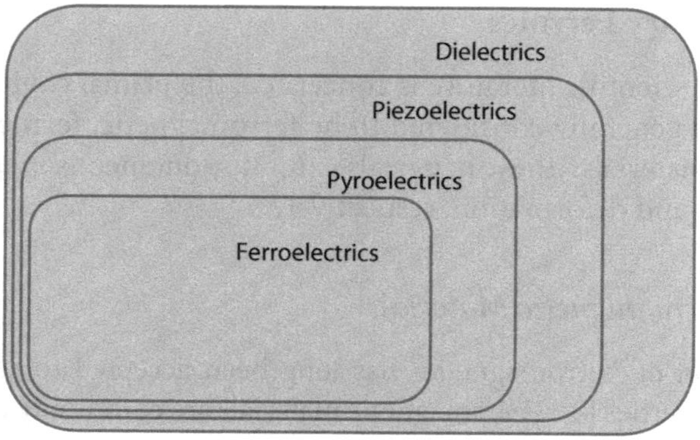

Figure 2.1 Ferroelectrics in the dielectric class Venn diagram [6].

largest among ferroic materials and finds applications in various fields such as mechanical switches and piezoelectric sensors. While ferroelasticity might appear to have little bearing on the search for multiferroic materials, it has a big impact on how ferroelectric and ferromagnetic properties are coupled. Strain and ferroelasticity have been found to mediate much of this coupling. As with other ferroic properties, ferroelasticity is characterized by its hysteresis. Ferroelastic materials have a broad range of applications in technology, including in the development of shape memory alloys, which are materials that can be programmed to return to a specific shape after deformation [7].

2.2.2 Secondary Ferroics

In addition to these primary types, several other types of ferroic materials exhibit multiple types of ferroic behavior, including Multiferroics, Ferroelastoelectric, Ferromagnetoelastic, and Ferromagnetoelectric materials [8]. These secondary types of ferroic materials attract a lot of interest due to their potential applications in advanced technologies, such as spintronic devices, energy conversion devices, and sensors.

2.2.2.1 Multiferroics

Multiferroic materials have attracted considerable attention due to their potential for multifunctional applications, as they display at least two ferroic orders such as ferroelectric, ferromagnetic, or ferroelastic. The magnetoelectric (ME) effect is one of many unique phenomena that might result from the coupling relationship between these various order parameters.

While this effect has been observed in some natural materials at low temperatures, there is growing interest in developing synthetic multiferroic materials that exhibit stronger coupling and the ME effect at higher temperatures. The ME effect has a wide range of possible uses, including in spintronics, multiple-state memory, and information storage [9]. For instance, the ME effect can be utilized to create more sensitive magnetic field sensors or to develop memory devices that can be switched between different states using electric or magnetic fields.

2.2.2.2 *Ferroelastoelectric Materials*

Ferroelastoelectric materials are a subclass of multiferroic materials that exhibit three distinct ferroic orders: ferroelectric, ferromagnetic, and ferroelastic. These materials can generate an electric polarization in response to mechanical stress or strain, and vice versa. This coupling between the mechanical and electrical properties of these materials makes them highly desirable for many different uses. In these materials, the coupling between the electric, magnetic, and elastic degrees of freedom can lead to the emergence of unique phenomena such as strain-mediated magnetoelectric coupling. Ferroelasticity refers to the ability of a material to undergo a reversible deformation, such as stretching or compression, and maintain its original shape when the stress is removed [10].

2.2.2.3 *Ferromagnetoelastic Materials*

Ferromagnetoelastic materials are a type of multiferroic material that exhibits two ferroic orders: ferromagnetic and ferroelastic. These materials possess spontaneous magnetization and are capable of undergoing reversible deformations (such as stretching or compression) while maintaining their magnetic properties. The coupling between the magnetic and elastic degrees of freedom in Ferromagnetoelastic materials can lead to the emergence of novel phenomena, such as magnetoelastic coupling, which refers to the interaction between the magnetic and elastic degrees of freedom in a material [10].

When a ferromagnetoelastic material undergoes deformation, it can cause a change in its magnetic properties, such as the direction of its magnetization or its magnetic susceptibility. Ferromagnetoelastic materials have received attention due to their potential for various applications, such as in magnetic field sensing, magnetic energy conversion, and actuation. By utilizing the coupling between the magnetic and elastic degrees of freedom, ferromagnetoelastic materials offer a unique and promising avenue for the development of next-generation magnetic devices.

2.2.2.4 *Ferromagnetoelectric Materials*

Ferromagnetoelectric materials are a type of multiferroic material that exhibits two ferroic orders: ferromagnetic and ferroelectric. These materials possess spontaneous magnetization and electric polarization, which can be coupled through a phenomenon known as magnetoelectric coupling. The coupling of the magnetic and electric degrees of freedom in a material is referred to as magnetoelectric coupling. When applied to ferromagnetoelectric materials, an external magnetic field can result in an electric polarization or electric field, while an applied electric field can result in an electric field or a magnetic field [10]. This coupling between the electric and magnetic degrees of freedom can be utilized for various applications, such as in magnetic field sensing, energy conversion, and data storage.

Ferromagnetoelectric materials have received significant attention due to their potential for multifunctional applications. They provide a viable route for the creation of future gadgets including magnetoelectric sensors, actuators, and memory systems. By utilizing the coupling between the magnetic and electric degrees of freedom, Ferromagnetoelectric materials can provide enhanced performance, lower power consumption, and improve functionality.

2.3 Past of Ferroic Materials

Ferroic materials belong to a class of substances that spontaneously organize electrically or magnetically and can switch between the two states when an external electric or magnetic field is applied. The term "ferroic" comes from the Latin word "ferrum," which means iron, as many of the first materials to exhibit ferroic behavior were iron-based alloys. With several applications in technology and engineering, the study of ferroic materials has been a significant topic of research in materials science and physics [11].

A lengthy and fascinating history of ferroic materials may be traced back to the early discoveries of electricity and magnetism. In this chapter, we will explore the history of ferroic materials, starting with the discovery of magnetism and electricity, and following the development of the understanding of ferroic behavior over time.

2.3.1 Discovery of Magnetism and Electricity

The study of ferroic materials began with the discovery of magnetism and electricity. The ancient Greeks were aware of the magnetic properties of certain stones, yet William Gilbert, a physician to Queen Elizabeth I, did not discover this until the 16th century, and published a comprehensive study of magnetism, which included the first use of the term "magnetic pole." In the 18th century, Benjamin Franklin performed experiments on electricity, leading to the discovery of positive and negative charges. In the early 19th century, electromagnetic induction was discovered by Michael Faraday, which set the foundation for the evolution of electric motors and generators [12].

2.3.2 Discovery of Ferromagnetism

The discovery of ferromagnetism began in the early 19th century with the work of Hans Christian Oersted and André-Marie Ampère, who discovered the connection between electricity and magnetism. In 1825, William Sturgeon created the first electromagnet by wrapping a wire around an iron core, demonstrating the magnetic properties of iron. In 1845, Michael Faraday discovered that certain materials, such as iron, nickel, and cobalt, exhibit a much stronger magnetic field than other materials when subjected to an external magnetic field. This behavior, called ferromagnetism, was named by William Whewell in 1847 [13].

2.3.3 Discovery of Ferroelectricity

Pierre Curie and his brother Jacques Curie's research, which followed their discovery of the piezoelectric effect in 1880, marked the beginning of the study of ferroelectricity in the late 19th century. The piezoelectric effect is the ability of certain materials, such as quartz, to generate an electric field when subjected to mechanical stress [14]. In 1921, Joseph Valasek discovered ferroelectricity in Rochelle salt, a natural mineral that exhibits spontaneous electric polarization when cooled below a certain temperature.

2.3.4 Discovery of Ferroelasticity

The discovery of ferroelasticity began in the mid-20th century with the work of James S. Kasper and his colleagues, who first identified the shape-memory property of a gold-cadmium alloy. The shape-memory effect is the ability of certain materials to retain their original shape after being deformed when heated above a certain temperature. In the 1960s, the first ferroelastic material, barium titanate, was discovered by W. P. Mason and his colleagues.

In a recent analysis, Cross and Newnham identified numerous significant dates in the development of ferroelectricity over the previous seven decades. These events include the discovery of Rochelle salt in its early stages, J. Valasek's initial comparison of magnetism to ferromagnetism, C. B. Sawyer's first feasible use of a ferroelectric for bimorph "benders" and "twisters" and their implementation in tube-type phonographs, and H. Mueller's elegant phenomenological description, which was later used by A. F. [15]. These dates are important for understanding the historical development and early applications of ferroelectric materials. However, the specific years or time periods associated with these dates are not specified in the statement.

During 10 years that was arguably the most exciting in the history of ferroelectricity, 24 new ferroelectrics were discovered. This increase in activity was probably influenced by the friendly rivalry between Berndt Matthias and his coworkers at Bell Labs and Ray Pepinsky and his group at Penn State. It was at that moment when many of the practical materials used today, such as $LiNb0_3$, $KNb0_3$, $PbNb_20_6$, and SbSI, were discovered. These 10 years were likely a crucial period in the development of the field.

The "age of high science" in the 1960s and 1970s created the groundwork for our current comprehension of ferroelectrics. This period saw the discovery of numerous new ferroelectrics, including SbSI and others, and the development of key theoretical and experimental tools, such as the soft-mode theory, Raman spectroscopy, inelastic neutron scattering, etc. In addition to establishing the basis for ferroelectric behavior in significant materials such as solid solutions of PbTi03 and PbTi03:PbZr03, the entire symmetry classification for all ferroics was also developed.

However, by the middle of the 1970s, interest in fundamental research on ferroelectricity began to wane, as the focus shifted toward the feasible applications of ferroelectric materials in industries such as electroceramics. While theoreticians continued to explore more esoteric phenomena, electroceramics focused on making useful materials for applications such as dielectric, piezoelectric, pyroelectric, and electro-optic materials. This practical evolution of ferroelectric materials is the subject of ongoing research and development, with new and improved materials being developed for a wide range of applications [4].

2.4 Present of Ferroic Materials

Today, ferroic materials are used in a broad range of applications, from electronics to energy storage to medical devices [16]. The creation of multiferroics, or materials with both ferroelectric and ferromagnetic order, is one recent development in the study of ferroic materials. These materials have potential applications in new types of data storage and computing devices, as well as sensors and actuators [16].

The creation of novel piezoelectric materials is another field of ongoing ferroic material research. Piezoelectric materials, which can be found in anything from petrol lighters to ultrasound scanners, transform mechanical energy into electrical energy. Researchers are exploring new materials that exhibit enhanced piezoelectric properties, such as lead-free materials that are more environmentally friendly than traditional lead-based piezoelectrics [17].

Ferroic materials are also being used in energy storage applications. Lithium-ion batteries, which are used in everything from cell phones to electric vehicles, rely on materials that can store and release electrical energy. Due to their high energy density and quick charging times, ferroic compounds like lithium niobate and lithium tantalate are being investigated as promising materials for use in next-generation batteries.

In the field of medicine, ferroic materials are being used to develop new types of medical devices. For example, ferromagnetic materials are used in magnetic resonance imaging (MRI) machines to create high-resolution images of the body. Ferroelectric materials are being explored as potential materials for use in implantable devices, such as pacemakers and sensors [18].

Here are some of the present developments in ferroic materials:

Multiferroic materials: These materials exhibit more than one ferroic property, such as ferroelectricity and ferromagnetism, making them useful for a broad range of applications such as data storage, sensing, and energy harvesting [9].

Biocompatible ferroelectric materials: Ferroelectric materials are being developed for biomedical applications due to their biocompatibility, such as for drug delivery, biosensors, and tissue engineering [18].

Nanoscale ferroic materials: The development of nanoscale ferroic materials has opened up new avenues for their use in high-density memory storage, sensing, and energy conversion.

Ferroelectric polymers: Polymers with ferroelectric properties are being developed for use in flexible electronics, such as sensors and energy harvesting devices.

Ferroelectric thin films: thin films of ferroelectric materials are being developed for use in microelectronics, such as non-volatile memory and microelectromechanical systems (MEMS).

Ferroic materials for energy applications: Ferroic materials are being explored for use in energy applications, such as piezoelectric energy harvesting and solid oxide fuel cells [1].

Artificial intelligence-guided discovery: Machine learning algorithms are being used to accelerate the discovery of new ferroic materials with desirable properties.

2.5 Properties of Ferroic Materials

Ferroic materials exhibit a range of unique properties that make them highly desirable for various technological applications. In this essay, we will discuss some of the properties of ferroic materials, including ferroelectricity, ferromagnetism, ferroelasticity, and their applications. The ability of a material to have an impulsive electric polarization that can be retained by an external electric field is known as ferroelectricity [17].

Ferroelectric materials are used in a broad range of applications, including capacitors, piezoelectric sensors, actuators, and memory devices. The electric polarization in these materials arises from the displacement of ions from their equilibrium positions under the influence of an electric field. The most commonly used ferroelectric materials are perovskite oxides, such as lead zirconate titanate (PZT), which exhibits high piezoelectric coefficients and is widely used in medical ultrasound imaging [18].

A material's ability to maintain a magnetic field even when there is no external magnetic field is known as ferromagnetism [19]. Ferromagnetic materials are materials that have a permanent magnetization that can be aligned in a specific direction by an external magnetic field. The magnetic moments of the atoms in the material are aligned in the same direction, resulting in a net magnetization [20]. Ferromagnetic materials are used in a variety of applications, including data storage, sensors, and motors. The magnetic moment in these materials arises from the alignment of the spins of unpaired electrons in the material. The most commonly used ferromagnetic materials are iron, cobalt, and nickel and their alloys [21].

Ferroelasticity is a property that describes the ability of a material to undergo a reversible deformation under the influence of an external stimulus, such as temperature, electric or magnetic fields, or stress. Ferroelastic materials are materials that exhibit a reversible change in shape or deformation in response to external stress or strain [7]. The deformation is caused by the reorientation of structural domains in the material, which can be switched by external stress or strain. Ferroelastic materials are used in a variety of applications, including microelectromechanical systems (MEMS), sensors, actuators, and energy harvesting devices. The most commonly used ferroelastic materials are shape memory alloys, such as nickel-titanium (NiTi), which exhibit large deformation and recovery strains and are used in medical implants and robotics.

Ferroic materials can also exhibit multiferroic behavior, which describes the coexistence of ferroelectricity and ferromagnetism in the same material. Multiferroic materials are highly desirable for various technological applications, including data storage, sensors, and spintronics. The most commonly used multiferroic materials are bismuth ferrite ($BiFeO_3$), which exhibits high ferroelectric and antiferromagnetic ordering temperatures and is used in various memory and logic devices [22]. Overall, ferroic materials have a wide range of applications in electronics, data storage, sensors, and other fields. Grasping the properties of these materials is essential for the evolution of new technologies and devices.

These properties make them highly desirable for various technological applications, including medical imaging, data storage, robotics, and energy harvesting. As researchers continue to explore the properties of ferroic materials and push the boundaries of their applications, we can expect to see even more innovative and transformative developments in the future.

In conclusion, ferroic materials exhibit a range of unique and desirable properties, including ferroelectricity, ferromagnetism, ferroelasticity, and multiferroicity. In addition to the above properties, ferroic materials also exhibit other unique properties, including hysteresis, which is the dependence of the material's response on the history of the applied field, and domain structure, which is the arrangement of ferroic domains within the material.

2.6 Scaling of Properties

Ferroic materials exhibit a variety of interesting and useful properties, including ferromagnetism, ferroelectricity, and ferroelasticity. These properties are known to exhibit scaling behavior with temperature, magnetic field, and electric field. In this essay, we will discuss the scaling of properties of ferroic materials, including how they change with external stimuli, and how researchers use this knowledge to optimize these materials for various applications [4].

Scaling of ferroic properties is an essential aspect of understanding these materials' behavior. It describes how the properties of the material change concerning external stimuli, such as temperature, magnetic field, and electric field. The scaling behavior of ferroic properties can be expressed in terms of power laws, which describe the rate at which the properties change as a function of the external stimuli. The scaling behavior of ferroic properties is dependent on the nature of the interactions between the constituent atoms or molecules in the material and can be understood through the study of phase transitions.

The scaling behavior of ferroic properties has important implications for the optimization of these materials for various applications. For example, the scaling of ferroelectric properties with temperature is essential for designing high-temperature piezoelectric sensors and actuators. The scaling of ferromagnetic properties with the magnetic field is crucial for developing high-density magnetic storage devices and spintronics [23].

In addition to external stimuli, the scaling behavior of ferroic properties can also be influenced by the size of the material. The properties of ferroic materials change as the size of the material decreases, which is known as size scaling. This phenomenon is essential for developing nanoscale devices and optimizing their performance. For example, the size scaling of ferroelectric properties is used to develop nanoscale capacitors and memory devices [24].

Another important aspect of scaling ferroic properties is the scaling of domain sizes. The domains are regions of the material that have a uniform orientation of polarization or magnetization. The domain size of ferroic materials is important because it affects the materials' performance and stability. The domain size scales with the external stimuli, and researchers can use this knowledge to control the domain size and optimize the properties of the material for various applications [25].

In conclusion, the scaling of properties of ferroic materials is an essential aspect of understanding and optimizing these materials for various applications. The scaling behavior of ferroic properties with external stimuli, size, and domain size has important implications for developing high-performance ferroic devices, including sensors, actuators, and memory devices. As researchers continue to explore the scaling of ferroic properties, we can expect to see more innovative and transformative developments in the future [4].

2.7 Recent Advances in Ferroic Materials

Ferroic materials are a class of materials that exhibit ferromagnetic, ferroelectric, or ferroelastic behavior, and their unique properties make them useful for a wide range of technological applications. Significant advancements have been made recently in the study and development of ferroic materials, including the discovery of new materials, advances in understanding their properties, and the development of new applications [22].

One significant recent advance in the field of ferroic materials is the discovery of new materials that exhibit unique ferroic properties. For example, the discovery of multiferroic materials, which exhibit both ferromagnetic and ferroelectric behavior, has opened up new avenues for research and applications. One promising example of a multiferroic material is bismuth ferrite, which has both ferromagnetic and ferroelectric properties and can be used for magnetic storage and spintronics applications.

Another recent advance in ferroic materials is the development of new techniques for controlling their properties. For example, researchers have developed methods for controlling the domain structure of ferroelectric materials, which has important implications for developing high-performance ferroelectric devices such as sensors and actuators [26]. The use of external stimuli such as electric fields, magnetic fields, and strain is effective in controlling the domain structure of ferroelectric materials.

In addition to developing new materials and techniques for controlling their properties, researchers have also made significant advances in understanding the fundamental physics of ferroic materials. For example, advances in computational modeling and simulation have allowed researchers to study the behavior of ferroic materials at the atomic and molecular level, which has led to a better understanding of the mechanisms that drive their properties. Additionally, new experimental techniques such as scanning probe microscopy and X-ray diffraction have allowed researchers to observe the behavior of ferroic materials with high spatial and temporal resolution, providing insights into their behavior at the nanoscale.

The development of new applications for ferroic materials is another significant recent advance in the field. For example, ferroelectric materials are being developed for use in energy harvesting applications, where they can be used to convert mechanical energy into electrical energy. Additionally, multiferroic materials are being developed for use in spintronics and magnetic memory devices, where they can be used to store and manipulate data using magnetic fields and electrical currents.

In conclusion, recent advances in ferroic materials have led to significant improvements in our understanding of their properties and the development of new materials and applications. The discovery of new materials, the development of new techniques for controlling their properties, advances in computational modeling and simulation, and the development of new applications have all contributed to the rapid progress in this field. As researchers continue to explore the potential of ferroic materials, we can expect to see further advances in the coming years, which will have important implications for a wide range of technological applications.

2.8 Conclusion

In conclusion, ferroic materials are a class of materials that exhibit spontaneous electric or magnetic ordering, which can be shifted with the application of an external electric or magnetic field. They may be categorized into primary ferroics, such as ferromagnetic, ferroelectric, and ferroelastic materials, and secondary ferroics, which exhibit multiple types of ferroic behavior. Ferroic materials have been the subject of extensive research for several decades, leading to the discovery of many new materials and technological applications, such as piezoelectric actuators, sensors, and memory devices. Their properties can be tuned by varying their composition, structure, and processing conditions, and their scaling can be controlled by manipulating the size, shape, and orientation of the materials. Recent advancements in ferroic materials include the development of new synthesis methods, the discovery of materials with enhanced properties, and the exploration of novel applications. Ongoing research and development in this field are expected to continue to yield new materials and applications, as well as the improvement of existing technologies.

References

1. Wadhawan, V.K., Ferroic materials: A primer. *Resonance*, 7, 15, 2002.
2. Gao, W., Brennan, R., Hu, Y., Wuttig, M., Yuan, G., Quandt, E., Ren, S., Energy transduction ferroic materials. *Mater. Today*, 21, 771, 2018.
3. Martin, L.W. and Rappe, A.M., Thin-film ferroelectric materials and their applications. *Nat. Rev. Mater.*, 2, 2016. p.(1-14).
4. Cross, L.E., Ferroic materials and composites: Past, present and future, in: *Advanced Ceramics III*, S. Sōmiya, (Ed.), pp. 71–102, Springer Netherlands, Dordrecht, 1990.
5. Goldman, A., *Handbook of modern ferromagnetic materials*, Springer, 1999, New York.
6. Kim, T.Y., Kim, S.K., Kim, S.W., Application of ferroelectric materials for improving output power of energy harvesters. *Nano Converg.*, 5, 2018, p.30.
7. Salje, E.K.H., Ferroelastic materials. *Annu. Rev. Mater. Res.*, 42, 265, 2012.
8. Newnham, R.E. and Cross, L.E., Symmetry of secondary ferroics. I. *Mater. Res. Bull.*, 9, 927, 1974.

9. Nan, C.W., Bichurin, M.I., Dong, S., Viehland, D., Srinivasan, G., Multiferroic magnetoelectric composites: Historical perspective, status, and future directions, in: *Progress in Advanced Dielectrics*, pp. 191–293, World Scientific, Singapore, 2020.
10. Wadhawan, V.K., Ferroic materials, in: *Smart Structures*, V.K. Wadhawan, (Ed.), pp. 75–118, Oxford University Press, Walton Street, Oxford, 2008.
11. Redfern, S.A.T., Wang, C., Hong, J.W., Catalan, G., Scott, J.F., Elastic and electrical anomalies at low-temperature phase transitions in $BiFeO_3$. *J. Phys. Condens. Matter.*, 20, 452205, 2008.
12. Watson, W., XXXI. An account of Mr. Benjamin Franklin's treatise, lately published, intituled, Experiments and observations on electricity, made at Philadelphia in America. *Philos. Trans. R. Soc. Lond.*, 47, 202, 1752.
13. Coey, J.M.D., *Magnetism and magnetic materials*, Cambridge University Press, Cambridge, England, 2010.
14. Lines, M.E. and Glass, A.M., Applications of ferroelectrics, in: *Principles and Applications of Ferroelectrics and Related Materials*, M.E. Lines, and A.M. Glass, (Eds.), pp. 559–607, Oxford University Press, Walton Street, Oxford, 2010.
15. Salje, E.K.H., Hayward, S.A., Lee, W.T., Ferroelastic phase transitions: Structure and microstructure. *Acta Crystallogr. Sect. A Found. Crystallogr.*, 61, 3, 2005.
16. Fukunaga, M., Takesada, M., Onodera, A., Ferroelectricity in layered perovskites as a model of ultra-thin films. *World J. Condens. Matter Phys.*, 06, 224, 2016.
17. Scott, J.F., Applications of modern ferroelectrics. *Sci. (80-.)*, 315, 954, 2007.
18. Saxena, A. and Bhalla, A.S., Bioferroics and biomultiferroics: Ferroic behaviour of biological systems. *Mater. Res. Innov.*, 17, 440, 2013.
19. Fiebig, M., Revival of the magnetoelectric effect. *J. Phys. D. Appl. Phys.*, 38, R123, 2005.
20. Spaldin, N.A., *Magnetic materials: Fundamentals and applications*, Cambridge University Press, Cambridge, England, 2010.
21. Müllner, P., *Shape memory alloys*, K. Otsuka and C.M. Wayman (Eds.), p. 284, +xiv, Cambridge University Press, ISBN 0-521-44487-X (hardcover), *Mater. Sci. Eng. A.*, 268, 246, 1999, Cambridge, England.
22. Ramesh, R. and Spaldin, N.A., Multiferroics: Progress and prospects in thin films. *Nat. Mater.*, 6, 21, 2007.
23. Eric Cross, L., Relaxor ferroelecirics. *Ferroelectrics*, 76, 241, 1987.
24. Tsatskis, I., Salje, E.K.H., Heine, V., Pattern formation during phase transitions: Kinetics of partially conserved order parameters and the role of gradient energies. *J. Phys. Condens. Matter.*, 6, 11027, 1994.
25. Eerenstein, W., Mathur, N.D., Scott, J.F., Multiferroic and magnetoelectric materials. *Nature*, 442, 759, 2006.
26. Setter, N., Damjanovic, D., Eng, L., Fox, G., Gevorgian, S., Hong, S., Kingon, A., Kohlstedt, H., Park, N.Y., Stephenson, G.B., Stolitchnov, I., Taganstev, A.K., Taylor, D.V., Yamada, T., Streiffer, S., Erratum: Ferroelectric thin films: Review of materials, properties, and applications (Journal of Applied Physics (2006) 100 (051606)). *J. Appl. Phys.*, 100, 2006. p. 1-41.

3

Future Perspectives of Ferroic/ Multiferroic Materials

Sajida Shamsher*, Maria Wasim, Aneela Sabir, Muhammad Sahfiq, Abdur Rehman Mushtaq Ahmad† and Ishna Adeefa

Institute of Polymer and Textile Engineering, University of the Punjab, Lahore, Pakistan

Abstract

The chapter "Future Perspectives of Ferroic/Multiferroic Materials" looks at the bright future of these two types of solid-state materials. The introduction to the chapter lays the foundation for understanding the fundamentals of ferroic and multiferroic materials, as well as their types. The characterization techniques are evaluated to investigate them at the atomic and nanoscale levels. Additionally, a variety of applications are discussed. These include photovoltaic multiferroic solar cells, magnetic recording read heads, multi-state memory, multiferroic microwave phase shifters, and magnetoelectric devices. It also discusses the difficulties encountered in the field of these materials, providing insights into potential future directions and also demonstrating how these materials can be incorporated into existing technologies, highlighting the potential for opening up new features.

In conclusion, this chapter offers an in-depth evaluation of the state of these materials currently and their promising future possibilities. This chapter offers a greater understanding of the possible improvements and applications of ferroic and multiferroic materials in a variety of industries, making it an invaluable resource for researchers and scientists.

Keywords: Ferroics, multiferroics, magnetoelectric devices, photovoltaic multiferroics, multi-state memories, perovskite ferroelectrics, ferroelastic materials, random access memory

**Corresponding author*: sajida95sajo@gmail.com

†*Corresponding author*: abdurrehmanmushtaqahmad@gmail.com

Inamuddin, Tariq Altalhi and Mohammad Abu Jafar Mazumder (eds.) Ferroic Materials-Based Technologies, (35–54) © 2024 Scrivener Publishing LLC

3.1 Introduction

Ferroic and multiferroic materials have experienced a rise in popularity over the past 15 years due to their intriguing physical characteristics and enormous potential for technological application. In recent years, advances in materials synthesis, characterization, and theory have led to a deeper understanding of the underlying physics governing these materials. This has enabled the development of novel ferroic and multiferroic materials with improved performance and new functionalities. The future perspectives of these materials include the exploration of novel materials systems and the development of new device concepts based on their properties. Additionally, the integration of these materials into advanced electronic devices holds great promise for the development of new technologies.

Ferroic materials, such as ferroelectric, ferromagnetic, and ferroelastic materials, have been used in different technological applications, including information storage, sensing, and energy conversion. Multiferroic materials, wherein two or more ferroic order parameters are present at once, have the potential to enable new device architectures and functionalities. They can be utilized to give materials new properties, such as the ability to electrically control ferromagnetism at ambient temperature. Multiferroic materials are perfect candidates for a new generation of technological applications and could potentially generate billions of dollars. The study of ferroic/multiferroic materials is an active and exciting area of research with application possibilities in various fields [1]. Figure 3.1 describes the types of ferroic materials.

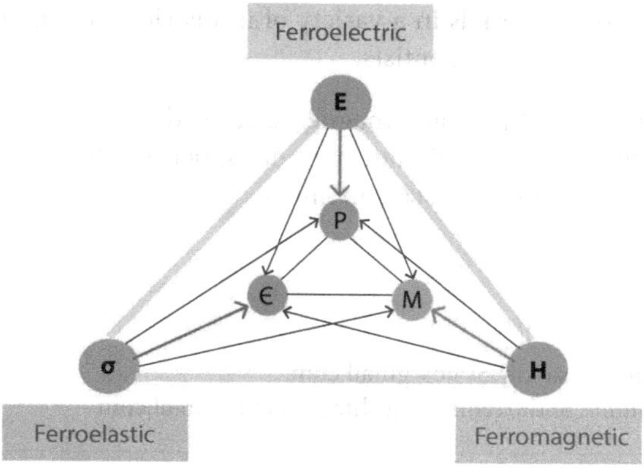

Figure 3.1 Types of ferroic materials.

3.2 Ferroic and Multiferroic Materials and Types

3.2.1 Ferroic Materials

Ferroic materials are a group of substances that show spontaneous ordering of their electric, magnetic, and/or elastic dipoles or domains, which give rise to their unique and useful properties. There are three main types of ferroic materials:

a) Ferroelectric Materials
Ferroelectrics are materials with spontaneously switchable polarization. The emergence of a hysteresis loop, the defining characteristic of ferroelectricity, is caused by the electric field-induced transition from one polarization state to another.
In various modes of operation, ferroelectrics can be utilized to harvest energy from a variety of sources, including solar, thermal, magnetic, wind, and mechanical vibrations [2].

b) Ferromagnetic Materials
Ferromagnetic materials are those that, even in the absence of an external magnetic field, exhibit spontaneous net magnetization at the atomic level. Ferromagnetic materials become strongly magnetized in the field's direction when exposed to an external magnetic field. Ferromagnetic materials are attracted to a magnet very strongly. Even after the external magnetizing field is withdrawn, these materials will continue to be magnetic for a while. Hysteresis is the name of this quality.
Ferromagnetic materials are commonly used to store nonvolatile data on cassettes, hard drives, magnetic random access memory, and other media [3].

c) Ferroelastic Materials
Ferroelastic materials are ferroic materials that exhibit spontaneous strain and can be switched between two or more stable states by applying external stress. Ferroelectricity and ferromagnetism in ferroic materials are mechanically equal to ferroelasticity. When a ferroelastic material is stressed, it undergoes a transition from one stable phase to another with an alternative crystal structure or orientation. This stress-induced phase change causes the material to spontaneously strain. Ferroelasticity manifests itself in the shape memory effect and superelasticity. At room temperature, depending on the circumstances, the typical ferroelastic alloy nitinol (nickel-titanium) can display either superelasticity.

Under ambient conditions, ferroelastic materials have significant potential for use in switches, force sensors, and nonvolatile memory devices [4].

3.2.2 Multiferroic Materials

A group of solid-state chemicals known as a multiferroic material has a minimum of two order states, such as the piezo-elastic, electric, or magnetic phases. They have potential applications in next-generation memory and sensing devices, as well as in energy conversion. Recent advances include the finding of novel multiferroic materials as well as the development of techniques for controlling their properties.

There are two main types of multiferroic materials:

a) Single-Phase Multiferroic Materials

These are the materials that are naturally isotropic and homogeneous, containing both magnetic order states and electric order states at all times and in all parts. Single-phase multiferroics may exhibit elastic/piezo-elastic properties. These are extremely rare in nature and extremely difficult to synthesize artificially.

Examples of Single-Phase Multiferroics: Cr_2O_3, $Cr_3B7O_{13}Cl$,$TbPO_4$, YIG, $BiFeO_3$, $BiMnO_3$, $RMnO_3$, etc.

There are two further types of single-phase multiferroics: The fundamental distinction between these two classes is in the physical mechanics of electric polarization. Figure 3.2 shows the relationship among multiferroic/ferromagnetic/ferroelectric/magnetoelectric materials.

i. Type I Single-Phase Multiferroics

This type exhibits compounds containing both magnetic and electric phases, but these phases are essentially independent

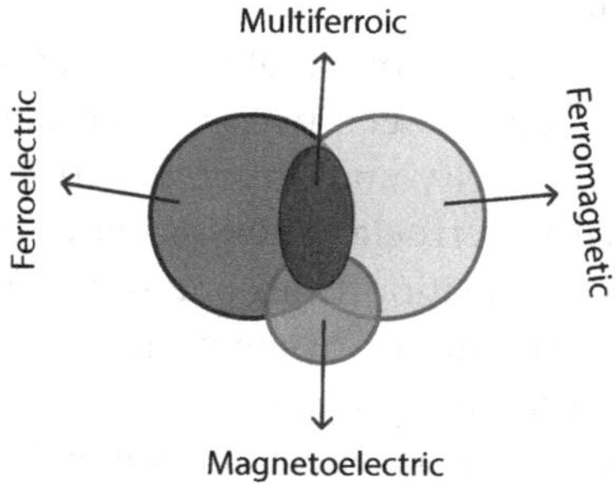

Figure 3.2 The relationship among multiferroic/ferromagnetic/ferroelectric/magnetoelectric materials.

of one another and each has a unique microscopic origin. These materials are important due to their potential use in applications such as spintronics, where the magnetic and electric characteristics being coupled can be harnessed to create new types of electronic devices, e.g., $BiFeO_3$.

ii. Type II Single-Phase Multiferroics
This is a class of materials in which electric polarization is thought to be generated by a specific magnetic order, resulting in an extraordinary magnetoelectric response. Orthorhombic RMnO3 is one such material, where the magnetic ordering of the Mn ions is thought to induce electric polarization. Because of their potential for use in spintronics and other advanced electronic applications, these materials are of great interest.

b) Multiferroic Composites
Multiferroic composites are materials that have both magnetic and electric order states as well as piezo order states. However, composite multiferroics are different from single-phase multiferroics because the phases of the order are physically split from one another. The electric order phase of composites is a dielectric insulator, whereas the magnetic phase can be either a dielectric insulator or a metallic conductor. As a result of the availability of a greater variety of constituent materials, the fabrication of composite multiferroics is made considerably more versatile by this potent characteristic. Composite multiferroics have indirect magneto-electric coupling.

Examples:

$CoFe_2O_4$ in $BaTiO_3$matrix, $BaTiO_3$-$CoFe_2O_4$ composite fiber, $NiFe_2O_4$/PZT laminated layers, etc. [1].

3.3 Emerging Ferroic and Multiferroic Materials

3.3.1 Introduction to Emerging Ferroic and Multiferroic Materials

The development of emerging ferroic and multiferroic materials is an exciting area of research, and these materials have the potential to revolutionize various fields, including electronics, energy harvesting, and computer technologies. These materials have gained significant attention in recent years due to their unique properties, which are not observed in conventional materials. These materials have several advantages over conventional materials, for example, sensitivity to external stimuli, low energy consumption, potential for miniaturization, and low cost. They have the potential to reduce the size and increase the efficiency of electronic devices by combining magnetic and electric properties in a single material.

3.3.2 Examples of Emerging Ferroic and Multiferroic Materials

There are several examples of emerging ferroic/multiferroic materials. Some of the most prominent examples are flexoelectric materials, metamaterials, magnetoelectric materials, perovskite ferroelectrics, hybrid perovskite materials, topological ferroics, multiferroic metal-organic frameworks (MOFs), lead zirconate titanate (PZT), magnetostrictive-piezoelectric composites, etc.

a) Flexoelectric Materials
Flexoelectric materials are a type of material that can generate an electric field in response to a mechanical strain gradient. Examples of flexoelectric materials include ZnO, $BaTiO_3$, and $LiNbO_3$.

Researchers have demonstrated the existence of flexoelectricity in several 2D materials, including graphene, MoS_2, and BN, which could potentially be used in flexible and stretchable electronic devices [5].

b) Magnetoelectric Materials
These are materials that exhibit both magnetic and electric properties and can convert magnetic fields into electrical signals or vice versa. Examples of magnetoelectric materials include $BiFeO_3$, $Pb(Zr,Ti)O_3$ (PZT), and Cr_2O_3 [6].

c) Perovskite Ferroelectrics
Perovskite ferroelectrics are a class of materials with the ABO3 crystal structure and oxygen octahedra that share a corner. Some of the most commonly used piezoelectric and ferroelectric ceramics are in this group. These include strontium titanate (STO), barium titanate (BaTiO3), lead titanate (PbTiO3), and barium strontium titanate (BST), among others. Due to their unique qualities, such as high dielectric constants and high piezoelectric coefficients, these materials have been studied a lot to find out how they could be used in electronics, energy storage, and sensing [7].

d) Topological Ferroics
Topological materials are a class of materials that have unique electronic and magnetic properties due to their non-trivial topology. Recently, researchers

have discovered topological ferroics, which are materials that exhibit both ferroelectricity and non-trivial topology. These materials have potential applications in spintronics and quantum computing [8].

e) Hybrid Perovskite Materials
Hybrid perovskites are a class of materials that consist of both organic and inorganic components and exhibit unique electronic and optical properties. Recently, researchers have discovered ferroelectricity in hybrid perovskites, which has potential applications in optoelectronics and photovoltaics [9].

f) Multiferroic Metal-Organic Frameworks (MOFs)
MOFs are a group of porous materials made up of metal ions and organic ligands. Recently, researchers have discovered multiferroic properties in MOFs, which have potential applications in information storage and spintronics [10].

3.4 Introduction to Advances in Characterization Techniques of Ferroic/Multiferroic Materials

To understand these properties and optimize the performance of these materials, it is essential to develop advanced characterization techniques. In recent years, there have been significant advances in characterization techniques of ferroic and multiferroic materials, which have enabled researchers to study the materials at the atomic and nanoscale levels.

Researchers have come up with and used a wide range of advanced characterization methods to learn more about the properties and behavior of ferroic and multiferroic materials. Techniques like scanning tunneling microscopy (STM) and atomic force microscopy (AFM) have been used for high-resolution imaging and handling of these materials at the nanoscale. X-ray and neutron scattering methods, such as X-ray diffraction (XRD), and polarized neutron scattering (PNS), have been applied to the investigation structure of the crystal and magnetic properties of these materials. Researchers have been able to learn more about the basic physics of these materials and how they could be used in different areas of technology thanks to these advanced characterization methods.

3.4.1 Scanning Probe Microscopy

Scanning probe microscopy techniques including STM and AFM have found widespread usage in nanoscale imaging and help resolve the local structural, electronic, and magnetic properties of ferroic and multiferroic materials. The mechanical and electrical properties of these materials, as well as their surface morphology and topography, can be investigated with these methods [11].

3.4.2 X-Ray Diffraction and Scattering

The crystal structure and magnetic characteristics of ferroic and multiferroic materials have been widely studied using X-ray diffraction (XRD) and small-angle X-ray scattering (SAXS). The crystal structure of these materials can be determined by XRD, while the magnetic domain structure and magnetic ordering can be determined by SAXS [12].

3.4.3 Neutron Scattering

Polarized neutron scattering (PNS) has been used to study the magnetic properties of ferroic and multiferroic materials. Neutrons are highly sensitive to magnetic structures, making them an ideal tool for studying the magnetic ordering in these materials.

3.4.4 Raman Spectroscopy

Raman spectroscopy is incredibly helpful for analyzing the lattice dynamics involved in the phase transitions of ferroic materials, and it should shed some light on the dynamics involved in the phase transitions of multiferroics [13].

3.5 Applications

Ferroic and multiferroic materials have unique characteristics that make them suitable for a variety of uses. However, different uses need different characteristics from a material. For most applications, there are common requirements that must be met ideally at ambient temperatures, such as strong magneto-electric interaction phenomena, easy production using cheap chemical elements and compounds, suitability for mass manufacturing and integration at the chip/wafer level, and the ability to keep working at micro- and nano-scale sizes. For the creation of multiferroic materials and devices to go well, these rules must be strictly followed [14]. Figure 3.3 represents the applications of ferroic/multiferroic materials.

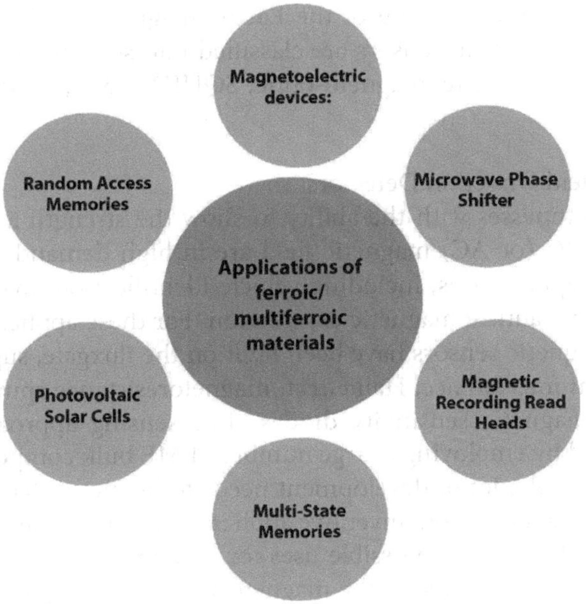

Figure 3.3 Applications of ferroic/multiferroic materials.

3.5.1 Magnetoelectric Devices

Magnetoelectric (ME) materials have the property of coupling magnetic and electric fields. These materials have undergone substantial research due to their potential for low power consumption, high sensitivity, and fast response.

Magnetic Field Sensors

Using the phenomenon of magnetically induced magneto-electric, multiferroic materials may be utilized to construct magnetic field sensors. Without physical contact, magnetic sensors can detect the emergence or modification of magnetic fields in their immediate environment. From these changes, they can infer data regarding characteristics of both the magnetic field and other objects' auxiliary characteristics, e.g., presence, angle, rotation, electrical currents, temperature, speed, and direction. Magnetic sensor output signals require signal processing to be translated into the appropriate parameter.

In addition to other uses, magnetic field sensors are used in biomedical applications, medical imaging, positioning and speed measurement, security systems, metal and weapon detection, imaging using a microscope, cosmic objects, the mapping of the Earth's magnetic field, and geopositioning. Magnetic field sensors are classified into several types, including inductive coil, flux gate, magneto-optic, SQUID, AMR, GMR, and CMR [1].

Magnetic Field Direction Detectors
Magnetic compasses with the ability to show the strength and/or orientation of a DC (or AC) magnetic field are in high demand for a variety of practical applications, including vehicle identification, marine navigation, and aeronautical magnetic exploration. For these applications, many different magnetic sensors have been built on the fluxgate, superconducting quantum interference, Hall effect, magnetoresistance spin valves, and enormous magnetic sensitivity diodes. This sensing approach is easily implemented by employing a large number of ME bulk composites.

There are still a lot of development needs to be met in terms of device design, physical operation, inventive application, compatible modules, etc. There are still numerous possible uses for magnetoelectric devices based on ME bulk materials due to the magnetostriction principle and the coupling mechanism. Large application potential is demonstrated by the likely remarkable qualities of bulk magnetoelectric devices, including a lower volume, lighter weight, and a higher coupling effect [15].

3.5.2 Multiferroic Microwave Phase Shifter

Microwave phase shifters hold significant importance in various fields such as phased array antenna systems, telecommunications, and radar applications. Phase shifters commonly employ semiconductor, ferroelectric, and ferrite materials. Ferroelectric phase shifters have quick electric tunability and use less power than magnetic ferrite phase shifters. However, they are associated with significant loss at frequencies exceeding 5 GHz. The integration of ferrite materials and ferroelectric materials can result in the creation of a multiferroic composite, which can serve as a dual-phase adjustable microwaves system that operates at frequencies exceeding 5 GHz.

3.5.3 Multiferroic Magnetic Recording Read Heads

Ferroic/Multiferroic materials have gained significant attention in recent years due to their unique properties, which allow for the development of new and innovative technologies.

One such application of multiferroic materials is the development of multiferroic magnetic recording read heads. These read heads use a combination of ferroelectric and ferromagnetic materials to create a device that is both sensitive to magnetic fields and capable of producing an electric signal.

The use of multiferroic materials in magnetic recording read heads has several advantages over traditional read heads. It allows for the reduction in the size of the read head, as both the magnetic and electric signals can be generated from the same device. This can lead to the development of higher-density storage devices, which are increasingly in demand as data storage requirements continue to grow.

In addition, the use of multiferroic materials can also lead to improvements in read head sensitivity and speed, as the coupling between the ferroelectric and ferromagnetic orders can lead to a more efficient transfer of signals between the two. This can result in faster read and write times, as well as greater accuracy in data storage and retrieval.

3.5.4 Multi-State Memories and Multiferroic Random Access Memories

A crucial part of computer memories and consumer electronics is random access memory (RAM). The volatility of RAM is contingent upon the technology employed, as it may be either volatile or non-volatile. The term "Volatile RAM" refers to the characteristic of computer chips to retain stored information only as long as a continuous supply of electric power is maintained. Upon deactivation of power, the stored information becomes inaccessible. Random access memory (RAM), despite its volatile nature, provides significantly quicker data retrieval rates compared to traditional non-volatile data storage. In conjunction with the microprocessor, RAM is largely used for the storage and execution of operating system code, operating software, and any other related memory processes.

The proposition of ferroelectric-based non-volatile RAM technology (FRAM) was initially introduced by a graduate student from MIT in 1952. Non-volatile RAM could replace volatile RAM to avoid data loss and speed up computer startup without having to wait for the software to load. Magneto-resistive or magnetic random access memory (MRAM) is another non-volatile RAM method that uses magnetic states and the magneto-resistance property to store data bits in multilayers.

The coexistence of magnetic order and electric polarization in a multiferroic material has the potential to integrate the benefits of both MRAM and FRAM and a permanent random access memory with the capacity to store numerous memory states is created.

3.5.5 Photovoltaic Multiferroic Solar Cells

Solar radiation is widely considered the most viable option for renewable energy acquisition. Electric current can only be generated by photons that possess energy greater than the band gap. Therefore, the engineering of semiconductors possessing a narrow band gap results in greater absorption of photons and subsequently generates a higher current, albeit with a lower voltage output. On the contrary, semiconductors possessing a significant band gap yield greater voltages; however, the current is restricted due to a substantial proportion of unabsorbed solar photons. Today's semiconductors' solar power technology can capture a minor proportion of the solar spectrum, which is generally limited to visible photons of approximately 2 eV. This phenomenon is alternatively referred to as the solar power dilemma.

The dilemma of photovoltaic semiconductor solar cells can be avoided by employing the bulk photovoltaic effect, an alternative technology. The aforementioned phenomenon is observed in ferroelectric substances, specifically in thin films made of perovskite ferroelectric oxides such as LiNbO3, PZT, and BaTiO3. Photons having energies that are greater than the band gap between the dielectric and ferroelectric structures are absorbed in a polarized ferroelectric crystal with internal polarization P, resulting in the creation of holes and electrons in the majority of the ferroelectric crystal.

In ferroelectric materials, the internal polarization generates an internal electric field, causing charge separation within the crystal. This leads to a photovoltaic effect that is fundamentally unique compared to the typical interfacial influence seen in semiconductors. Ferroelectrics exhibit a bulk effect known as the photovoltaic effect, with the internal field E originating from the crystal's internal remanent polarization. Thus, the internal field exists throughout rather than being restricted to a tiny interfacial region, the ferroelectric crystal's bulk portion [1].

3.6 Challenges and Future Directions for Ferroic and Multiferroic Materials

There are still some challenges of ferroic and multiferroic materials that need to be addressed to realize their full potential. Some of the major challenges and future directions for ferroic and multiferroic materials are as follows.

3.6.1 Stability and Reliability

One of the major challenges facing the development of ferroic and multiferroic materials is the issue of stability and reliability. These materials frequently have complex microstructures and domain structures that respond to environmental conditions such as temperature, humidity, and mechanical stress. As a result, the performance and dependability of certain materials may deteriorate over time due to the progressive erosion of their properties.

Recent research has focused on the stability and reliability of these materials. One method is to develop innovative materials with greater intrinsic stability, such as those with high Curie or Néel temperatures. Rare-earth transition-metal alloys have been explored for a variety of reasons, including their high magnetic anisotropy and thermal stability. The usage of oxide materials with strong oxygen ion mobility improves ferroelectric domain stability even further [16].

3.6.2 Integration with Existing Technologies

Creating interfaces and device architectures that work well with ferroic and multiferroic materials is one of the biggest problems that need to be solved before these materials can be used in current technologies. These materials often have unique qualities that require them to be processed and put together in a certain way.

Recent studies have focused on developing new strategies for their integration with existing technologies. The simplest method to do this is to look into the use of hybrid device architectures, which blend the unique properties of these materials with the functions of traditional materials. For example, ferroelectric materials in non-volatile memory devices have been investigated like FeRAMs and RRAMs, which offer improved performance and energy efficiency compared to conventional memory technologies.

Another approach is to accomplish new methods, like 3D printing and making microelectromechanical systems (MEMS), which are capable of making complex device architectures and interfaces. For example, researchers have explored the use of inkjet printing to fabricate ferroelectric thin films on flexible substrates, enabling the development of flexible and wearable electronics [17].

In addition to developing new integration techniques, future research directions will also need to address the issue of compatibility with existing manufacturing processes and standards. For example, the use of ferroelectric materials in existing semiconductor technologies will require the development of new deposition and processing techniques that are compatible with current semiconductor manufacturing processes. Furthermore, the use of multiferroic materials in magnetic storage and spintronics devices will require the development of new fabrication techniques and device architectures that can take advantage of their unique properties [18].

3.6.3 Scalability

The scalability of ferroic and multiferroic materials is a critical challenge for their integration into practical applications. These materials often exhibit unique properties that are promising for a range of technological applications, but their scalability is limited by several factors such as material synthesis, processing, and characterization.

Recent research has focused on developing scalable synthesis and processing techniques for ferroic and multiferroic materials. One approach is to explore the use of high-throughput screening techniques and computational materials design to accelerate the discovery and optimization of these materials. For example, researchers have used combinatorial synthesis techniques to rapidly screen a large number of materials and identify promising candidates for ferroelectric and multiferroic applications.

Another approach is to explore the use of scalable processing techniques, such as roll-to-roll deposition and printing, to fabricate large-area films and devices. For example, researchers have explored the use of inkjet printing and aerosol jet printing to fabricate ferroelectric thin films on flexible substrates, enabling the development of flexible and stretchable electronics.

In addition to developing scalable synthesis and processing techniques, future research directions will also need to address the issue of characterization and standardization. As these materials are scaled up, it becomes increasingly important to develop standardized characterization techniques that can accurately measure their properties and ensure their reproducibility. Furthermore, the integration of these materials into practical applications will require the development of standardization and quality control measures to ensure their performance and reliability [19].

3.6.4 Novel Applications

The unique properties of these materials have led to a wide range of novel applications in areas such as energy conversion, data storage, and sensing.

However, there remain many challenges and opportunities for the development of new applications for these materials.

Nowadays, ferroelectric materials have been used for energy harvesting applications. Ferroelectric materials can convert mechanical energy into electrical energy through the piezoelectric effect, making them promising candidates for a range of applications, such as powering small electronic devices and sensors. Researchers are investigating the use of novel ferroelectric materials, such as lead-free perovskites and thin films, which offer improved performance and reliability compared to traditional materials [20].

Another area of research is the use of multiferroic materials for magnetic and electric field sensing. These materials can exhibit strong coupling between electric and magnetic properties, enabling the development of novel sensors that can detect both electric and magnetic fields simultaneously. For example, researchers have demonstrated the use of multiferroic materials for magnetic field sensing in high-temperature environments, which could have important applications in aerospace and industrial settings [21].

3.7 Integration of Ferroic and Multiferroic Materials into Current Technology

3.7.1 Integration of Multiferroic Materials into Memory Devices

Multiferroic materials exhibit strong coupling between magnetic and ferroelectric properties, making them promising candidates for use in novel memory devices that combine both magnetic and electric switching.

One approach to integrating multiferroic materials into memory devices is through the use of spin-transfer torque (STT) switching. STT memory devices use the spin of electrons to control the magnetization of a ferromagnetic layer, allowing for non-volatile data storage. Researchers have investigated the use of multiferroic materials as the reference layer in STT devices, which can offer improved energy efficiency and reduced write error rates compared to traditional devices [22].

Another approach is the use of magnetoelectric (ME) coupling in multiferroic memory devices. ME coupling enables the control of magnetic properties through an applied electric field and vice versa, allowing for electric field control of magnetization and readout of stored data. Researchers have investigated the use of multiferroic materials such as BiFeO3 and TbMnO3 in ME memory devices, which have shown promising results in terms of high storage density, low power consumption, and quick operation.

The integration of multiferroic materials into memory devices is still in the early stages of development, and several challenges need to be overcome, such as improving the scalability and reliability of these devices. However, with continued research and development, it is expected that multiferroic memory devices will play an important role in next-generation memory technologies [23].

3.7.2 Integration of Ferroelectric Materials into Energy Harvesting Devices

The integration of multiferroic materials into energy harvesting devices has the potential to significantly improve their performance and efficiency. Multiferroic materials can convert mechanical energy into electrical energy through the magnetostrictive and piezoelectric effects, making them promising candidates for use in energy harvesting devices.

One approach to integrating multiferroic materials into energy harvesting devices is through the use of magnetoelectric (ME) composites. ME composites consist of a piezoelectric material and a magnetostrictive material, which are mechanically coupled to each other through a thin interlayer. When subjected to mechanical stress or vibration, the piezoelectric material generates an electrical voltage, which can be enhanced by the magnetic field induced by the magnetostrictive material [24].

Researchers have investigated the use of multiferroic materials such as Terfenol-D and Pb(Zr,Ti)O3 (PZT) in ME composites for energy harvesting applications. These materials have shown promising results in terms of high power density, broad frequency response, and low energy consumption.

The integration of multiferroic materials into energy harvesting devices is still in the early stages of development, and several challenges need to be addressed, such as improving the efficiency and reliability of these devices. However, with continued research and development, it is expected that multiferroic energy harvesting devices will play an important role in meeting the growing demand for sustainable and efficient energy sources [25].

3.7.3 Integration of Ferroelectric Materials into Sensors

With their remarkable properties, multiferroic materials have captured the attention of scientists worldwide. In recent years, the integration of multiferroics into sensors has shown tremendous promise as one of the potential applications of these materials. Using the properties of multiferroic materials, researchers intend to create sensors with enhanced sensitivity, precision, and energy efficiency for a variety of sensing applications.

Multiferroic sensors are ideal for biomedical sensing, environmental monitoring, and industrial process control because they can detect magnetic fields and electric charges with high precision and accuracy. In response to the increasing interest in this field, researchers are investigating various methods for incorporating multiferroic materials into sensors, such as the use of magneto-electric (ME) sensors and multiferroic nanosensors. Multiferroic sensors are anticipated to revolutionize the sensing industry, with improvements in sensitivity, selectivity, and overall performance, despite the challenges that lay ahead [26].

3.7.4 Integration of Ferromagnetic Materials into Spintronic Devices

Because of their ability to manipulate magnetism via electric fields, multiferroic materials have received interest in the field of spintronics. This property, known as magnetoelectric coupling, has allowed for the incorporation of multiferroic materials into spintronic devices, resulting in the development of innovative devices with enhanced functionality.

The use of multiferroic tunnel junctions (MTJs) for magnetic memory is one example of the use of multiferroic materials in spintronic devices. MTJs are made up of two ferromagnetic layers separated by a thin multiferroic layer that acts as a tunnel barrier. Using multiferroic tunnel barriers can result in considerable improvements in MTJ performance, such as lower power consumption and greater thermal stability. For example, using a BiFeO3 tunnel barrier in an MTJ can result in a tenfold reduction in power consumption and a fourfold gain in thermal stability over traditional MTJs.

Furthermore, the incorporation of multiferroic materials into spintronic devices has permitted the creation of innovative devices such as multiferroic spin valves (MSVs). MSVs are made up of a multiferroic layer sandwiched between two ferromagnetic layers, and the multiferroic layer can be utilized to regulate the magnetization of the ferromagnetic layers using electric fields. According to research, the utilization of MSVs can lead to the development of devices with enhanced functionality, such as low-power magnetic memory and reconfigurable logic [27].

3.7.5 Integration of Multiferroic Materials into Microwave Devices

Multiferroic materials are regarded as good candidates for the construction of microwave devices such as tunable resonators, filters, and phase shifters due to their ability to adjust the electrical and magnetic properties of the material with an external electric field. One example of a multiferroic material used in microwave devices is bismuth ferrite (BiFeO3), which has high ferroelectric polarization and magnetization, making it an ideal material for use in the microwave frequency range.

Recent research has shown that the inclusion of multiferroic materials into microwave devices can greatly increase their performance. For example, researchers have reported the development of a tunable bandpass filter using a multiferroic BiFeO3 thin film. The filter exhibited a high-quality factor and a wide tuning range with low insertion loss, indicating the excellent performance of the device. Another example of the integration of multiferroic materials is the development of a phase shifter using a lead zirconate titanate (PZT) thin film. The device showed high tunability and low loss in the microwave frequency range, making it an ideal candidate for use in communication systems.

However, challenges such as processing techniques and compatibility with existing microwave devices need to be addressed to fully integrate multiferroic materials into microwave devices. Research is ongoing to find ways to optimize the properties of multiferroic materials for use in microwave devices and to develop new processing techniques to achieve better integration. Overall, the incorporation of multiferroic materials into microwave devices has the potential to revolutionize microwave technology, resulting in the development of more efficient and high-performance devices for a wide range of applications [28].

3.8 Conclusion

In conclusion, ferroic and multiferroic materials have the potential to revolutionize modern technology with their unique properties such as ferroelectricity, ferromagnetism, and magnetoelectricity. This book chapter has highlighted the advances made in this field, including the discovery and characterization of new materials, advancements in characterization techniques, and the development of new applications such as magnetoelectric devices, spintronics, sensors and actuators, energy harvesting, and data storage.

However, there are still several challenges that need to be addressed, such as scalability and sustainability. The integration of ferroic and multiferroic materials into current technology also presents new opportunities and challenges.

By addressing these challenges and pursuing these opportunities, the field of ferroic and multiferroic. Shortly, exciting advancements in the field of smart materials and structures might be anticipated.

References

1. Vopson, M.M., Fundamentals of multiferroic materials and their possible applications. *Crit. Rev. Solid State Mater. Sci.*, 40, 223, 2015.
2. Gupta, S., *Introduction to ferroelectrics and related materials*, in *Ferroelectric Materials for Energy Harvesting and Storage*. p. 1-41, Woodhead Publishing Series, Elsevier, 2021.
3. Marghussian, V., Magnetic properties of nano-glass ceramics, in: *Nano-Glass Ceramics*, V.B.T.-N.-G.C. Marghussian, (Ed.), pp. 181–223, William Andrew Publishing, Oxford, 2015.
4. Wadhawan, V.K., Ferroic materials: A primer. *Resonance*, 7, 15, 2002.
5. Shu, L., Liang, R., Rao, Z., Fei, L., Ke, S., Wang, Y., Flexoelectric materials and their related applications: A focused review. *J. Adv. Ceram.*, 8, 153, 2019.
6. Bichurin, M., Petrov, V., Priya, S., Bhalla, A., Multiferroic magnetoelectric composites and their applications. *Adv. Condens. Matter Phys.*, 2012, 129794, 2012.
7. Ranjan Sahu, D., Introductory chapter: Basics and importance of ferroelectric materials for applications, in: *Multifunctional Ferroelectric Materials*. Ch. 1, D.R. Sahu, (Ed.), IntechOpen, Rijeka, 2021.
8. Stamps, R.L. and H. Schultheib, *Recent Advances in Topological Ferroics and Their Dynamics*.
9. Chen, Q., De Marco, N., Yang, Y., Song, T.B., Chen, C.C., Zhao, H., Hong, Z., Zhou, H., Yang, Y., Under the spotlight: The organic-inorganic hybrid halide perovskite for optoelectronic applications. *Nano Today*, 10, 355, 2015.
10. Hou, J., Ashling, C.W., Collins, S.M., Krajnc, A., Zhou, C., Longley, L., Johnstone, D.N., Chater, P.A., Li, S., Coulet, M.V., Llewellyn, P.L., Coudert, F.X., Keen, D.A., Midgley, P.A., Mali, G., Chen, V., Bennett, T.D., Metal-organic framework crystal-glass composites. *Nat. Commun.*, 10, 41, 2019.
11. Guo, H. and J. Zhang, *Scanning probe microscopy of epitaxial oxide thin films*. Epitaxial Growth of Complex Metal Oxides, p. 331-367, Woodhead Publishing Series, 2022..
12. Devashankar, S., Mariappan, L., Sureshkumar, P., Rathnakumari, M., Growth and characterization of tetramethyl ammonium tetrachloro zincate II: A ferroic crystal. *J. Cryst. Growth*, 311, 4207, 2009.
13. Ghosh, A., Sahu, J.R., Bhat, S.V., Rao, C.N.R., A Raman study of multiferroic $LuMnO_3$. *Solid State Sci.*, 11, 1639, 2009.

14. Kumar, M., Shankar, S., Kumar, A., Anshul, A., Jayasimhadri, M., Thakur, O.P., Progress in multiferroic and magnetoelectric materials: Applications, opportunities and challenges. *J. Mater. Sci. Mater. Electron.*, 31, 19487, 2020.
15. Mao, Q., Wu, J., Hu, Z., Xu, Y., Du, Y., Hao, Y., Guan, M., Wang, C., Wang, Z., Zhou, Z., Dong, S., Ren, W., Liu, M., Jiang, Z., Magnetoelectric devices based on magnetoelectric bulk composites. *J. Mater. Chem. C*, 9, 5594, 2021.
16. Sheng, K., Yuan, H.K., Wang, Z.Y., Intrinsic ferromagnetism in 2D h-CrC semiconductors with strong magnetic anisotropy and high Curie temperatures. *J. Mater. Chem. C*, 9, 16495, 2021.
17. Huang, Y., Nie, M., Li, B., Yin, K., Sun, L., An integrated MEMS device for *in-situ* four-probe electro-mechanical characterization of PT nanobeam, in: *Proceedings of the IEEE International Conference on Micro Electro Mechanical Systems (MEMS)*, pp. 771–774, 2023.
18. Luo, Z.D., Yang, M.M., Liu, Y., Alexe, M., Emerging opportunities for 2D semiconductor/ferroelectric transistor-structure devices. *Adv. Mater.*, 33, 2005620, 2021.
19. Zi, W., Jin, Z., Liu, S., Xu, B., Flexible perovskite solar cells based on green, continuous roll-to-roll printing technology. *J. Energy Chem.*, 27, 971, 2018.
20. Shi, Z., Guo, J., Chen, Y., Li, Q., Pan, Y., Zhang, H., Xia, Y., Huang, W., Lead-free organic–inorganic hybrid perovskites for photovoltaic applications: Recent advances and perspectives. *Adv. Mater.*, 29, 1605005, 2017.
21. Halpin, J. and Keeney, L., Naturally layered aurivillius phases: Flexible scaffolds for the design of multiferroic materials. *OAJ Mater. Devices*, 5, 49, 2021.
22. Tang, D.D. and Pai, C., Spin-Torque-Transfer (STT) MRAM engineering, in: *Magnetic Memory Technology*, pp. 151–203, 2020.
23. Cho, J.-H. and Jo, W., Progress in the development of single-phase magnetoelectric multiferroic oxides. *Ceramist*, 24, 228, 2021.
24. Chen, G., Jin, Z., Chen, J., A review: Magneto-optical sensor based on magnetostrictive materials and magneto-optical material. *Sens. Actuators Rep.*, 5, 100152, 2023.
25. Sharma, S., Mishra, S.S., Kumar, R., Yadav, R.M., Recent progress on polyvinylidene difluoride-based nanocomposites: Applications in energy harvesting and sensing. *New J. Chem.*, 46, 18613, 2022.
26. Singh, S. and Melnik, R., Coupled multiphysics modelling of sensors for chemical, biomedical, and environmental applications with focus on smart materials and low-dimensional nanostructures. *Chemosensors*, 10, 157, 2022.
27. Jose, J.K. and Balakrishnan, R., Multiferroics for spintronic applications, in: *Nanotechnology in Electronics*, pp. 301–316, 2023.
28. Tomina, E.V., Perov, N.S., Mittova, I.Y., Alekhina, Y.A., Stekleneva, O.V., Kurkin, N.A., Microwave synthesis and magnetic properties of bismuth ferrite nanopowder doped with cobalt. *Russ. Chem. Bull.*, 69, 941, 2020.

4

Basic Principles and Measurement Techniques of Electrocaloric Effect in Ferroelectric Materials

Madhushree P. [1], N. S. Kiran Kumar[1], P. Saidi Reddy[2] and K. C. Sekhar[1]*

[1]*Department of Physics, School of Basic and Applied Sciences, Central University of Tamil Nadu, Thiruvarur, India*

[2]*Department of Applied Sciences, National Institute of Technology, Goa, Farmagudi, Ponda, Goa, India*

Abstract

The development of innovative refrigeration technologies becomes essential for the reduction of hazardous gas emissions in conventional cooling systems. In this context, electrocaloric effect (ECE)-based cooling technology seems to be a promising alternative to the conventional vapor compression-based cooling technology due to its unique advantages such as high efficiency, easy miniaturization and integration, eco-friendly, lightweight, and low cost. Therefore, in this chapter, first we discuss the working principle, a brief history, the mechanism, different theoretical approaches, and different direct and indirect measurement techniques for quantifying ECE. Further, we focus on the recent developments and different strategies that are adopted to enhance ECE in different lead-based and lead-free ferroelectrics that include barium titanate-based, sodium potassium niobate-based, and nano-scale binary oxide (ZrO_2, HfO_2)-based materials.

Keywords: Electrocaloric effect, ferroelectric materials, perovskites, solid-state refrigeration, binary oxides

**Corresponding author*: sekhar.koppole@gmail.com

Inamuddin, Tariq Altalhi and Mohammad Abu Jafar Mazumder (eds.) Ferroic Materials-Based Technologies, (55–98) © 2024 Scrivener Publishing LLC

4.1 Introduction

Over the decades, classical vapor compression-based technology is widely used as cooling technology in refrigerators, air conditioners, etc. In this technology, the gases like hydrochlorofluorocarbons, freon, and chlorofluorocarbons are commonly used as refrigerants, which are deleterious to the environment. The emission of these gases into the atmosphere is a threat to life on Earth. Hence, the development of innovative refrigeration technologies has become a necessity in order to reduce the emission of hazardous gases and produce more efficient cooling devices [1–4]. In this context, the magnetic and electrocaloric (EC) cooling techniques based on magnetocaloric effect (MCE) and electrocaloric effect (ECE), respectively, seem to be promising alternatives to conventional vapor compression-based cooling techniques [1, 2, 4]. The phenomenon of temperature variation in a magnetic material when exposed to a magnetic field (MF) is called MCE [1, 2]. When an external MF is applied to a ferromagnetic material, the magnetic dipoles get oriented in the direction of the field. This orientation of dipoles lowers the energy in terms of entropy and increases the material's temperature. Conversely, upon the removal of MF, magnetic dipoles return to the initial state, thus increasing entropy and consequently lowering the material's temperature. MCE-based cooling systems have some advantages such as higher efficiency, low cost, environment friendly, etc., over classical vapor compressor systems [1]. However, there is a demand for giant permanent magnets to trigger the desired changes in MF, which is the main drawback for the magnetic refrigeration technology, because of the high cost of permanent magnets and the magnetocaloric materials [1, 2].

Recent studies in cooling technology have shown that EC cooling has the potential to overcome these limitations and to create highly efficient, ecological, budget-friendly cooling devices [2, 4, 5]. The ECE in ferroelectric (FE) materials can be expressed as the isothermal entropy (ΔS) change and adiabatic temperature (ΔT) change by the influence of an external electric field (EF), which is an electrical analogue of MCE [1, 3, 4, 6–9]. Further, ECE-based cooling systems are highly efficient like MCE-based cooling systems, but are also compact in size, lightweight, and easily integrable in devices. Therefore, EC materials are actively considered for the future solid-state refrigeration devices [1, 2, 4].

Back in 1930s, Kobeko and Kurtschatov experimentally investigated ECE in Rochelle salt, but the quantification was not reported in their studies [1, 10–12]. Later, many anomalies were observed in the published results due to lack of standardized transliteration of Cyrillic letters and many of the investigations did not come into light before the influx of computer search engines. In 2011, J.F. Scott has discussed the detailed anomalies in the citation [12]. The discovery of a giant ECE in anti-ferroelectric $PbZr_{0.95}Ti_{0.05}O_3$ (PZT) thin film by Mischenko *et al.* in 2006 caught the attention of the EC refrigeration field. The ECE with ΔT of 12 K was observed at temperature T_{EC} = 226°C, above the phase transition temperature of 222°C for a field of ΔE = 480 kV/cm in the ~350 nm PZT thin film [13]. In 2008, Zhang *et al.* from Penn State announced a giant ECE in FE P(VDF-TrFE) polymer thin films [14]. In these two innovative works, high fields were achieved with low operating voltages due to lower thickness of films, and thus, a giant EC response was noted. Further, continuous efforts have been made to create a material that can satisfy the practical requirements of solid-state refrigeration devices.

Therefore, in this chapter, we first discuss the principle, a brief history, theoretical approaches, and quantifying techniques of ECE, its applications and then the recent advances in electrocaloric materials that includes lead (Pb) based, Pb-free ferroelectric and nanoscale binary oxides in detail.

4.2 Electrocaloric Effect (ECE)

The physical phenomenon ECE is commonly found in the FE materials, and it is manifested in terms of two parameters, ΔT or ΔS, which arise due to variation in external EF [5]. Recently, the ECE has emerged as an appealing phenomenon with potential applications in a wide range of fields, including new generation cooling and heating systems that are more environmental friendly and efficient [15, 16].

4.2.1 Brief History of ECE

In 1878, W. Thomson theoretically predicted the ECE in his work on thermo-elastic, thermo-magnetic, and pyroelectric properties of matter. Based on the theory of pyroelectricity, he proposed that a pyroelectric material experiences a heating or cooling effect when exposed to an EF [17]. After 50 years, Kobeko and Kurtschatov in 1930 first observed EC response in Rochelle salt without quantification of temperature change [1, 9–11]. In 1943, Hautzenlaub quantitatively measured the ECE in Rochelle salt [1]. Later in 1963, Wiseman and Kuebler did the measurements in Rochelle salt and obtained a ΔT of 0.0036 K for a field of 1.4 kV/cm at temperature

22.2°C [12, 18]. This temperature change was insufficient for the practical applications. Thus, the scientific community was forced by the industry to develop new materials that can be used for applications in the real world. As soon as the innovation of superconductivity and its application started to blow up in the science field, the realm of ECE really began to take off [1]. Later in 1977, Radebaugh discussed the feasibility of EC refrigeration in the 4 K to 15 K temperature range. EC refrigeration was studied in the low-temperature limit, where the refrigerator absorbed heat from a load (4 K) and transferred to a reservoir (15 K) [12, 19]. The refrigerator was developed and tested based on $SrTiO_3$ and $KTaO_3$. The largest observed ΔT in $SrTiO_3$ is 0.3 K at 10 K under an EF of 20 kV/cm. In $KTaO_3$ single crystal (SC), a ΔT of 0.2 K at 10 K was observed when EF was 15.26 kV/cm [19].

During this period, only few research groups have dedicated their research to solid-state ECE-based cooling systems due to the low ΔT [1]. This attitude towards the ECE-based cooling system has changed in 2006 after reporting a huge ΔT of nearly 12 K in $PbZr_{0.95}Ti_{0.05}O_3$ through an indirect method [13]. Later in 2010, the first approach has been made to measure the ECE in a thin film via a direct measurement method [20]. Figure 4.1 shows the timeline of the major events and innovation of ECE [1].

4.2.2 Working Principle

Under the action of EF, the dipolar state in a polar material becomes more ordered from a less ordered state. Also, on removing the field, the dipolar state returns to the less ordered state [21]. Although the microscopic mechanism of ECE is still not completely known to the scientific community, on the macroscopic scale, ECE is explained based on the change in entropy between two polar states under the change in EF [1, 7, 21]. Figure 4.2 schematically represents an EC cooling cycle. *A* is the initial state, where the temperature is T_1, which is taken as room temperature (RT), and the field is zero ($E_1 = 0$). The system has large entropy and the dipoles are randomly oriented.

- Step 1 (A →B): The system is exposed to an EF. The dipoles reorient along the direction of the field. There is no heat transfer to the surrounding during the transition from A to B, and this transition is an adiabatic process. Since $S(T_1, E_1) = S(T_2, E_2)$, the entropy change is zero. The temperature increases by an amount ΔT ($T_2 = T_1 + \Delta T$).

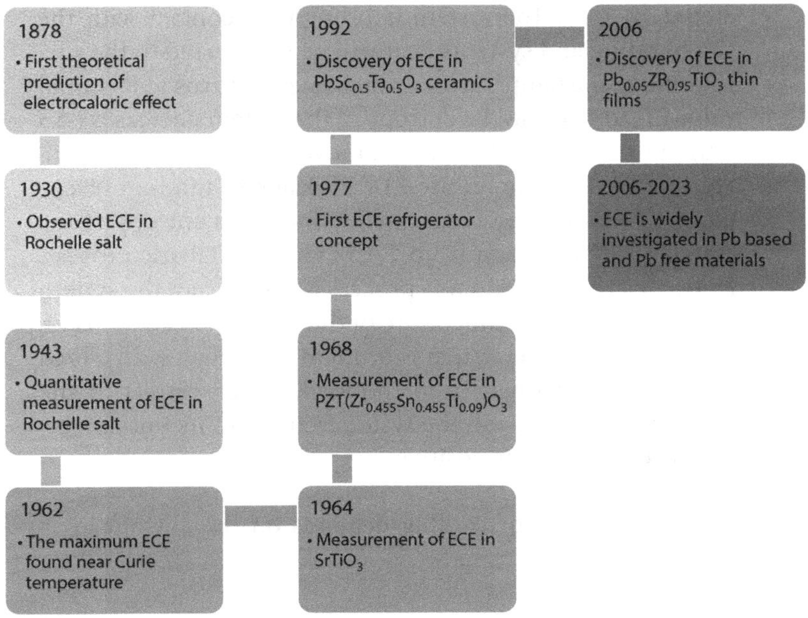

Figure 4.1 Timeline of innovations/breakthrough in electrocaloric effect [1].

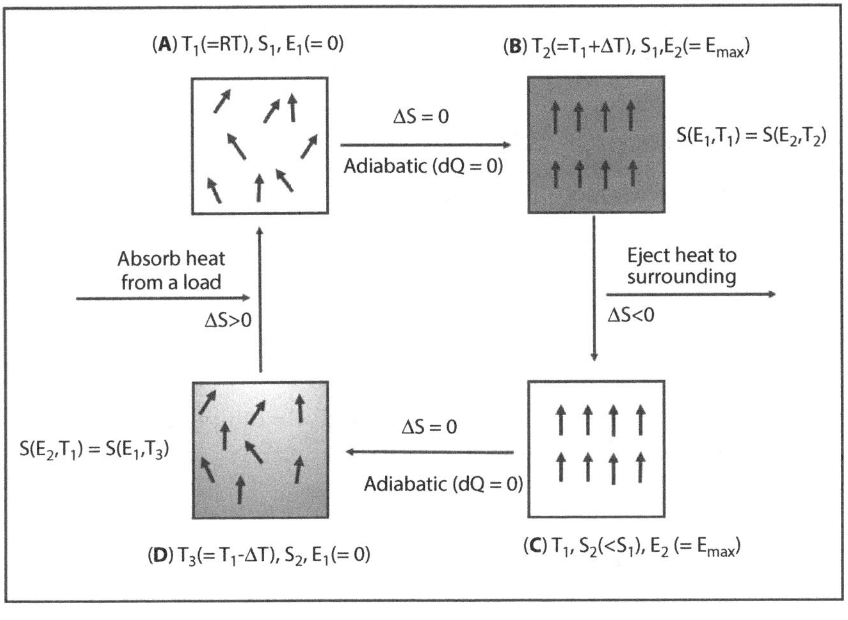

Figure 4.2 Schematic representation of the EC heating/cooling cycle [1,21].

- Step 2 (B →C): The system is brought in contact with the surrounding and system's heat will be transferred to the surrounding. The temperature of the system returns to its initial value (T_2 to T_1), and the entropy of the system decreases ($\Delta S < 0$) in this process. The system moves from state B to C.
- Step 3 (C →D): The removal of EF leads to random orientation of dipoles and consequently increases entropy of the system. The transition from C to D is an adiabatic process. Hence heat required in this process is taken from the system and it reduces the temperature of the system ($T_3 = T_1 - \Delta T$).
- Step 4 (D →A): The system is brought in contact with a load and heat is absorbed from the load, which results in the cooling of the load and the system returns to its initial temperature. Here, ΔS is positive.

Hence, the EC heating or cooling depends on the entropy change as shown in Figure 4.2 [21].

4.2.3 Theory

ECE is established as a thermodynamic phenomenon and the conventional thermodynamic relations guide through the explanation of the same. In general, two approaches such as Maxwell and Landau phenomenological approaches are proposed to describe ECE [7].

4.2.3.1 *Maxwell Approach*

In this approach, Maxwell thermodynamic relations are used to calculate the ECE in dielectric materials. The Gibbs free energy (G) in differential form is [7]

$$dG = -SdT - \epsilon_i X_i - P_i dE_i \tag{4.1}$$

where S = entropy, E = electric field, ϵ = strain , T = temperature, X = stress, and P = dielectric polarization.

From Equation 4.1,

$$S = -\left(\frac{\partial G}{\partial T}\right)_{X,P} \quad \epsilon = -\left(\frac{\partial G}{\partial X_i}\right)_{T,P} \quad P_i = -\left(\frac{\partial G}{\partial E_i}\right)_{T,X} \tag{4.2}$$

The standard Maxwell relation for variable pairs (*S*,*T*) and (*P*,*E*) is

$$\left(\frac{\partial S}{\partial E_i}\right)_{T,X} = \left(\frac{\partial P_i}{\partial T}\right)_{E,X} \tag{4.3}$$

The change in entropy during the adiabatic process at constant stress *X* is

$$dS = \left(\frac{\partial S}{\partial E}\right)_T dE + \left(\frac{\partial S}{\partial T}\right)_E dT = 0 \tag{4.4}$$

Equation 4.4 can be re-written as follows:

$$\left(\frac{\partial T}{\partial E}\right)_S = \frac{T}{C_E}\left(\frac{\partial P}{\partial T}\right)_E = \frac{T}{C_E} P_E \tag{4.5}$$

Here, $C_E = \rho c_E$ = heat capacity per unit volume $\left(\mathrm{JK^{-1}m^{-3}}\right)$ and $P_E = \left(\frac{\partial P}{\partial T}\right)_E$ = pyroelectric coefficient at constant *E*.

ΔT can be calculated by integrating Equation 4.5 as given below:

$$\Delta T = -\int_{E1}^{E_2} \left(\frac{T(E)}{C_E(T)}\right)\left(\frac{\partial P}{\partial T}\right)_E dE \tag{4.6}$$

To estimate ΔT using Equation 4.6, the experimental values of temperature and EF-dependent $C_E(T)$ and $P(E,T)$ are used. This procedure for the estimation of ΔT through the Maxwell approach is known as the indirect method [7].

4.2.3.2 *Landau Phenomenological Approach*

The EC materials typically consist of dipoles and weakly polarizable lattices [7]. Therefore, the entropy of a system $S(E,T)$ is the sum of the entropy of weakly polarizable lattice, ($S_l(T)$) and that of dipoles ($S_d(E,T)$) as follows

$$S(E,T) = S_l(T) + S_d(E,T) \tag{4.7}$$

For an adiabatic process, the overall change in entropy is zero. Hence, Equation 4.7 implies

$$\Delta S(E,T) = \Delta S_l(T) + \Delta S_d(E,T) = 0 \tag{4.8}$$

Assuming that $S_l(T)$ is independent of E, the entropy change, $\Delta S_l(T)$ from T_1 to T_2 can be written as,

$$\Delta S_l = \int_{T_1}^{T_2} \frac{C_l(T)}{T} dT \cong C_l(T_1) log\left(\frac{T_2}{T_1}\right) \tag{4.9}$$

where $C_l(T)$ = lattice heat capacity. $\Delta S_d(E,T)$ can be expressed as the difference of entropy between final and initial points (E_1, T_1) and (E_2, T_2).

By substituting Equation 4.9 in Equation 4.8, we get

$$\left[S_d(E_2,T_2) - S_d(E_1,T_1)\right] + C_l(T_1) log\left(\frac{T_2}{T_1}\right) = 0 \tag{4.10}$$

Equation 4.10 can be re-written as follows:

$$T_2 = T_1 exp\left[\frac{S_d(E_1,T_1) - S_d(E_2,T_2)}{C_l(T_1)}\right] \tag{4.11}$$

Landau free energy of the dipolar subsystem is given by

$$F = F_0 + \frac{1}{2}aP^2 + \frac{1}{4}bP^4 + \frac{1}{6}cP^6 + \ldots - EP \tag{4.12}$$

where $P = P(E,T)$ is the field and temperature-dependent dielectric polarization and F_0 is the field-independent configurational part.

a is a temperature-dependent parameter and it is related to the system under consideration. Other $b, c, d, ..$ parameters are assumed to be independent of temperature [7].

$S_d(E,T)$ is calculated from Landau free energy. In equilibrium $\left(\frac{\partial F}{\partial P}\right)_T = 0$, so S_d can be written as

$$S_d\left(E,T\right) = -\left(\frac{\partial F}{\partial T}\right)_E \tag{4.13}$$

$$S_d = S_0 + S_1\left(P\right) \tag{4.14}$$

where S_0 = configurational entropy of dipoles and it is given by $S_0 = -\left(\frac{\partial F_0}{\partial T}\right)$ and $S_1\left(P\right)$ is given by

$$S_1\left(P\right) = -\frac{1}{2}a_1P^2 - \frac{1}{4}b_1P^4 - \frac{1}{6}c_1P^6 - \ldots \tag{4.15}$$

where $a_1 = \left(\frac{da}{dT}\right), b_1 = \left(\frac{db}{dT}\right), c_1 = \left(\frac{dc}{dT}\right), \ldots$

since b, c, ... are temperature independent and b_1, c_1, ... become zero. Therefore, for $E_2 = E$ and $E_1 = 0$, T_2 can be written as

$$T_2 = T_1 exp\left\{\frac{1}{2C_l\left(T_1\right)}\left[a_1\left(T_2\right)P^2\left(E,T_2\right) - a_1\left(T_1\right)P^2\left(0,T_1\right)\right]\right\} \tag{4.16}$$

The solution of Equation 4.15 can be obtained by iteration or numerical method. When $\Delta T = T_2 - T_1$, ΔT is much smaller than T_1 and T_2, i.e., $\Delta T \ll T_1$ and $\Delta T \ll T_2$, we can write $T_1 = T, T_2 = T + \Delta T$. Equation 4.16 becomes

$$\Delta T \cong \frac{T}{2C_l\left(T\right)}\left[a_1\left(T+\Delta T\right)P^2\left(E,T+\Delta T\right) - a_1\left(T\right)P^2\left(0,T\right)\right] \tag{4.17}$$

Equation 4.17 is a self-consistent equation for ΔT.

4.3 Direct and Indirect Measurement Techniques

A concise review of direct and indirect techniques to measure ECE is provided in this section. Various direct methods are widely investigated by researchers in past years, such as differential scanning calorimetry, scanning thermal microscopy, fast infrared (IR) photometry, modified high-resolution calorimetry, thermocouple, thermometer, and specifically designed calorimeters [22–25]. Maxwell's thermodynamic relation-based indirect method is also included.

4.3.1 Direct Methods for Measurement of ECE

Differential scanning calorimetry, IR photometry, and scanning thermal microscopy are discussed in this section. A brief description about the method and typical examples of ECE obtained from direct measurement techniques are given.

4.3.1.1 Differential Scanning Calorimetry (DSC)

Primarily, the thermal properties of the materials, such as melting point, glass transition temperature, crystallization behavior, heat capacity, thermal stability, phase transitions, and reaction kinetics are determined using DSC. The DSC is preferable due to advantages like quick operation with commercially available instruments and software, relatively good sensitivity for detecting enthalpy changes, and ability to find the thermal properties of small samples [7]. The primary challenge in the DSC method is incorporation of modifications that are required to apply large electric field in commercially available instruments [7].

Figure 4.3a depicts the schematic diagram of a modified DSC setup [26]. Sanlialp and co-workers assessed the ECE in $0.65Ba(Zr_{0.2}Ti_{0.8})O_3$-$0.35(Ba_{0.7}Ca_{0.3})TiO_3$ (0.65BZT-0.35BCT) ceramics using the following DSC setup [26]. The sample cell has two spring loaders and thin varnished constantan wires that were specially constructed for imparting field to the sample. Alumina crucibles are used as sample holders, in which silver was coated on the inner surface as an insulator for the DSC sensors. The

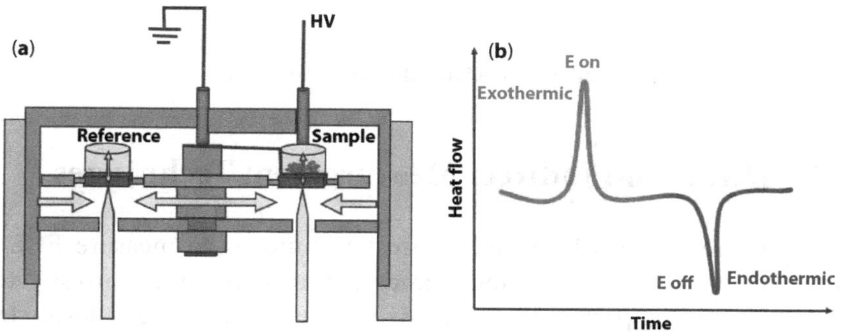

Figure 4.3 (a) Schematic diagram of DSC setup [22], (b) typical DSC heat flow curve [22].

connection between sample and sensor is assured by spring loaders, which produces thermal loss during the measurement [26]. Figure 4.3b is a typical DSC curve, where exothermic peak corresponds to the release of heat to the surrounding and endothermic peak is the absorption of heat from the surrounding. The on and off conditions of EF result in exothermic peak and endothermic peak, respectively [26]. When field is applied, the heat flow first increases and then decreases, due to a change from a disordered to an ordered state. The heat flow decreases and then increases upon the removal of field, which indicates the change from an ordered to a disordered state [27].

Figure 4.4a demonstrates the heat flow (H) measurements of 0.65BZT-0.35BCT ceramics [26]. Electric field is applied at constant temperatures (isothermal conditions) and H is marked as a function of time [7, 26, 28, 29]. The ECE measurements were performed in the temperature range 20°C to 100°C at an interval of 10°C. At each temperature (20, 30, 40°C), the sample is kept for 30 min to stabilize the temperature. ECE is measured at three electric fields 10, 20, and 30 kV/cm, where field is applied and removed with a time duration of 100 s, respectively. The heat flow measurements at 60°C are shown in Figure 4.4a. The area swept by endothermic and exothermic peak gives a specific heat exchange value ΔQ. From Equation 4.18, ΔT is calculated and shown in Figure 4.4c where C_p = specific heat capacity, K = thermal loss factor, and m_{ECE} and m_c are masses of sample at ECE and at C_p measurement, respectively [26].

$$\Delta T = \frac{\Delta Q m_{ECE}}{C_p m_c} K \tag{4.18}$$

Specific heat capacity is measured using the DSC [26, 30]. During the measurement, a certain amount of heat is lost, which arises due to the alumina

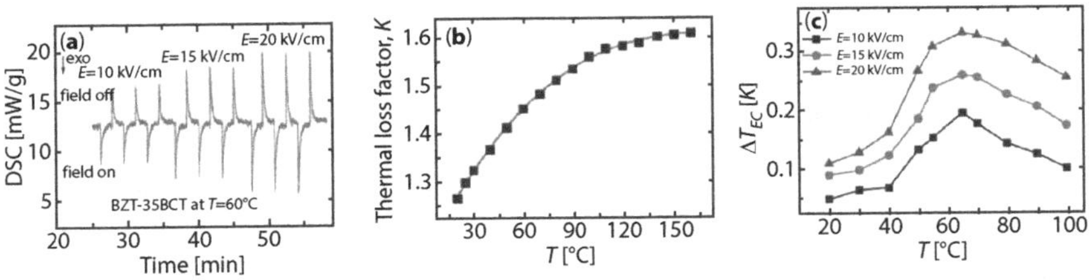

Figure 4.4 (a) Heat flow measurements of 0.65BZT-0.35BCT ceramics at different EF using DSC, (b) variation in thermal loss factor with temperature, (c) temperature dependence of ΔT observed in 0.65BZT-0.35BCT ceramics. (Reprinted with permission from [26].)

crucible's reduced thermal conductivity. To compensate for this, heat flow is corrected by a thermal loss factor. The reference measurements are conducted on in-built known resistance values in the relevant temperature and heat flow ranges. The variation of thermal loss factor with temperature is shown in Figure 4.4b. The thermal loss factor is almost linearly dependent on the temperature up to 100°C, and above that, it becomes saturated [26, 30].

4.3.1.2 *Fast Infrared Photometry*

Fast infrared photometry refers to the temporal and spatial measurement of ECE by capturing thermal image of the sample. This technique is faster and compact compared to other direct techniques like DSC [25]. There are two variants of this method based on the type of sensor used [7,23]: IR sensor [31] and fast IR camera [22]. The schematic diagrams of the two variants are shown in Figures 4.5a and 4.6a.

Lu *et al.* investigated ECE in P(VDF-TrFE) 68/32 mol% copolymers using the IR sensor [31]. The P(VDF-TrFE) 68/32 mol% films are coated using the solution-cast method and then five times uniaxially stretched. The irradiated polymer films are coated with aluminum (Al) electrodes having a diameter of 6 mm using thermal evaporator. Later, the films were stuck on the Kapton film of thickness 12 μm. A thin Al film with resistance R of 30 Ω was coated on the Kapton film that acts as a heater. The experimental setup is shown in Figure 4.5a. A voltage V is applied to the heater, which produces Joule heat. This heat alters the sample temperature and the temperature change is detected using the IR sensor [31]. Figure 4.6a shows a schematic diagram of ECE measurement using the IR camera [25]. The camera has an

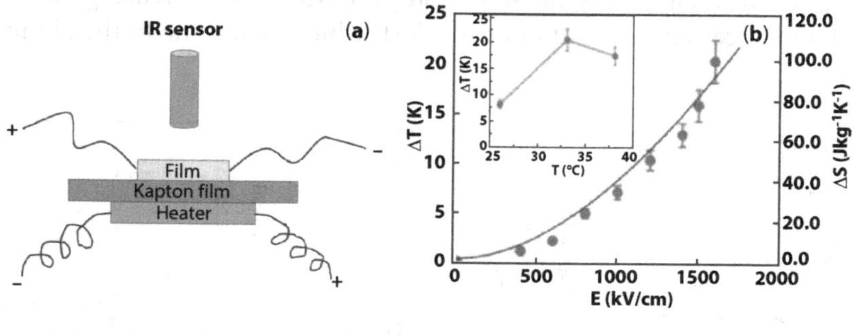

Figure 4.5 (a) Illustrative diagram of fast infrared photometry setup with IR sensor [31], (b) directly measured ECE for irradiated copolymer [P(VDF-TrFE) 68/32%]. Temperature dependence of ΔT under 160 MV/m, measured at different temperatures around RT, is given in the inset. (Reprinted with permission from [31].)

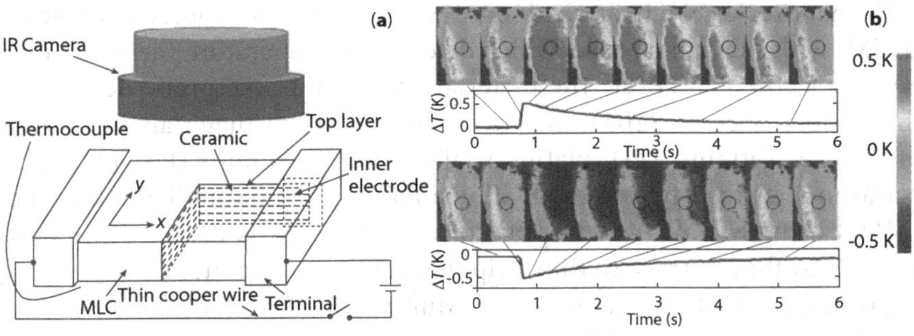

Figure 4.6 (a) Graphical illustration of ECE measurement setup using IR camera. (Reprinted with permission from [25].) (b) ΔT measured upon the application of field (upper image) and removal of field (lower image) in $BaTiO_3$-based MLC by IR camera. The circles mark the location where ΔT is measured. The color code gives the largest $\Delta T = 0.50$ K. (Reprinted with permission from [22].)

array of lenses with a semiconductor InSb sensor. The camera captures the thermal radiation emitted from the surfaces and the thermal images have a good spatial resolution of 6.5 μm and about 25 mK thermal sensitivity [22, 25]. Emissivity of the surface depends on its roughness. Hence, to acquire high accuracy, the sample surface should be flat [22].

Liu *et al.* used an IR camera to image/record the EC response of a commercial multilayer capacitor (MLC). It consists of around 150 layers of $BaTiO_3$-based thick films (11 μm thick) and each layer is separated by a 2 μm thick Ni electrode [25]. The MLC is connected to a generator through a thin copper wire. Tin was coated on the MLC terminal to ensure the electrical connection with the copper wire, and thereby with the generator. To measure the temperature of the MLC, the MLC terminal was connected to a thermocouple. The voltage was applied and removed from the MLC in cycles. The sample gets heated and cooled according to the applied voltage. The InSb sensors detect the photons emitted from the MLC and corresponding IR images are produced. Figure 4.6b is an IR image of the $BaTiO_3$-based MLC recorded in different locations on the MLC. The color code given along with Figure 4.6b shows that $\Delta T = 0.50$ K is observed in the MLC [22].

4.3.1.3 *Scanning Thermal Microscopy (SThM)*

Scanning thermal microscopy (SThM) was developed based on the scanning probe microscopy (SPM) [24]. SThM is a more apt method for studying ECE in thin films with substrates because it allows thermal measurements at the nanoscale [22]. In a broad sense, SThM is an atomic force microscope (AFM) with a thermal probe, which detects the temperature variation using a piezoelectric cantilever. The material used in thermal probe determines the performance of SThM and the commonly used materials in thermal probe are platinum (Pt), platinum alloy, and palladium (Pd) [24]. Kar-Narayan *et al.* investigated the ECE in a $BaTiO_3$-based multilayer capacitor (MLC) using SThM. A conceptual picture of SThM is depicted in Figure 4.7 [22]. A thin Pt track (a thin strip that acts as a resistance thermometer) deposited on the tip of AFM, Veeco Multimode AFM, a cantilever holder, and Anasys Instruments GLA-5 tips are the major components of the SThM instrument. MLC consists of 200 $BaTiO_3$ layers, each with a thickness of 6.5 μm and each layer is separated by a 2 μm thick Ni electrode. The Ni electrodes allow the EC heat conduction between $BaTiO_3$ layers and terminals. A Wheatstone bridge is used to ensure the accuracy of the measurements. The track resistance R_t is one of the arms of the bridge. The location of the tip is fixed before each measurement. Also, the output voltage V_{out} is made zero by using variable resistance R_3 prior to each measurement.

Using a Pt-100 thermometer, the V_{out} measurement is calibrated. The experiment was performed with a tip-sample separation of 5 μm, to avoid electrical contact and destructive contact with the MLC. A variable voltage source is attached to the MLC terminal; 200 V is applied to MLC and observed EC heating of the sample. The increase in temperature was $\Delta T = 0.58 \pm 0.08$ K in 1 s and most of the heat gets dissipated to the piezo scanner in 20 s. Upon the removal of the voltage, EC cooling is observed in 1 s. Then, over 20 s, the MLC absorbed heat mainly from the scanner as depicted in Figure 4.8a. The ΔT value depends on the separation between the tip and sample as evidenced from Figure 4.8b.

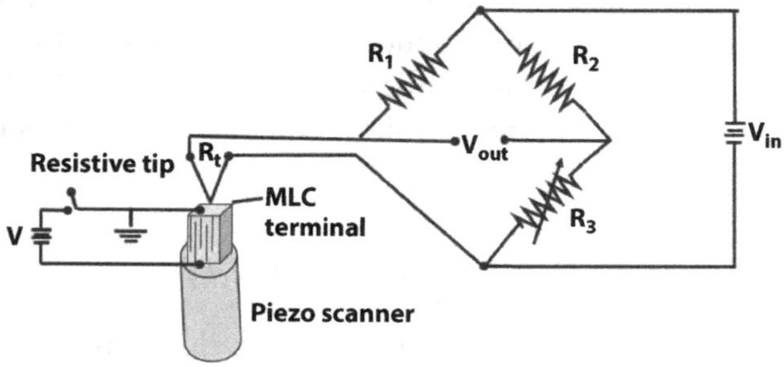

Figure 4.7 Schematic diagram of SThM [22].

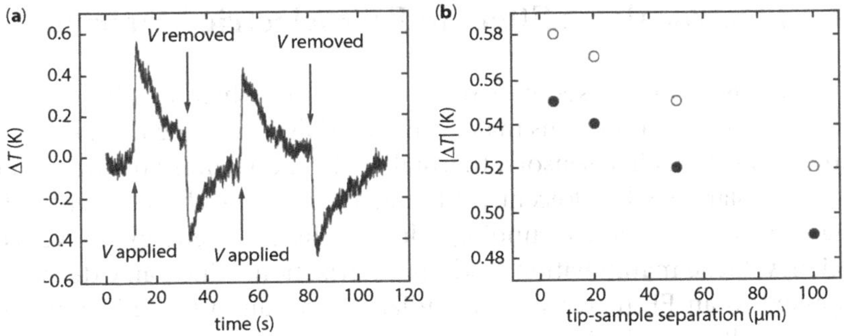

Figure 4.8 SThM measurements of ECE in $BaTiO_3$-based MLC. (a) ΔT vs. time, on applying and removing field, (b) EC heating and cooling as a function of tip-sample separation (open circles denoted the heating and closed circles denoted cooling). (Reprinted with permission from [22].)

4.3.2 Indirect Method

The expressions for ΔT and ΔS as derived from the Maxwell approach (Section 4.2.3.1) are

$$\Delta T = -\int_{E_2}^{E_1} \frac{T}{\rho C_E}\left(\frac{\partial P}{\partial T}\right) dE \tag{4.19a}$$

$$\Delta S = -\int_{E_2}^{E_1} \left(\frac{\partial P}{\partial T}\right) dE \tag{4.19b}$$

where E_1 and E_2 denote initial and final applied electric field and other variables are defined in Section 4.2.3.1. The polarization as a function temperature is derived from T-dependent P–E loops. Further, variation of the slope ($\partial P/\partial T$) of the curves with T is plotted at distinct EFs [7]. Maintaining the precision of P–T and ($\partial P/\partial T$) is not easy as the polarization is discontinuous near transition temperatures in first-order phase transition FE materials. In P–T curves, short temperature intervals are preferred to obtain very smooth isothermal P–T curves, which is a necessary condition to avoid abnormal shift in EC peak positions and errors in estimation of ΔT value. However, this method is most commonly used in the literature to evaluate EC performance of FE thin and thick films and bulk ceramics [21, 32–34].

4.4 Electrocaloric Effect in Ferroelectric Materials

FE materials show the spontaneous polarization, which can be switched by applying external EF. This makes FE materials promising candidates for many applications like sensors, thermal applications, non-volatile memories, and solar cells [5]. Recently, FE materials seem to have potential for solid-state refrigeration technology due to the large energy conversion efficiency, facile manipulation, and miniaturization. When an external EF is applied to an FE material under adiabatic conditions, the T increases, which is called adiabatic polarization heating. The removal of EF causes the reduction in temperature of the FE material, known as adiabatic depolarization cooling. The FE materials are classified into two categories: lead-based and lead-free ferroelectrics.

4.4.1 Lead-Based Ferroelectric Materials

This section discusses in detail the ECE in various lead (Pb)-based ceramics such as PZT ($Pb(Zr, Ti)O_3$), PLZT ($Pb(La, Zr, Ti)O_3$), and PMN-PT ($P(Mn, Nb, Ti)O_3$-$PbTiO_3$).

4.4.1.1 PZT-Based Normal Ferroelectrics

PZT ($Pb(Zr_{1-x}Ti_x)O_3$) is a solid solution of $PbZrO_3$ (orthorhombic)–$PbTiO_3$ (tetragonal), and the properties of PZT can be tuned by varying Zr/Ti ratio. A weak ECE with a ΔT of 0.19 K at 30 kV/cm in the vicinity of T_c (463 K) was reported in solid-state synthesized $Pb_{0.88}La_{0.12}(Zr_{0.65}Ti_{0.35})O_3$ ceramics [35]. In 2006, the discovery of large ECE in PZT thin film by Mischenko and co-workers was a major breakthrough in the field of EC materials [13]. A temperature change of 12 K was observed under an EF of 480 kV/cm at T= 222°C [13].

Recently, various approaches such as (i) doping [31, 36–39], (ii) electric field engineering [40, 41], (iii) misfit strain [42, 43], (iv) effect of domain walls [44], and (v) impact of polarization domains and charged defects [45] have been adopted to improve the ECE and to achieve low working temperature. The effect of doping, misfit strain, and electric field engineering on ECE are discussed in this section.

Doping Effect

Doping is one of the simplest strategies to improve EC response, and thus, different doping elements such as La^{3+} and Mn^{3+} into PZT thin films have been widely investigated. An increased pyroelectric coefficient of 3.52×10^{-4}

C/m^2 K was reported in 1 mol% Mn-doped PZT thin film deposited on the Pt/Ti/SiO_2/Si substrate, which is high compared to the value of 2.11×10^{-4} C/m^2 K obtained in undoped PZT [36, 46]. ECE is proportional to pyroelectric coefficient (Equation 4.6), which implies that Mn^{3+} doping can enhance the PZT thin film's ECE. Recently, Peng *et al.* showed tailoring of positive ECE into negative ECE by choosing an appropriate substrate [47]. Figure 4.9a–d shows the influence of different substrates on EC response of PZT thin films. A large ECE of $\Delta T = 44.5$ K at 313 K is found in Mn-doped PZT thin films deposited on Pt/TiO_2/SiO_2/Si substrates as shown in Figure 4.9a. By depositing the Mn-doped PZT thin film directly on the substrate (p-GaN, carrier concentration $n = 4 \times 10^{17}$), a pure negative ECE of $\Delta T = -23.5$ K is recorded with a broad working temperature of 150 K as seen in Figure 4.9c. Due to the modest Gibbs free energy difference between the phases, the field-induced structural phase change from tetragonal and/or orthorhombic to rhombohedral plays a critical role in realizing a substantial negative ECE [47].

Different compositions of La-doped PZT such as $Pb_{0.92}La_{0.08}(Zr_{0.65}Ti_{0.35})_{0.98}O_3$ [37] $(Pb_{0.865}La_{0.09})(Zr_{0.65}Ti_{0.35})O_3$ [38], $Pb_{0.9}La_{0.1}Zr_{0.65}Ti_{0.35}O_3$ [48], $Pb_{0.89}La_{0.11}(Zr_{0.7}Ti_{0.3})_{0.9725}O_3$, and $Pb_{0.93}La_{0.07}(Zr_{0.82}Ti_{0.18})O_3$ [39] have been studied by

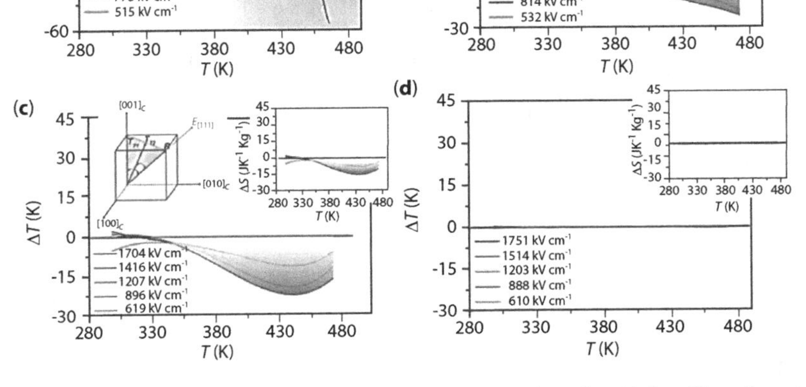

Figure 4.9 ΔT values as a function of temperature of Mn-doped PZT thin films deposited on (a) Pt/TiO_x/SiO_2/Si(100), (b) $LaNiO_3$/n-Si(100), (c) p-GaN, and (d) n-Si(100) substrates. Corresponding ΔS is given in insets. (Reprinted with permission from [47].)

different groups, and their performance is tabulated in Table 4.1. The $Pb_{0.92}La_{0.08}(Zr_{0.65}Ti_{0.35})_{0.98}O_3$ ceramics exhibited $\Delta T \sim 0.26$ K at unit field 1 kV/mm. The RT ECE is 0.06 K, which is larger (three times) than PZT [37]. In Si-doped PZT thin films, an enhanced ECE is observed for 1 mol% Si-doped PZT. This is attributed to the presence of multiple phases in Si-doped PZT. In this composition, orthorhombic and tetragonal phases equally coexist, which, in turn, provides a large number of polar states. Consequently, it provides a larger ECE than PZT [49].

Further, Roh and colleagues examined the effect of electric field and thickness on ECE in La^{3+}-doped PZT thin films [48]. The ΔT increases with increasing film thickness and becomes saturated above 840 nm thickness. This enhancement in ECE is attributed to the residual stress. In the current study, residual stress is reduced when film thickness is increased, which, in turn, increases the ECE. Maximum $\Delta T = 0.68$ K and EC strength $\Delta T/\Delta E = 3.8$ mK cm/kV is obtained for the PLZT film at a thickness of 1080 nm at a temperature ~ 300 K. The maximum ΔT increases with electric field and the T_{EC} shifts to higher temperature with field.

Table 4.1 EC parameters of different doped PZT ceramics.

Material	ΔT (K)	ΔE (kV/cm)	T_{EC} (K)	$\Delta T/\Delta E$ (10^{-6} K mV^{-1})	Reference
$(Pb_{0.88}Sr_{0.08})(Nb_{0.08}(Zr_{0.53}Ti_{0.47})_{0.92})O_3$	0.65	15	453	0.43	[50]
$Pb_{0.3}Sr_{0.7}TiO_3$	1.52	80	287.9	0.19	[51]
$Pb_{0.3}Ca_{0.05}Sr_{0.65}TiO_3$	1.71	80	292.5	0.21	[51]
$Pb_{0.3}Ca_{0.1}Sr_{0.6}TiO_3$	1.43	80	291	0.18	[51]
$Pb_{0.3}Ca_{0.15}Sr_{0.55}TiO_3$	1.49	80	281	0.19	[51]
$PbZr_{0.95}Ti_{0.05}O_3$- 1 mol % Si	8.5	390	203	-	[49]
$Pb_{0.92}La_{0.08}(Zr_{0.65}Ti_{0.35})O_3$	0.8	30	433	-	[37]
$(Pb_{0.865}La_{0.09})(Zr_{0.65}Ti_{0.35})O_3$	0.23	193	473	-	[38]
$Pb_{0.89}La_{0.11}(Zr_{0.7}Ti_{0.3})O_3$	2.21	70	423	0.453	[39]
$Pb_{0.93}La_{0.07}(Zr_{0.82}Ti_{0.18})O_3$	1.3	30	403	0.433	[39]

Recently, M. Zhang *et al.* fabricated $0.95Pb_{0.92}La_{0.08}(Zr_{0.70}Ti_{0.30})_{0.98}O_3$-$0.05BiFeO_3$ thin films ($BiFeO_3$-La co-doped PZT) by the sol-gel method [52]. The epitaxial films were coated on Nb-doped $SrTiO_3$ (Nb:STO) substrate with (111), (110), and (100) orientations. The addition of bismuth ferrite improved the polarization of the thin film and also stabilized the R3c phase. Alternatively, doped lanthanum lowered the T_c and increased the degree of the dielectric–permittivity–dispersion, and broadened the working temperature range. A large ECE of 20.6 K was observed near room temperature under the field of 1956 kV/cm for the films deposited on (111)-oriented Nb:STO. The self-induced poling that occurred in the process of preparation of thin film accounts for the large ECE observed in the (111)-oriented thin film. Besides, a broad operation temperature range of 100 K was observed, which makes $BiFeO_3$-La co-doped PZT a promising candidate for low-temperature EC cooling applications [52].

Influence of Misfit Strain

It is known that strain is one of the key parameters that influences the properties of FE thin films. Misfit strain in thin films occurs as a result of deviation of thermal expansion coefficients of substrate from that of film and their mismatched lattice parameters. The dependence of misfit strain on the ECE is theoretically calculated using Landau–Devonshire phenomenological theory [42, 53, 54]. It is shown that the values of ΔT and the T_{EC} increases with the increase of misfit strain. The compressive and tensile misfit strain has shown an opposite effect on EC parameters. With the increase of compressive misfit strain, T_{EC} shifts to higher temperatures, the magnitude of ΔT decreases and diminishes its dependence on T, the tensile misfit strain shifts T_{EC} to lower temperature and increases the magnitude of ΔT, and ΔT becomes more dependent on temperature [53]. The variation in EC response under the influence of misfit strain is studied in various FE thin films, and it infers that ECE can be enhanced by choosing an appropriate substrate via misfit strain [42, 43]. The misfit strain dependence on ECE of the epitaxial PZT thin film is theoretically investigated. According to the phenomenological Landau–Devonshire theory, the EC coefficient is highly influenced by the misfit strain. The epitaxial $Pb(Zr_{0.1}Ti_{0.9})O_3$ and $Pb(Zr_{0.5}Ti_{0.5})O_3$ thin films exhibit a large EC coefficient at RT [42]. The relationship between misfit strain and ECE has been discussed in various materials such as $SrTiO_3$, P(VDF-TrFE) polymer, $PbTiO_3$, and $Pb_{1-x}Sr_xTiO_3$ [54–57]. A theoretical study of the changes in RT ECE with composition and misfit strain in epitaxial $Pb_{1-x}Sr_xTiO_3$ thin film is conducted. The T_c shifts towards a lower-temperature region with the addition of Sr and it enhances the ECE. According to the temperature–misfit strain phase diagram, four different equilibrium phases may exist in the thin film $Pb_{1-x}Sr_xTiO_3$, and they are denoted as the paraelectric, $\alpha\alpha$, c, and γ phase. The EF is applied along the [001] direction. The paraelectric and $\alpha\alpha$ phases are unstable and transform into c and γ phases, respectively. In the $Pb_{1-x}Sr_xTiO_3$ thin film, c and γ phases are stable and c to γ phase transition is first order. Because of this, at phase boundary, ΔT is discontinuous. An optimum ECE of ΔT = 6 K at ΔE = 100 kV/cm is visible in $Pb_{0.5}Sr_{0.5}TiO_3$ composition. The thin films with tensile misfit strain possessed higher ΔT than films with compressive misfit strain, for all compositions. In conclusion, large ECE is tuned by changing the Sr content and misfit strain in $Pb_{1-x}Sr_xTiO_3$ thin films [57].

Electric Field Engineering

Usually, the ferroelectrics exhibited high ECE near the phase transitions. In PZT materials, the field-induced antiferroelectric (AFE) to ferroelectric (FE) phase transition can be easily achieved at low temperatures and high ECE near room temperature is highly advantageous for real-time applications. In this context, Qihan Peng *et al.* studied field-induced phase transitions and ECE in La-doped $Pb(Zr,Sn,Ti)O_3$ [58]. It is shown that a significant improvement in ΔT = 1.04 K at RT (30°C) was obtained in $(Pb_{0.94}La_{0.04})(Zr_{0.52}Sn_{0.30}Ti_{0.18})O_3$ bulk ceramic through AFE-to-FE transition [58]. Recently, Zhang *et al.* achieved giant ECE in $PbZr_{0.95}Ti_{0.05}O_3$/$PbZr_{0.52}Ti_{0.48}O_3$ bi-layer films by tuning the EF [59]. Bi-layer thin films exhibited a large ΔT = 24.8 K and entropy change of ΔS = 20.5 J K^{-1} kg^{-1} at an EF of 566 kV/cm around 125°C. The change in ΔT with EF is illustrated in Figure 4.10. The AFE/FE phase transition triggered by EF shows a crucial role in achieving an excellent ECE.

4.4.1.2 $Pb(Mg_{1/3}Nb_{2/3})O_3$- $PbTiO_3$ (PMN-PT) Relaxor Ferroelectrics

(1−x)$Pb(Mg_{1/3}Nb_{2/3})O_3$-x$PbTiO_3$ (PMN-PT) systems are typical relaxor ferroelectrics (RFE) that consist of polar nano regions (PNRs). The PNRs arise due to compositional fluctuations caused by different doping ions and induce a diffused phase transition and this led to a wide ECE peak in RFEs. The effect of PMN and PT composition on EC response of PMN-PT ceramics, single crystals, and thin films are investigated [60]. The team of L. Shaobo found the ECE in 0.9PMN-0.1PT and 0.85PMN-0.15PT ceramics for ferroelectric micro electro mechanical system (MEMS) micro cooler applications. The 0.85PMN-0.15PT showed a larger EC temperature change (1.71 K for a EF = 16 kV/cm) than 0.95PMN-0.15PT [61].

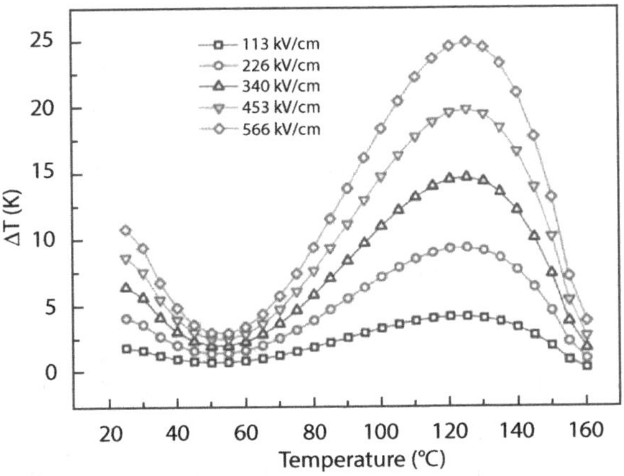

Figure 4.10 $PbZr_{0.95}Ti_{0.05}O_3/PbZr_{0.52}Ti_{0.48}O_3$ bilayer thin film's temperature depend on ΔT and corresponding electric fields. (Reprinted with permission from [59].)

In 2006, Daniel Guyomar *et al.* studied the ECE in 0.75PMN-0.25PT ceramic material by indirect and direct methods. The reversible heat change of 0.2 J/g for a field of 30 kV/cm is observed [62]. The EC and pyroelectric properties of the same composition (0.75PMN-0.25PT) were analyzed by Gael Sebald *et al.* [63]. A large pyroelectric coefficient (1300×10^{-6} $Cm^{-2}K^{-1}$) is found for SC, which is larger than that of ceramics, and a heat flow of 0.40 J/g under a field of 25 kV/cm is recorded [62]. Hagberg *et al.* explored the ECE in 0.87PMN-0.13PT ceramics via direct method. The maximum $\Delta T = 0.558$ K was observed at $T = 70°C$ for an applied field of 24 kV/cm. Although the operation T is nearer to the RT, the value of ΔT is small [64]. Q. J. Hua *et al.* investigated the dependence of composition on the ECE by using an eighth-order Landau–Devonshire theory in $(1-x)$ $Pb(Mg_{1/3}Nb_{2/3})O_3$–$xPbTiO_3$ SCs [65]. A decrease in the rhombohedral–tetragonal (R–T) and monoclinic–tetragonal (M_C–T) phase transition temperatures and a shift in T_c towards higher T are observed. A -Ve EC coefficient is obtained near M_C-T in $0.69Pb(Mg_{1/3}Nb_{2/3})O_3$-$0.31PbTiO_3$ SC, which is due to the increase of polarization with T. The formation and alignment of PNRs causes stable and high ECE over a wide range of T.

Effect of Crystal Orientation

It is established that dielectric and FE properties of SCs, thin films, and ceramics are strongly influenced by their orientation or texture and consequently affect the EC response of RFEs. J. Peräntie *et al.* measured the field-induced temperature change on the surface of (011)-oriented

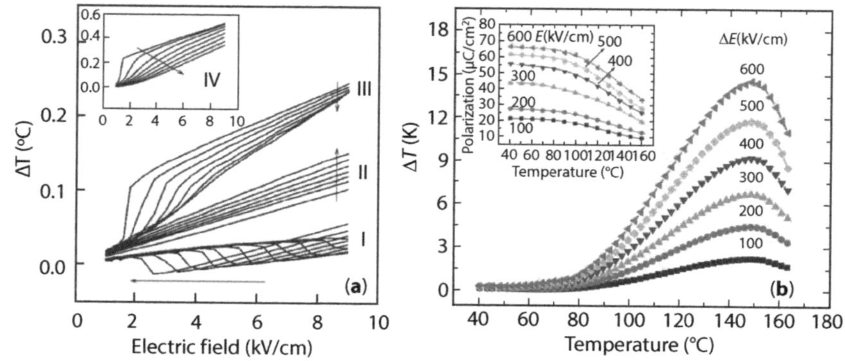

Figure 4.11 (a) Variation of ΔT with electric field in different ranges of temperatures such as 31–71°C (I), 73–97°C (II), 101–127°C (III), and 134–162°C (IV) of PMN-28PT single crystal. Arrow represents increasing direction of temperature. (Reprinted with permission from [66].) (b) ΔT as a function of T at different EF in PMN-32PT films. (Reprinted with permission from [69].)

$0.72PbMg_{1/3}Nb_{2/3}O_3$-$0.28PbTiO_3$ (PMN-28PT) through a thermocouple (chromel-alumel) which is connected to the sample [66]. Figure 4.11a depicts field-induced ΔT in (011)-oriented PMN-28PT in the T ranges of 31–71°C (I), 73–97°C (II), 101–127°C (III), and 134–162°C (IV). In range IV, (011)-oriented PMN-28PT SC displays orthorhombic to cubic phase transition and shows $\Delta T = 0.53$°C under EF = 9 kV/cm as seen in the inset of Figure 4.11a [66]. The (111)-oriented $0.9PbMg_{1/3}Nb_{2/3}O_3$–$0.1PbTiO_3$ (0.9PMN–0.1PT) SC displays $\Delta T = {\sim}1$K with only a change of 40 kV/cm field [67]. The RF magnetron sputtered $0.67PbMg_{1/3}Nb_{2/3}O_3$–$0.33PbTiO_3$ (PMN-33PT) films on $LaNiO_3$-coated SiO_2/Si substrates possess a (001)-oriented texture. The (001)-oriented PMN-33PT thin films exhibited a large ΔT = 14.5 K at 12 V as shown in Figure 4.11b [68]. It is due to the large change in polarization and entropy at high field during phase transition from a ferroelectric to a paraelectric state. Further, (100)-oriented $0.68PbMg_{1/3}Nb_{2/3}O_3$–$0.32PbTiO_3$ (PMN-32PT) thin films with a $Pb(Zr_{0.3}Ti_{0.7})O_3$/PbOx buffer layer demonstrated a $\Delta T = 13.4$ K at 15 V near the transition temperature of 146°C [69]. PMN-32PT films show slightly lower EC performance compared to PMN-33PT textured films.

4.4.2 Lead-Free Ferroelectric Materials

Although the Pb-based FE materials possess large ECE, the toxicity of Pb prevents them from being used in refrigeration devices. It is highly desirable to develop environmentally benign Pb-free materials with qualities comparable to Pb-based compounds. Several lead-free materials have emerged in this field including $BaTiO_3$, $K_{0.5}Bi_{0.5}TiO_3$, $Na_{0.5}Bi_{0.5}TiO_3$, and $(K_{0.5}Na_{0.5})NbO_3$-based compositions [21].

4.4.2.1 *$BaTiO_3$-Based Ceramics*

$BaTiO_3$ (BTO) is the first lead-free perovskite ceramic material that exhibited ferroelectricity. BTO exhibited high dielectric constant, Curie temperature above room temperature, and good piezoelectric and pyroelectric properties. It displays a series of phase transitions like rhombohedral↔orthorhombic (183K), orthorhombic↔tetragonal (278K), and tetragonal↔cubic phase (393K) [21, 70]. The ECE in BTO SCs, ceramics, and thin and thick films is widely explored. For instance, BTO SCs show ECE with $\Delta T = 1.6$ K and $\Delta S = 1.9$ J/kg K at a low EF of 10 kV/cm [71]. Similarly, Moya and fellow researchers studied the ECE in BTO SC at two different applied fields: 12 kV/cm ($\Delta T = 0.90$°C at 128°C) and 4 kV/cm ($\Delta T = 0.87$°C at 123°C) [72]. Later, Liu *et al.* studied barocaloric and pressure-mediated ECE in BTO SC [73]. The thermodynamic model estimated $\Delta T = 0.97$ K (8 kV/cm) and 1.03 K (12 kV/cm) at 0 GPa and is in agreement with the values obtained via a direct method. The hydrostatic pressure causes a shift in the ECE peak towards lower temperature and a slight reduction in ΔT values as shown in Figure 4.12a and b [73].

Compared to ceramics and SCs, thin films exhibit better EC response as larger fields can be obtained with small voltages. Cao and Li investigated the ECE in BTO thin films using a transverse Ising model and reported a large $\Delta T = 11.78$ K at 490 K [74]. Yang Bai *et al.* studied the relation between ECE and temperature, field intensity, and varying field rate in the

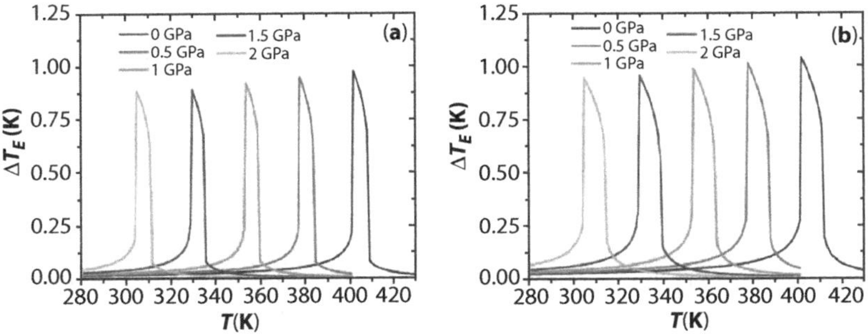

Figure 4.12 Temperature and hydrostatic pressure dependence of ΔT at (a) 8 kV/cm and (b) 12 kV/cm of $BaTiO_3$ single crystal. (Reprinted with permission from [73].)

BTO thick film [75]. The EF-induced reorientation of dipole and deformation in the lattice causes large ECE with $\Delta T = 7.1$ K and an entropy change of $\Delta S = 10.1$ J/kg K at 80°C under a field of 800 kV/cm. The ECE increases monotonically with EF [76].

By choosing the appropriate doping material, the phase transition temperature, micro-structure, dielectric and FE properties of BTO can be tuned. The EC response can be enhanced by selecting the right composition that obeys the invariant critical point (ICP) [77, 78]. At ICP, more than one phase coexist and a greater number of polar states are available, which, in turn, increase the polar phase entropy. In addition, the electric field required to switch between two polar states is small. Consequently, a significant ECE originates under low EF. The four phases of the BTO ceramic system can be unified into a single-phase transition by adding dopants like Zr, Sn, and/or Sr, which can induce a large ECE under low EFs [77–81]. Qian and co-workers investigated the influence of Zr doping into the B-site of barium titanate on EC response of $BaZr_xTi_{1-x}O_3$ ($x = 0.15, 0.2$, and 0.25) ceramics and reported a $\Delta T = 4.5$ K in $BaZr_{0.2}Ti_{0.8}O_3$ ceramics [80]. S. K. Upadhyay *et al.* showed that Sn doping in the Ti site of $BaTiO_3$ shifts cubic to tetragonal transition temperature towards low T [79]. The variation of EC strength with Sn doping at room temperature and phase transition is shown Figure 4.13. The $BaTi_{0.95}Sn_{0.05}O_3$ ceramics showed optimum EC strength ($\xi = 0.19$ K mm/kV) at RT under low EF (3.62 kV/cm).

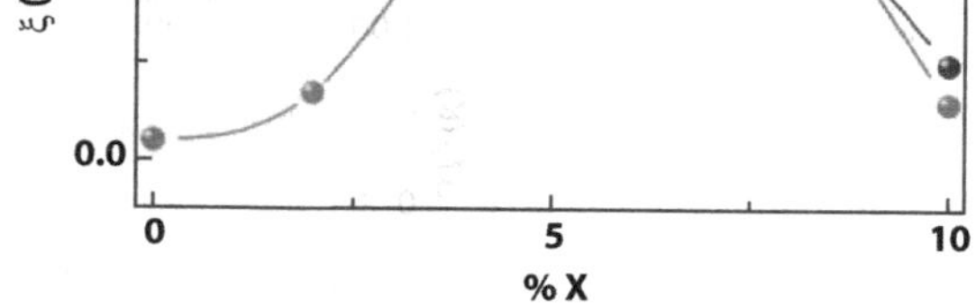

Figure 4.13 EC strength variation of $BaTi_{1-x}Sn_xO_3$ with Sn doping ($x = 0, 2, 5, 10\%$) at room temperature across transition, i.e., T_{C-T} for $x \leq 5\%$ and T_{C-R} for $x = 10\%$. (Reprinted with permission from [79].)

Further, Luo *et al.* reported ECE in $BaTi_{1-x}Sn_xTiO_3$ ceramics with $x = 0.08$, 0.105, and 0.14 near room temperature by using the indirect method. Enhanced $\Delta T = 0.61$ K and EC strength (0.31 K mm/kV) was reported for composition $x = 0.105$ near RT at a field of 20 kV/cm. The optimum EC response is due to coexistence of four phases, i.e., orthorhombic, rhombohedral, tetragonal, and cubic for the composition of $x = 0.105$ [82].

Further, A-site doping in B-site-doped $BaTiO_3$ shifted T_c towards low temperature [81]. The T_c of $Ba_{1-x}Sr_xTi_{0.9}Sn_{0.1}O_3$ ceramics slowly decreased with increasing Sr content. The T_c decreased to RT and is beneficial for near-RT refrigeration applications. The $Ba_{0.94}Sr_{0.06}Ti_{0.9}Sn_{0.1}O_3$ ceramics exhibited an EC strength of 0.31K mm/kV ($\Delta T = 0.46$ K) at 47°C. This demonstrates that $Ba_{1-x}Sr_xTi_{0.9}Sn_{0.1}O_3$ is an emerging candidate for eco-friendly RT refrigeration [81]. The EC properties of $BaTiO_3$-based materials are listed in Table 4.2.

4.4.2.2 *$Ba(Zr_{0.2}Ti_{0.8})O_3$–$(Ba_{0.7}Ca_{0.3})TiO_3$ (BZCT)-Based Ferroelectrics*

Recently, (1–x) $BaZr_{0.2}Ti_{0.8}O_3$-$xBa_{0.7}Ca_{0.3}TiO_3$ (BZCT) is considered as one of the potential lead-free materials, which can replace PZT, commonly used in many piezo-, pyro-, and ferroelectric-based devices [101–103]. It exhibited outstanding dielectric, piezo-, and ferroelectric properties due to simultaneous existence of multi-phases like tetragonal, rhombohedral, and cubic phases as shown in Figure 4.14a [103, 104]. Y. Bai *et al.* studied the effect of composition in (1–x) $BaZr_{0.2}Ti_{0.8}O_3$-$xBa_{0.7}Ca_{0.3}TiO_3$ ceramics. The composition $x = 0.3$ exhibited maximum $\Delta T = 0.30$°C (20 kV/cm) and the broad cooling peak due to the existence of the tri-critical point, i.e., coincidence of three phases of BZCT as seen in Figure 4.14a [104]. Y. Zhou *et al.* also studied the effect of composition in the range of $x = 0.3$ to 0.7 using the Maxwell relation-based indirect method. The composition $x = 0.4$ shows a monotonic +Ve ECE with $\Delta T = 0.58$°C (125°C) under a moderate EF of 28 kV/cm. At higher compositions ($x \geq 0.5$), ceramics shows simultaneous existence of an abnormal -Ve ECE and usual +Ve ECE [105]. Recently, E. V. Ramana and team examined the effect of growth rate (200–5 mm/h) on EC performance of $Ba(Ti_{0.8}Zr_{0.2})O_3$–$(Ba_{0.7}Ca_{0.3})TiO_3$ (BZCT) fiber crystals grown by laser-heated pedestal growth (LHPG) [106] and compared the performance with ceramics prepared via the solid-state reaction method. The fiber crystal grown with 5 mm/h and ceramics exhibited ΔT values of 0.95 and 1.3 K, respectively, under an EF of 30 kV/cm. The effect of MgO doping on ferroelectric and EC properties of (1–x)(0.6Ba($Ti_{0.8}Zr_{0.2}$)

Table 4.2 ECE parameters of some $BaTiO_3$ based ceramics.

Material	ΔT (K)	ΔE (kV/cm)	T_{EC} (K)	Reference
$BaTiO_3$	0.75	-	216	[83]
$BaTiO_3$	0.4	7.5	391	[72]
$BaTiO_3$	1.8	17.6	353	[84]
$BaTiO_3$	0.2	40	303	[85]
$BaTiO_3$	0.63	40	406	[85]
$BaTiO_3$	0.42	20	400	[86]
$BaTiO_3$	1.3	10	397	[87]
$BaTiO_3$	0.84	5	287	[88]
$BaSn_{0.11}Ti_{0.89}O_3$-2 mol% CuO	0.33	10	313	[89]
$Ba(Ti_{0.975}Y_{0.025})O_3$	1.26	50	375	[90]
$Ba(Ti_{0.9675}Y_{0.0375})O_3$	1.03	50	364	[90]
$Ba(Ti_{0.95}Y_{0.05})O_3$	1.15	50	346	[90]
$Ba(Ti_{0.9375}Y_{0.0625})O_3$	0.37	50	340	[90]
$BaSn_{0.11}Ti_{0.89}O_3$-5 mol% Y	1.4	50	328	[91]
$BaTiO_3$-2% Eu	0.57	12	400	[92]
BHT-70BST	0.46	30	334	[93]
BHT-50BST	0.39	30	341	[93]
BHT-40BST	0.34	30	337	[93]
$Ba(Ti_{0.99}In_{0.01})O_{2.995}$	0.22	10	290	[94]
$Ba(Ti_{0.99}In_{0.05})O_{2.975}$	0.42	36	314	[94]
$Ba(Ti_{0.99}In_{0.03})O_{2.985}$	0.29	26	324	[94]
$Ba_{0.9}Sr_{0.1}Ti_{0.98}Fe_{0.015}O_3$	1.06	25	358	[95]
$BaTi_{0.89}Sn_{0.11}O_3$	0.71	25	322	[96]
$Ba(Zr_{0.05}Ti_{0.95})O_3$	2.4	30	386	[97]

(*Continued*)

Table 4.2 ECE parameters of some $BaTiO_3$ based ceramics. (*Continued*)

Material	ΔT (K)	ΔE (kV/cm)	T_{EC} (K)	Reference
$Ba(Zr_{0.3}Ti_{0.7})O_3$	1.18	30	301	[97]
$Ba(Zr_{0.15}Ti_{0.85})O_3$	1.2	30	334	[97]
$Ba(Zr_{0.25}Ti_{0.75})O_3$	1.1	30	298	[97]
$Ba(Zr_{0.1}Ti_{0.9})O_3$	1.7	30	356	[97]
$Ba_{0.995}Ce_{0.005}Ti_{0.99}Mn_{0.01}O_3$	1.22	30	393	[98]
$Ba_{0.985}Ce_{0.015}Ti_{0.99}Mn_{0.01}O_3$	1.07	30	370	[98]
$Ba_{0.97}Ce_{0.030}Ti_{0.99}Mn_{0.01}O_3$	0.41	30	328	[98]
$Ba_{0.96}Ce_{0.040}Ti_{0.99}Mn_{0.01}O_3$	0.50	30	298	[98]
$Ba_{0.955}Ce_{0.045}Ti_{0.99}Mn_{0.01}O_3$	0.42	30	293	[98]
$Ba_{0.94}Sr_{0.06}Ti_{0.9}Sn_{0.1}O_3$	0.46	15	320	[81]
$Ba_{0.95}Ca_{0.05}Ti_{0.95}Sn_{0.05}O_3$	0.53	10.81	386	[99]
$0.8BaZr_{0.18}Ti_{0.82}O_3$-$0.2BaSn_{0.11}Ti_{0.89}O_3$	3.5	100	-	[78]
$Ba_{0.8}Ca_{0.2}(Zr_{0.1}Ti_{0.9})_{0.925}Sn_{0.075}O_3$	0.214	8	294	[100]
$BaTi_{0.895}Sn_{0.105}O_3$	0.61	20	301	[77]

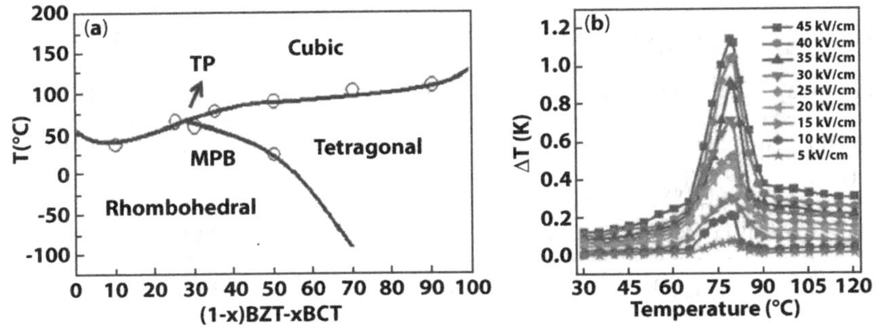

Figure 4.14 (a) Composition-dependent phase diagram of BZT-BCT. (Reprinted with permission from [104].) (b) ΔT as function temperature under different electric fields for $0.99(0.6Ba(Ti_{0.8}Zr_{0.2})O_3$-$0.4(Ba_{0.7}Ca_{0.3})TiO_3)$-$0.01MgO$ ceramics. (Reprinted with permission from [107].)

O_3–0.4($Ba_{0.7}Ca_{0.3}$)TiO_3)-xMgO ceramics by varying $x = 0–0.05$ obtained an optimum $\Delta T = 1.12$ K (45 kV/cm) for composition $x = 0.01$ in the vicinity of tetragonal–cubic transition temperature as depicted in Figure 4.14b [107]. Table 4.3 presents the EC parameters of (Ba,Ca)(Zr,Ti)O_3-based ceramics.

Table 4.3 Electrocaloric effect in few (Ba,Ca)(Zr,Ti)O_3-based ceramics.

Material	ΔT (K)	ΔE (kV/cm)	T (K)	Reference
$Ba_{0.98}Ca_{0.02}(Zr_{0.085}Ti_{0.915})O_3$	0.6	40	358	[108]
$Ba_{0.8}Ca_{0.2}ZrTi_{0.96}O_3$	0.27	7.95	386	[109]
$Ba_{0.8}Ca_{0.2}TiO_3$	0.12	7.95	398	[109]
$Ba_{0.92}Ca_{0.08}Zr_{0.05}Ti_{0.95}O_3$	0.37	15	398	[110]
$Ba_{0.85}Ca_{0.15}Zr_{0.1}Ti_{0.9}O_3$	0.41	21.5	360	[111]
$Ba_{0.95}Ca_{0.05}(Zr_{0.1}Ti_{0.9})_{0.925}Sn_{0.075}O_3$	0.211	8	312	[112]
$Ba_{0.9}Ca_{0.1}(Zr_{0.1}Ti_{0.9})_{0.925}Sn_{0.075}O_3$	0.205	8	297	[112]
$Ba_{0.85}Ca_{0.15}(Zr_{0.1}Ti_{0.9})_{0.925}Sn_{0.075}O_3$	0.214	8	294	[112]
$Ba_{0.8}Ca_{0.2}(Zr_{0.1}Ti_{0.9})_{0.925}Sn_{0.075}O_3$	0.176	8	323	[112]
$Ba_{0.85}Ca_{0.15}Zr_{0.1}Ti_{0.88}Sn_{0.02}O_3$	0.84	32	353	[113]
$Ba_{0.85}Ca_{0.15}Zr_{0.1}Ti_{0.89}Fe_{0.01}O_3$	0.86	37	345	[114]
$Ba_{0.85}Ca_{0.15}Zr_{0.10}Ti_{0.90}O_3$	0.109	6.65	363	[115]
$Ba_{0.786}Ca_{0.214}Zr_{0.006}Ti_{0.994}O_3$	1.05	10	397	[116]
$0.95Ba_{0.67}Sr_{0.33}TiO_3$-$0.05Ba_{0.9}Ca_{0.1}Ti_{0.9}Zr_{0.1}O_3$	0.74	30	308	[117]
$(Ba_{0.85}Ca_{0.15})(Ti_{0.9}Zr_{0.06}Sn_{0.04})O_3$	0.27	12	342	[118]
$Ba_{0.85}Ca_{0.15}Zr_{0.1}Ti_{0.9}O_3$	0.1513	6	375	[119]
$[(Ba_{0.85}Ca_{0.15})_{0.97}Bi_{0.02}](Zr_{0.1}Ti_{0.9})O_3$	0.91	80	380	[120]
$(Ba_{0.85}Ca_{0.075}Sr_{0.075})(Zr_{0.1}Ti_{0.88}Sn_{0.02})O_3$	1.5	33	303	[121]

4.4.2.3 (K, Na) NbO_3 (KNN)-Based Ceramics

$KNbO_3$ and $NaNbO_3$ easily form a solid solution at a whole range of composition and $K_{0.5}Na_{0.5}NbO_3$ (KNN) is the most studied composition for different applications because of their rich dielectric and FE properties induced by a polymorphic phase boundary [60]. In previous sections, it is observed that multiphase coexistence is an effective strategy to obtain a large ΔT in lead-based and $BaTiO_3$-based ferroelectrics. The reduction in energy barrier and polarization anisotropy due to multiphase existence facilitates polarization rotation and consequently enhances ECE of an electrocaloric material. The polymorphic phase boundary of KNN benefits ECE. However, the EC response of KNN is less studied compared to lead-based materials. Direct measurement of ECE in bulk Pb-free RFE $K_{0.5}Na_{0.5}NbO_3$-$SrTiO_3$ (KNN-STO) ceramics is found to be $\Delta T = 0.5°C$ at ferroelectric phase transition under a field of 80 kV/cm [122]. J. Koruza *et al.* found that 0.85KNN-0.15STO ceramics are able to sustain high field (160 kV/cm) and subsequently exhibited a large $\Delta T_{EC} = 1.9°C$ and $\Delta S = 3.6$ J/kg K (67°C) under an EF of 159 kV/cm as displayed in Figure 4.15a [123]. The 0.9KNN-0.1Sr($Sc_{0.5}Nb_{0.5}$)O_3 FE nanocrystalline ceramics showed $\Delta T = 0.28$ K near $T_c = 357$ K for electric field 25 kV/cm [124]. Further, it is shown that 0.9KNN-0.1STO nanocrystalline ceramics exhibited a positive and negative ECE depending on operating temperature as shown in Figure 4.15b [125]. Similarly, both +Ve and -Ve ECE was also observed in (1−x) KNN-x$LaNbO_3$. Further, (1−x) KNN-x$SrZrO_3$ showed ΔT= 1.19°C at 94°C (under EF = 35 kV/cm) for x = 0.1 [126]. The EC response of different KNN-based materials is shown in Table 4.4. The simultaneous existence

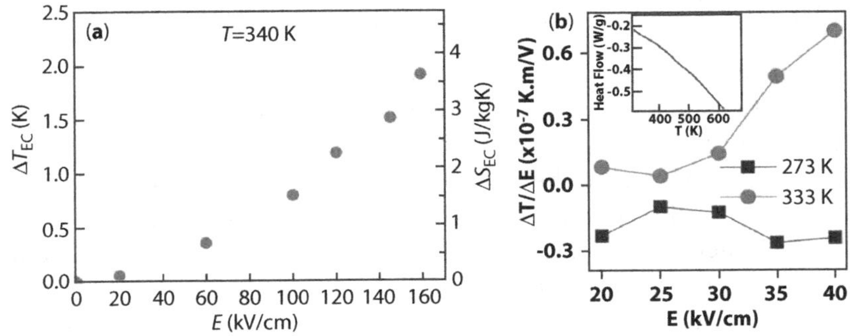

Figure 4.15 (a) ΔT_{EC} and ΔS_{EC} plot as a function of the electric field E at 340 K. (Reprinted with permission from [123].) (b) Electrocaloric responsivity coefficient ($\Delta T/\Delta E$) and DSC profile in the inset. (Reprinted with permission from [125].)

Table 4.4 ECE properties of some KNN based ceramics.

Material	ΔT (K)	ΔE (kV/cm)	T_{EC} (K)	Ref.
$K_{0.5}Na_{0.5}NbO_3$-$SrTiO_3$	0.3	60	305	[127]
$K_{0.5}Na_{0.5}NbO_3$-$SrTiO_3$	0.5	80	360	[122]
$0.85K_{0.5}Na_{0.5}NbO_3$-$0.15SrTiO_3$	1.9	159	340	[123]
$0.9(K_{0.5}Na_{0.5})NbO_3$-$0.1Sr(Sc_{0.5}Nb_{0.5})O_3$	0.28	25	257	[124]
$0.95(K_{0.48}Na_{0.52})(Nb_{0.95}Sb_{0.05})O_3$-$0.05Bi_{0.5}(Na_{0.82}K_{0.18})0.5ZrO_3$	0.48	40	300	[128]
$0.9(K_{0.5}Na_{0.5})NbO_3$-$0.1SrTiO_3$	0.43	40	330	[125]
$0.97K_{0.5}Na_{0.5}NbO_3$-$0.03LaNbO_3$	2.08	50	349	[126]
$0.97K_{0.5}Na_{0.5}NbO_3$-$0.03LiSbO_3$	3.33	40	345	[129]
$0.9(K_{0.5}Na_{0.5})NbO_3$-$0.1SrZrO_3$	1.19	35	367	[130]
$0.94(K_{0.5}Na_{0.5})NbO_3$-$0.06SrMnO_3$	−1.66	50	459	[131]
$0.9K_{0.5}Na_{0.5}NbO_3$-$0.1CaZrO_3$	0.27	40	402	[132]
$0.97K_{0.5}Na_{0.5}NbO_3$-$0.03Bi_{0.5}Na_{0.5}ZrO_3$	0.22	35	373	[133]

of +Ve and -Ve ECE enhances the cooling efficiency and this KNN-based ceramics become promising electrocaloric material [125].

4.4.2.4 *Hafnia and Zirconia-Based Ferroelectric Thin Films*

The interest in ferroelectrics was regained after the unexpected discovery of ferroelectricity at nanoscale in Si-doped HfO_2 and then both experimental and theoretical investigations were done in hafnia and zirconia film (HfO_2 and ZrO_2) to understand the origin of ferroelectricity [134, 135]. An initial wake-up pre-cycling is needed to induce ferroelectricity in these materials. Recently, different approaches such as doping, texture, defects induction, interface engineering, NH_3 plasma treatment, etc. are employed to achieve wake-up free ferroelectricity [135]. The existence of ferroelectricity at few nanometers scale is an advantage for nano refrigerators. The effect of Zr doping concentration on EC response of $Hf_{1-x}Zr_xO_2$ with x varying from 0.5 to 0.8 is shown in Figure 4.16b. Depending on the composition, it

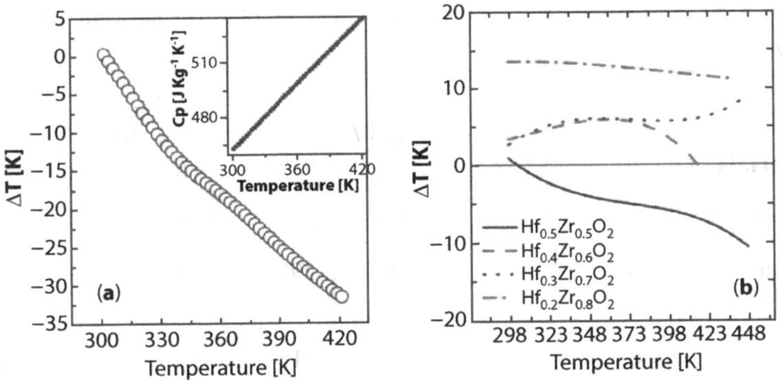

Figure 4.16 (a) Effect of Zr composition in $Hf_{1-x}Zr_xO_2$ films on its ΔT at different temperature. (Reprinted with permission from [140].) (b) Variation of ΔT with temperature exhibited by $Zr_{0.85}Ti_{0.15}O_2$ MIM configuration. (Reprinted with permission from [137].)

exhibited a positive and negative ECE effect. The $Hf_{0.5}Zr_{0.5}O_2$ film exhibited a negative ECE in the temperature range of 310–448 K, whereas other compositions exhibited a positive ECE and $Hf_{0.2}Zr_{0.8}O_2$ films showed optimum ΔT compared to other composition films [136, 137]. It is worthy to observe that $Hf_{0.2}Zr_{0.8}O_2$ exhibited good EC response even at room temperature, which make them attractive for real-time application. Further, the effect of different dopants (Zr, Si, Al, Gd, Y) on EC response of doped HfO_2 is summarized in Table 4.5. Among different dopants, yttrium-doped HfO_2 (Y:HfO_2) films exhibited optimum $\Delta T = 24.1$ K and EC responsivity ($\Delta T/\Delta E$) = 0.007 K.cm kV^{-1} at a temperature of 423 K. Although ZrO_2 and its AFE properties are suitable for ECE studies, the EC properties of ZrO_2 films have not been explored much yet. In 2020, B. Allouche *et al.* [138] reported the EC properties of an AFE ZrO_2 film experimentally for the first time. It exhibited negative ECE with $\Delta T = -31$ K under a field of 3.45 $MVcm^{-1}$ at a temperature of 413 K as seen in Figure 4.16a. The inset shows the variation of specific heat capacity with temperature. This high adiabatic temperature change is related to the crystalline disorder and AFE-FE phase transition. Later, Yu-Hua Liu *et al.* [139] found $\Delta T = -14.8$ K for an EF of 4000 kV/cm at 388 K for Ti-doped ZrO_2 thin films ($Zr_{0.85}Ti_{0.15}O_2$) in metal–insulator–metal (MIM) configuration. Therefore, HfO_2- and ZrO_2-based ferroelectrics are attractive for nanoscale solid-state refrigeration applications.

Table 4.5 ECE in HfO2-based thin films (cat% refers to the cationic percentage [137]).

Materials	ΔT (K)	ΔE (kV/cm)	T (K)	$\Delta T/\Delta E$ (K cm/kV)	Reference
$Hf_{0.2}Zr_{0.8}O_2$	13.4	3260	298	0.003	[136]
5.6 cat% Si-doped HfO_2	9.5	3330	298	0.003	[141]
3.8 mol% Si-doped HfO_2	4	1000	413	0.004	[141]
4.3 mol% Si-doped HfO_2	7.4	2330	393	0.003	[141]
5.7 cat% Al-doped HfO_2	−7.4	3900	80	−0.002	[142]
6.9 cat% Al-doped HfO_2	5.7	3900	295	0.001	[142]
3.9 mol% Gd-doped HfO_2	3.1	4000	295	0.0008	[142]
$Hf_{0.5}Zr_{0.5}O_2$	−10.8	3260	448	−0.003	[143]
Si-doped HfO_2	4.4	3000	353	-	[141]

4.5 Summary

The recent advances in the electrocaloric effect in ferroelectrics make them a promising solution for eco-friendly solid-state refrigeration technology. This chapter discussed different theoretical approaches of the electrocaloric effect, various direct methods to estimate adiabatic temperature change (ΔT), and different types of ferroelectric materials that include lead-based, barium titanate (BTO)-based, sodium/potassium niobate-based (KNN), and zirconia and hafnia-based ferroelectrics. In most of the literatures, ΔT was estimated from the indirect method, which may give error in estimation if temperature interval is not small, particularly in first-order phase transition ferroelectrics. Hence, direct measurement methods are preferable. Due to environmental issues, lead-free ferroelectric materials are essential. In this context, $BaTiO_3$-based materials exhibited better ECE with giant ΔT and responsivity ($\Delta T/\Delta E$) under low EF. However, they suffer from low dielectric breakdown, thus limiting their practical applications. Compared to BT-based materials, KNN-based materials exhibited high dielectric breakdown strength, but their ΔT and $\Delta T/\Delta E$ values are low, which need to be improved. Both KNN-based and $Hf_{1-x}Zr_xO_2$ exhibited a positive and a negative ECE depending on composition and operating temperature, which make them attractive for efficient and faster cooling. Therefore, we need to find better ways to improve dielectric breakdown strength in BT-based materials and EC response of other lead-free materials. Binary oxides (ZrO_2/HfO_2) seem to be attractive for the practical applications because of their large ΔT values, yet the direct measurement of EC on these materials is still lacking. Even though most of the materials exhibited giant ECE near the phase transition temperature, it is essential to bring phase transition temperature near RT for the development of efficient solid-state refrigerators.

References

1. Correia, T., Zhang, Q., Electrocaloric Effect: An Introduction. In: *Engineering Materials*. pp. 1–15. Springer, Berlin, Heidelberg, 2014.
2. Ožbolt, M., Kitanovski, Tušek, J., Poredoš, A., Electrocaloric vs. magnetocaloric energy conversion. *Int. J. Refrig.*, 37, 16, 2014.
3. Steven Brown, J. and Domanski, P.A., Review of alternative cooling technologies. *Appl. Therm. Eng.*, 64, 252, 2014.
4. Kumar, A., Thakre, Jeong, D.Y., Ryu, J., Prospects and challenges of the electrocaloric phenomenon in ferroelectric ceramics. *J. Mater. Chem. C*, 7, 6836, 2019.
5. Zhang, J., Hou, X., Zhang, Y., Tang, G., Wang, J., Electrocaloric effect in ferroelectric materials: From phase field to first-principles based effective Hamiltonian modeling. *Mater. Rep.: Energy*, 1, 100050, 2021.
6. Lu, S.G. and Zhang, Q., Large electrocaloric effect in relaxor ferroelectrics. *J. Adv. Dielectr.*, 02, 1230011, 2012.
7. Kutnjak, Z., Rožič, B., Pirc, R., Electrocaloric Effect: Theory, Measurements, and Applications. In: *Wiley Encyclopedia of Electrical and Electronics Engineering*. pp. 1–19. John Wiley & Sons, Inc. River Street, Hoboken, NJ, 2015.
8. Rose, M.C. and Cohen, R.E., Giant electrocaloric effect around T_c. *Phys. Rev. Lett.*, 109, 187604, 2012.
9. Wu, H.H., Zhu, J., Zhang, T.Y., Double hysteresis loops and large negative and positive electrocaloric effects in tetragonal ferroelectrics. *Phys. Chem. Chem. Phys.*, 17, 23897, 2015.

10. Scott, J.F., Electrocaloric materials. *Annu. Rev. Mater. Res.*, 41, 229, 2011.
11. Shebanovs, L., Borman, K., Lawless, W.N., Kalvane, A., Electrocaloric effect in some perovskite ferroelectric ceramics and multilayer capacitors. *Ferroelectr.*, 273, 137, 2002.
12. Lu, S.G., Zhang, Q.M., Kutnjak, Z., The electrocaloric effect (ECE) in ferroelectric polymer films, in: *Thin Film Growth*, pp. 364–383, Elsevier, Woodhead Publishing, Cambridgeshire, United Kingdom, 2011.
13. Mischenko, A.S., Zhang, Q., Scott, J.F., Whatmore, R.W., Mathur, N.D., Giant electrocaloric effect in thin-film $PbZr_{0.95}Ti_{0.05}O_3$. *Science*, 311, 1270, 2006.
14. Neese, B., Chu, B., Lu, S.G., Wang, Y., Furman, E., Zhang, Q.M., Large electrocaloric effect in ferroelectric polymers near room temperature. *Science*, 321, 821, 2008.
15. Valant, M., Electrocaloric materials for future solid-state refrigeration technologies. *Prog. Mater. Sci.*, 57, 980, 2012.
16. Kholkin, A.L., Kopyl, S., Tselev, A., Suchaneck, G., Electrocaloric-based applications: Challenges and perspectives, in: *The Electrocaloric Effect*, pp. 407–425, Elsevier, Woodhead Publishing, Cambridgeshire, United Kingdom, 2023.
17. Thomson, W., On the thermoelastic, thermomagnetic, and pyroelectric properties of matter. *Lond. Edinb. Dublin Philos. Mag. J. Sci.*, 5, 4, 1878.
18. Wiseman, G.G. and Kuebler, J.K., Electrocaloric effect in ferroelectric rochelle salt. *Phys. Rev.*, 131, 2023, 1963.
19. Radebaugh, R., Lawless, W.N., Siegwarth, J.D., Morrow, A.J., Feasibility of electrocaloric refrigeration for the 4–15 K temperature range. *Cryogenics*, 19, 187, 1979.
20. Lu, S.G., Rožič, B., Zhang, Q.M., Kutnjak, Z., Pirc, R., Lin, M., Li, X., Gorny, L., Comparison of directly and indirectly measured electrocaloric effect in relaxor ferroelectric polymers. *Appl. Phys. Lett.*, 97, 202901, 2010.
21. Zhang, Q. and Peng, B., Electrocaloric effect in lead-free ferroelectric perovskites, in: *The Electrocaloric Effect*, pp. 143–173, Woodhead Publishing, Cambridgeshire, United Kingdom, 2023.
22. Kar-Narayan, S., Crossley, S., Moya, X., Kovacova, Abergel, V.J., Bontempi, A., Baier, N., Defay, E., Mathur, N.D., Direct electrocaloric measurements of a multilayer capacitor using scanning thermal microscopy and infra-red imaging. *Appl. Phys. Lett.*, 102, 032903, 2013.
23. Kutnjak, Z. and Rožič, B., Indirect and direct measurements of the electrocaloric effect, in: *Electrocaloric Materials*, pp. 147–182, Springer, Berlin, Heidelberg, 2014.
24. Zhang, Y., Zhu, W., Hui, F., Lanza, M., Borca-Tasciuc, T., Muñoz Rojo, M., A review on principles and applications of scanning thermal microscopy (SThM). *Adv. Funct. Mater.*, 30, 1900892, 2020.
25. Liu, Y., Dkhil, B., Defay, E., Spatially resolved imaging of electrocaloric effect and the resultant heat flux in multilayer capacitors. *ACS Energy Lett.*, 1, 521, 2016.

26. Sanlialp, M., Shvartsman, V.V., Acosta, M., Dkhil, B., Lupascu, D.C., Strong electrocaloric effect in lead-free $0.65Ba(Zr_{0.2}Ti_{0.8})O_3$-$0.35(Ba_{0.7}Ca_{0.3})TiO_3$ ceramics obtained by direct measurements. *Appl. Phys. Lett.*, 106, 062901, 2015.
27. Biao, L., XinHua, W., ZhenHua, T., Bo, L., Tao, T., ZhiWei, X., TianFu, Z., XinGui, T., Yong, X., Jie, L., ShengGuo, L., Large electrocaloric effect in $BaTiO_3$ based multilayer ceramic capacitors. *Sci. China Technol. Sci.*, 59, 1054, 2016.
28. Liu, Y., Scott, J.F., Dkhil, B., Direct and indirect measurements on electrocaloric effect: Recent developments and perspectives. *Appl. Phys. Rev.*, 3, 031102, 2016.
29. Guyomar, D., Sebald, G., Guiffard, B., Seveyrat, L., Ferroelectric electrocaloric conversion in $0.75(PbMg_{1/3}Nb_{2/3}O_3)$–$0.25(PbTiO_3)$ ceramics. *J. Phys. D: Appl. Phys.*, 39, 4491, 2006.
30. Goupil, F.L., Berenov, A., Axelsson, A.K., Valant, M., Alford, McN.N., Direct and indirect electrocaloric measurements on 001-$PbMg_{1/3}Nb_{2/3}O_3$-$30PbTiO_3$ single crystals. *J. Appl. Phys.*, 111, 124109, 2012.
31. Lu, S.G., Rožič, B., Zhang, Q.M., Kutnjak, Z., Li, X., Furman, E., Gorny, L.J., Lin, M., Malič, B., Kosec, M., Blinc, R., Pirc, R., Organic and inorganic relaxor ferroelectrics with giant electrocaloric effect. *Appl. Phys. Lett.*, 97, 162904, 2010.
32. Smail, S., Benyoussef, M., Taïbi, K., Bensemma, N., Manoun, B., Marssi, M.E., Lahmar, A., Structural, dielectric, electrocaloric and energy storage properties of lead free $Ba_{0.975}La_{0.017}(ZrxTi_{0.95-x})Sn_{0.05}O_3$ (x=0.05; 0.20) ceramics. *Mater. Chem. Phys.*, 252, 123462, 2020.
33. Zhang, C., Du, Q., Li, W., Su, D., Shen, M., Qian, X., Li, B., Zhang, H., Jiang, S., Zhang, G., High electrocaloric effect in barium titanate-sodium niobate ceramics with core-shell grain assembly. *J. Materiomics*, 6, 618, 2020.
34. Chen, H., Ren, T.L., Wu, X.M., Yang, Y., Liu, L.T., Giant electrocaloric effect in lead-free thin film of strontium bismuth tantalite. *Appl. Phys. Lett.*, 94, 182902, 2009.
35. Ra, Yoo, C.M., J.H., Chol, S.H., Lee, J.Y., Electrocaloric effect of inorganic low temperature sintering $(Pb_{0.88}La_{0.08})(Zr_{0.65}Ti_{0.35})O_3$ ceramics. *IOP Conf. Ser.: Mater. Sci. Eng.*, 87, 012084, 2015.
36. Zhang, Q. and Whatmore, R.W., Improved ferroelectric and pyroelectric properties in Mn-doped lead zirconate titanate thin films. *J. Appl. Phys.*, 94, 5228, 2003.
37. Selvamani, R., Singh, G., Tiwari, V.S., Electro-caloric effect in PLZT (8/65/35) ceramic. *AIP Conf. Proc.*, 1447, 1281–1282, 2012.
38. Roh, I.J., Kwon, B., Gyu, H., Moon, Kim, J.S., Kang, C.Y., Electrocaloric effect in $Pb_{0.865}La_{0.09}(Zr_{0.65}Ti_{0.35})O_3$ thin film. *J. Sens. Sci. Technol.*, 23, 224, 2014.
39. Lu, B., Li, P., Tang, Z., Yao, Y., Gao, X., Kleemann, W., Lu, S.G., Large electrocaloric effect in relaxor ferroelectric and antiferroelectric lanthanum doped lead zirconate titanate ceramics. *Sci. Rep.*, 7, 45335, 2017.

40. Qiu, J.H. and Jiang, Q., Effect of electric field on electrocaloric effect in $Pb(Zr_{1-x}Ti_x)O_3$ solid solution. *Phys. Lett. A*, 372, 7191, 2008.
41. Zhang, T., Li, W., Cao, W., Hou, Y., Yu, Y., Fei, W., Giant electrocaloric effect in PZT bilayer thin films by utilizing the electric field engineering. *Appl. Phys. Lett.*, 108, 162902, 2016.
42. Qiu, J.H. and Jiang, Q., Misfit strain dependence of electrocaloric effect in epitaxial $Pb(Zr_{1-x}Ti_x)O_3$ thin films. *J. Appl. Phys.*, 103, 084105, 2008.
43. Zuo, Z., Chen, B., Wang, B., Yang, H., Zhan, Q., Liu, Y., Wang, J., Li, R.W., Strain assisted electrocaloric effect in $PbZr_{0.95}Ti_{0.05}O_3$ films on $0.7Pb(Mg_{1/3}Nb_{2/3})O_3$-$0.3PbTiO_3$ substrate. *Sci. Rep.*, 5, 16164, 2015.
44. Karthik, J. and Martin, L.W., Effect of domain walls on the electrocaloric properties of $Pb(Zr_{1-x}Ti_x)O_3$ thin films. *Appl. Phys. Lett.*, 99, 032904, 2011.
45. Weyland, F., Bradeško, A., Ma, Y.B., Koruza, J., Xu, B.X., Albe, K., Rojac, T., Novak, N., Impact of polarization dynamics and charged defects on the electrocaloric response of ferroelectric $Pb(Zr,Ti)O_3$ ceramics. *Energy Technol.*, 6, 1519, 2018.
46. Zhang, Q. and Whatmore, R.W., Sol-gel PZT and Mn-doped PZT thin films for pyroelectric applications. *J. Phys. D: Appl. Phys.*, 34, 2296, 2001.
47. Peng, B., Wang, T., Liu, L., Chen, X., Li, J., Zhang, Q., Yang, R., Sun, W., Wang, Z.L., P-GaN-substrate sprouted giant pure negative electrocaloric effect in Mn-doped $Pb(Zr_{0.3}Ti_{0.7})O_3$ thin film with a super-broad operational temperature range. *Nano Energy*, 86, 106059, 2021.
48. Roh, I.J., Kwon, B., Baek, S.H., Kim, S.K., Kim, J.S., Kang, C.Y., Thickness dependent electrocaloric effect in $Pb_{0.9}La_{0.1}Zr_{0.65}Ti_{0.35}O_3$ films grown by sol–gel process. *J. Electron. Mater.*, 45, 1057, 2016.
49. Wang, J., Jiang, G., Huang, W., Liu, D., Yang, B., Cao, W., Enhanced electrocaloric effect in a Si-doped $PbZr_{0.95}Ti_{0.05}O_3$ film deposited on FTO substrate. *Appl. Phys. Lett.*, 115, 053901, 2019.
50. Chen, C., Liang, R., Liu, Z., Yan, S., Nie, X., Zhou, Z., Dong, X., Coexistence of negative and positive electrocaloric effect in Sr and Nb Co-doped $Pb(Zr,Ti)O_3$ ferroelectric ceramics. *Mater. Lett.*, 189, 303, 2017.
51. Liu-Yang, H., Shao-Bo, G., Shi-Guang, Y., MIENSDenis, R., Gen-Shui, W., Xian-Lin, D., Electrocaloric effect in $Pb_{0.3}CaxSr_{0.7-x}TiO_3$ ceramics near room temperature. *Inorg. Mater.*, 36, 1011–1014, 2019.
52. Zhang, M., Liu, L., Yang, R., Yu, P., Zhang, Q., Peng, B., Giant electrocaloric effect in $BiFeO_3$ and La codoped $PbZr_{0.7}Ti_{0.3}O_3$ epitaxial thin films in a broad temperature range. *J. Materiomics*, 8, 156, 2022.
53. Akcay, G., Alpay, S.P., Rossetti, G.A., Scott, J.F., Influence of mechanical boundary conditions on the electrocaloric properties of ferroelectric thin films. *J. Appl. Phys.*, 103, 024104, 2008.
54. Qiu, J.H. and Jiang, Q., Effect of misfit strain on the electrocaloric effect in epitaxial $SrTiO_3$ thin films. *Eur. Phys. J. B*, 71, 15, 2009.

55. Qiu, J.H., Ding, J.N., Yuan, N.Y., Wang, X.Q., Yang, J., Effect of misfit strain on the electrocaloric effect of P(VDF-TrFE) copolymer thin films. *Eur. Phys. J. B*, 84, 25, 2011.
56. Li, B., Zhang, X., Wang, J.B., Zhong, X.L., Wang, F., Zhou, Y.C., Giant electrocaloric effect of $PbTiO_3$ thin film tuned in a wide temperature range by the anisotropic misfit strain. *Mech. Res. Commun.*, 55, 40, 2014.
57. Qiu, J.H., Ding, J.N., Yuan, N.Y., Wang, X.Q., Composition and misfit strain dependence of electrocaloric effect of $Pb_{1-x}Sr_x TiO_3$ thin films. *Chin. Phys. B*, 22, 017701, 2013.
58. Peng, Q., Kandula, K.R., Xia, Z., Zhang, C., Zhang, Y., Yang, Y., Zhang, Q., Jiang, S., Xu, J., Zhang, G., Electric field induced phase transitions and electrocaloric effect of La_{3+} doped $Pb(Zr,Sn,Ti)O_3$ ceramics. *Ceram. Int.*, 47, 13939, 2021.
59. Zhang, T., Li, W., Cao, W., Hou, Y., Yu, Y., Fei, W., Giant electrocaloric effect in PZT bilayer thin films by utilizing the electric field engineering. *Appl. Phys. Lett.*, 108, 162902, 2016.
60. Suchaneck, G., Electrocaloric effect in relaxor ferroelectrics, in: *The Electrocaloric Effect*, pp. 175–204, Elsevier, Woodhead Publishing, Cambridgeshire, United Kingdom, 2023.
61. Shaobo, L. and Yanqiu, L., Research on the electrocaloric effect of PMN/PT solid solution for ferroelectrics MEMS microcooler. *Mater. Sci. Eng. B*, 113, 46, 2004.
62. Guyomar, D., Sebald, G., Guiffard, B., Seveyrat, L., Ferroelectric electrocaloric conversion in $0.75(PbMg_{1/3}Nb_{2/3}O_3)$–$0.25(PbTiO_3)$ ceramics. *J. Phys. D: Appl. Phys.*, 39, 4491, 2006.
63. Sebald, L., Seveyrat, D., Guyomar, L., Lebrun, B., Guiffard, Pruvost, S., Electrocaloric and pyroelectric properties of $0.75Pb(Mg_{1/3}Nb_{2/3})O_3$–$0.25PbTiO_3$ single crystals. *J. Appl. Phys.*, 100, 124112, 2006.
64. Hagberg, J., Uusimäki, A., Jantunen, H., Electrocaloric characteristics in reactive sintered $0.87Pb(Mg_{1/3}Nb_{2/3})O_3$–$0.13PbTiO_3$. *Appl. Phys. Lett.*, 92, 132909, 2008.
65. Qiu, J.H., Wang, X.Q., Yuan, N.Y., Ding, J.N., Composition dependence of electrocaloric effect in $(1-x)Pb(Mg_{1/3}Nb_{2/3})O_3$–$xPbTiO_3$ single crystals. *Commun. Theor. Phys.*, 64, 113, 2015.
66. Peräntie, J., Hagberg, J., Uusimäki, A., Jantunen, H., Electric field induced dielectric and temperature changes in a ⟨011⟩ Oriented Pb $(Mg_{1/3}Nb_{2/3})O_3$-$PbTiO_3$ single crystal. *Phys. Rev. B*, 82, 134119, 2010.
67. Luo, L., Chen, H., Zhu, Y., Li, W., Luo, H., Zhang, Y., Pyroelectric and electrocaloric effect of <111> oriented 0.9PMN–0.1PT single crystal. *J. Alloys Compd.*, 509, 8149, 2011.
68. Feng, Z., Shi, D., Dou, S., Large electrocaloric effect in highly (001) oriented $0.67PbMg_{1/3}Nb_{2/3}O_3$–$0.33PbTiO_3$ thin films. *Solid State Commun.*, 151, 123, 2011.

69. Feng, Z., Shi, D., Zeng, R., Dou, S., Large electrocaloric effect of highly (100)-oriented $0.68PbMg_{1/3}Nb_{2/3}O_3$–$0.32PbTiO_3$ thin films with a $Pb(Zr_{0.3}Ti_{0.7})O_3/PbO_x$ buffer layer. *Thin Solid Films*, 519, 5433, 2011.
70. Acosta, M., Novak, N., Rojas, V., Patel, S., Vaish, R., Koruza, J., Rossetti, G.A., Rödel, J., $BaTiO_3$ based piezoelectrics: Fundamentals, current Status, and perspectives. *Appl. Phys. Rev.*, 4, 041305, 2017.
71. Bai, Y., Ding, K., Zheng, G.P., Shi, S.Q., Qiao, L., Entropy change measurement of electrocaloric effect of $BaTiO_3$ single crystal. *Phys. Status Solidi A*, 209, 941, 2012.
72. Moya, X., Taulats, E.S., Crossley, S., González-Alonso, D., Kar-Narayan, S., Planes, A., Mañosa, L., Mathur, N.D., Giant electrocaloric strength in single-crystal $BaTiO_3$. *Adv. Mater.*, 25, 1360, 2013.
73. Liu, Y., Wei, J., Janolin, P.E., Infante, I.C., Lou, X., Dkhil, B., Giant room temperature barocaloric effect and pressure-mediated electrocaloric effect in $BaTiO_3$ single crystal. *Appl. Phys. Lett.*, 104, 162904, 2014.
74. Cao, H.X. and Li, Z.Y., Electrocaloric effect in $BaTiO_3$ thin films. *J. Appl. Phys.*, 106, 094104, 2009.
75. Bai, Y., Zheng, G., Shi, S., Direct measurement of giant electrocaloric effect in $BaTiO_3$ multilayer thick film structure beyond theoretical prediction. *Appl. Phys. Lett.*, 96, 192902, 2010.
76. Bai, Y., Zheng, G.P., Ding, K., Qiao, L., Shi, S.Q., The giant electrocaloric effect and high effective cooling power near room temperature for $BaTiO_3$ thick film. *J. Appl. Phys.*, 110, 094103, 2011.
77. Luo, Z., Zhang, D., Liu, Y., Zhou, D., Yao, Y., Liu, C., Dkhil, B., Ren, X., Lou, X., Enhanced electrocaloric effect in lead free $BaTi_{1-x}Sn_xO_3$ ceramics near room temperature. *Appl. Phys. Lett.*, 105, 102904, 2014.
78. Hou, Y., Yang, L., Qian, X., Zhang, T., Zhang, Q.M., Electrocaloric response near room temperature in Zr- and Sn-doped $BaTiO_3$ systems. *Phil. Trans. R. Soc. A*, 374, 20160055, 2016.
79. Upadhyay, S.K., Reddy, V.R., Bag, P., Rawat, R., Gupta, S.M., Gupta, A., Electrocaloric effect in lead-free Sn doped $BaTiO_3$ ceramics at room temperature and low applied fields. *Appl. Phys. Lett.*, 105, 112907, 2014.
80. Qian, X.S., Ye, H.J., Zhang, Y.T., Gu, H., Li, X., Randall, C.A., Zhang, Q.M., Giant electrocaloric response over a broad temperature range in modified $BaTiO_3$ ceramics. *Adv. Funct. Mater.*, 24, 1300, 2014.
81. Qi, S., Zhang, G., Duan, L., Zeng, T., Cao, J., Electrocaloric effect in Pb-free Sr doped $BaTi_{0.9}Sn_{0.1}O_3$ ceramics. *Mater. Res. Bull.*, 91, 31, 2017.
82. Yao, Y., Zhou, C., Lv, D., Wang, D., Wu, H., Yang, Y., Ren, X., Large piezoelectricity and dielectric permittivity in $BaTiO_3$-$XBaSnO_3$ system: The role of phase coexisting. *EPL*, 98, 27008, 2012.
83. Wang, L., Wang, J., Li, B., Zhong, X., Wang, F., Song, H., Zeng, Y., Huang, D., Zhou, Y., Enhanced room temperature electrocaloric effect in barium titanate thin films with diffuse phase transition. *RSC Adv.*, 4, 21826, 2014.

84. Lu, S.G. and Zhang, Q., Electrocaloric materials for solid-state refrigeration. *Adv. Mater.*, 21, 1983, 2009.
85. Patel, S. and Kumar, M., Influence of grain size on the electrocaloric and pyroelectric properties in non-reducible $BaTiO_3$ ceramics. *AIP Adv.*, 10, 085302, 2020.
86. Sun, Z., Wang, Q.M., Slaughter, W.S., A solid-state refrigeration based on electrocaloric effect: Device and its analytical model. *J. Appl. Phys.*, 124, 064503, 2018.
87. Han, X., Bai, Y., Qiao, L.J., The electrocaloric effect of $BaTiO_3$ ceramics using hydrothermal synthesized nano-sized starting powders. *A.M.R.*, 624, 138, 2012.
88. Bai, Y., Han, X., Ding, K., Qiao, L., Electrocaloric refrigeration cycles with large cooling capacity in barium titanate ceramics near room temperature. *Energy Technol.*, 5, 703, 2017.
89. Wang, Y., Li, J., Yuan, R., Gao, H., Lv, J., Xue, D., Hao, X., Yang, Y., Lookman, T., Dkhil, B., Lou, X., Enhanced electrocaloric effect in BaSn/TiO3 ceramics by addition of CuO. *J. Alloys Compd.*, 851, 156772, 2021.
90. Zhao, Y., Wang, Q., Lv, J.H., Zhao, X., Wang, C.M., Dielectric properties and electrocaloric effect of Yttrium-modified $BaTiO_3$ ceramics. *Ceram. Int.*, 47, 18610, 2021.
91. Qiang, H., Fu, S., Xu, Z., Electrocaloric effect in Sn- and Y-codoped $BaTiO_3$ ceramics. *Phys. Status Solidi A*, 218, 2100090, 2021.
92. Gwizd, P., Sitko, D., Jankowska-Sumara, I., Krupska-Klimczak, M., The electrocaloric effect in $BaTiO_3$:Eu ceramics determined by an indirect method. *Phase Transitions*, 94, 192, 2021.
93. Zhang, B., Lou, X., Zheng, K., Xie, X., Shi, P., Guo, M., Zhu, X., Gao, Y., Liu, Q., Kang, R., Enhanced electrocaloric effect in the Sm and Hf co-doped $BaTiO_3$ ceramics. *Ceram. Int.*, 47, 1101, 2021.
94. Zhao, Y., Liu, X.Q., Wu, S.Y., Chen, X.M., Effect of phase transition on electrocaloric effect in indium substituted $BaTiO_3$ ceramics. *J. Alloys Compd.*, 822, 153632, 2020.
95. Kumar, P. and Prakash, C., Enhanced electrocaloric effect in lead free $Ba_{0.90}Sr_{0.10}Ti_{1-3x/4}Fe_xO_3$ ceramics. *J. Alloys Compd.*, 839, 155461, 2020.
96. Merselmiz, S., Merselmiz, S., Hanani, Z., Mezzane, D., Spreitzer, M., Bradeško, A., Fabijan, D., Vengust, D., Amjoud, M., Hajji, L., Abkhar, Z., Razumnaya, A.G., Rožič, B., Luk'yanchuk, I.A., Kutnjak, Z., High energy storage efficiency and large electrocaloric effect in lead-free $BaTi_{0.89}Sn_{0.11}O_3$ ceramic. *Ceram. Int.*, 46, 23867, 2020.
97. Jian, X.D., Lu, B., Li, D.D., Yao, Y.B., Tao, T., Liang, B., Guo, J.H., Zeng, Y.J., Chen, J.L., Lu, S.G., Direct measurement of large electrocaloric effect in $Ba(Zr_xTi_{1-x})O_3$ ceramics. *ACS Appl. Mater. Interfaces*, 10, 4801, 2018.
98. Liu, S., Xie, Q., Zhang, L., Zhao, Y., Wang, X., Mao, P., Wang, J., Lou, X., Tunable electrocaloric and energy storage behavior in the Ce, Mn hybrid doped $BaTiO_3$ ceramics. *J. Eur. Ceram.*, 38, 4664, 2018.

99. Upadhyay, S.K., Fatima, I., Reddy, V.R., Study of electro-caloric effect in Ca and Sn co-doped $BaTiO_3$ ceramics. *Mater. Res. Express*, 4, 046303, 2017.
100. Kaddoussi, H., Lahmar, A., Gagou, Y., Dellis, J.L., Khemakhem, H., Marssi, M.E., Electro-caloric effect in lead-free ferroelectric Ba1−Ca($Zr_{0.1}Ti_{0.9}$)0.925$Sn_{0.075}O_3$ ceramics. *Ceram. Int.*, 41, 15103, 2015.
101. Singh, G., Bhaumik, I., Ganesamoorthy, S., Bhatt, R., Karnal, A.K., Tiwari, V.S., Gupta, P.K., Electro caloric effect in $0.45BaZr_{0.2}Ti_{0.8}O_3$-$0.55Ba_{0.7}Ca_{0.3}TiO_3$ single crystal. *Appl. Phys. Lett.*, 102, 082902, 2013.
102. Zhou, Y., Lin, Q., Liu, W., Wang, D., Compositional dependence of electrocaloric effect in lead-free $(1-x)Ba(Zr_{0.2}Ti_{0.8})O_{3-x}(Ba_{0.7}Ca_{0.3})TiO_3$ ceramics. *RSC Adv.*, 6, 14084, 2016.
103. Lu, S., Chen, G., Zhang, Y., Zhao, Z., Li, F., Lv, Z., Ma, Z., Wang, D., Lu, C., Li, S., Electrocaloric effect in lead-free $0.5Ba(Zr_{0.2}Ti_{0.8})O_3$-$0.5(Ba_{0.7}Ca_{0.3})TiO_3$ ceramic measured by direct and indirect methods. *Ceram. Int.*, 44, 21950, 2018.
104. Bai, Y., Han, X., Qiao, L., Optimized electrocaloric refrigeration capacity in leadfree $(1-x)BaZr_{0.2}Ti_{0.8}O_3$-$XBa_{0.7}Ca_{0.3}TiO_3$ ceramics. *Appl. Phys. Lett.*, 102, 252904, 2013.
105. Zhou, Y., Lin, Q., Liu, W., Wang, D., Compositional dependence of electrocaloric effect in lead-free $(1-x)Ba(Zr_{0.2}Ti_{0.8})O_3-x(Ba_{0.7}Ca_{0.3})TiO_3$ ceramics. *RSC Adv.*, 6, 14084, 2016.
106. Ramana, E.V., Ferreira, N.M., Mahajan, A., Tobaldi, D.M., Bdikin, I., Rožič, B., Kutnjak, Z., Valente, M.A., Processing mediated enhancement of fFerroelectric and electrocaloric properties in $Ba(Ti_{0.8}Zr_{0.2})O_3-(Ba_{0.7}Ca_{0.3})TiO_3$ lead-free piezoelectrics. *J. Eur. Ceram.*, 41, 6424, 2021.
107. Kumar, N.S.K., Jayakrishnan, A.R., Silva, J.P.B., Sekhar, K.C., Effect of MgO doping on energy storage and electrocaloric properties of ferroelectric $0.6Ba(Zr_{0.2}Ti_{0.8})O_3-0.4(Ba_{0.7}Ca_{0.3})TiO_3$ ceramics. *Mater. Today Commun.*, 35, 105754, 2023.
108. Wang, J., Yang, T., Chen, S., Li, G., Zhang, Q., Yao, X., Nonadiabatic direct measurement electrocaloric effect in lead-free Ba,Ca(Zr,Ti)O_3 ceramics. *J. Alloys Compd.*, 550, 561, 2013.
109. Asbani, B., Dellis, J.L., A., J.L., Lahmar, J.L., Courty, M., Amjoud, M., Gagou, Y., Djellab, K., Mezzane, D., Kutnjak, Z., Marssi, M.E., Lead-free $Ba_{0.8}Ca_{0.2}(Zr_xTi_{1-x})O_3$ ceramics with large electrocaloric effect. *Appl. Phys. Lett.*, 106, 042902, 2015.
110. Singh, G., Tiwari, V.S., Gupta, P.K., Electro-caloric effect in $(Ba_{1-x}Ca_x)(Zr_{0.05}Ti_{0.95})O_3$ a lead-free ferroelectric material. *Appl. Phys. Lett.*, 103, 202903, 2013.
111. Patel, S., Chauhan, A., Vaish, R., Electrocaloric behavior and temperature dependent scaling of dynamic hysteresis of $Ba_{0.85}Ca_{0.15}Ti_{0.9}Zr_{0.1}O_3$ ceramics. *Int. J. Appl. Ceram. Technol.*, 12, 899, 2015.

112. Kaddoussi, H., Lahmar, A., Gagou, Y., Dellis, J.L., Khemakhem, H., Marssi, M.E., Electro caloric effect in lead-free ferroelectric Ba1–Ca($Zr_{0.1}Ti_{0.9}$)0.925$Sn_{0.075}O_3$ ceramics. *Ceram. Int.*, 41, 15103, 2015.
113. Patel, S., Sharma, P., Vaish, R., Enhanced electrocaloric effect in $Ba_{0.85}Ca_{0.15}Zr_{0.1}Ti_{0.9-x}Sn_xO_3$ ferroelectric ceramics. *Phase Transitions*, 89, 1062, 2016.
114. Valant, M., Electrocaloric materials for future solid-state refrigeration technologies. *Prog. Mater. Sci.*, 57, 980, 2012.
115. Hanani, Z., Mezzane, D., Amjoud, M., Razumnaya, A.G., Fourcade, S., Gagou, Y., Hoummada, K., Marssi, M., Gouné, M., Phase transitions, energy storage performances and electrocaloric effect of the lead-free $Ba_{0.85}Ca_{0.15}Zr_{0.10}Ti_{0.90}O_3$ ceramic relaxor. *J. Mater. Sci.: Mater. Electron.*, 30, 6430, 2019.
116. Liu, D., Li, Q., Yan, Q., Electro-caloric effect in a BCZT single crystal. *CrystEngComm*, 20, 1597, 2018.
117. Xu, Z., Qiang, H., Chen, Y., Liu, G., Room temperature electrocaloric effect in (1–x)$Ba_{0.67}Sr_{0.33}TiO_{3-X}Ba_{0.9}Ca_{0.1}Ti_{0.9}Zr_{0.1}O_3$ ceramics under moderate electric field. *J. Mater. Sci.: Mater. Electron.*, 29, 7227, 2018.
118. Belkhadir, S., Neqali, A., Amjoud, M., Mezzane, D., Alimoussa, A., Choukri, E., Gagou, Y., Raevski, I., Marssi, M., Luk'yanchuk, I.A., Rožič, B., Kutnjak, Z., Structural, dielectric and electrocaloric properties of ($Ba_{0.85}Ca_{0.15}$)($Ti_{0.9}Zr_{0.1-x}Sn_x$)O_3 ceramics elaborated by sol–gel method. *J. Mater. Sci.: Mater. Electron.*, 30, 14099, 2019.
119. Mezzourh, H., Belkhadir, S., Mezzane, D., Amjoud, M., Choukri, E., Lahmar, A., Gagou, Y., Kutnjak, Z., Marssi, M., Enhancing the dielectric, electrocaloric and energy storage properties of lead-free $Ba_{0.85}Ca_{0.15}Zr_{0.1}Ti_{0.9}O_3$ ceramics prepared via sol-gel process. *Phys. B: Condens. Matter*, 603, 412760, 2021.
120. Wei, F. *et al.*, Structure, dielectric, electrostrictive and electrocaloric properties of environmentally friendly Bi-substituted BCZT ferroelectric ceramics. *Ceram. Int.*, 47, 34676, 2021.
121. Patel, S. and Kumar, M., Electrocaloric properties of Sr and Sn doped BCZT lead-free ceramics. *Eur. Phys. J. Appl. Phys.*, 91, 20905, 2020.
122. Rožič, B., Koruza, J., Kutnjak, Z., Cordoyiannis, G., Malič, B., Kosec, M., The electrocaloric effect in lead-free $K_{0.5}Na_{0.5}NbO_3$-$SrTiO_3$ ceramics. *Ferroelectr.*, 446, 39, 2013.
123. Koruza, J., Rožič, B., Cordoyiannis, G., Malič, B., Kutnjak, Z., Large electrocaloric effect in lead-free $K_{0.5}Na_{0.5}NbO_3$-$SrTiO_3$ ceramics. *Appl. Phys. Lett.*, 106, 202905, 2015.
124. Kumar, R. and Singh, S., Enhanced electrocaloric effect in lead-free 0.9($K_{0.5}Na_{0.5}$)NbO_3 0.1Sr($Sc_{0.5}Nb_{0.5}$)O_3 ferroelectric nanocrystalline ceramics. *J. Alloys Compd.*, 723, 589, 2017.
125. Gupta, A., Kumar, R., Singh, S., Coexistence of negative and positive electrocaloric effect in lead-free 0.9($K_{0.5}Na_{0.5}$)NbO_3-0.1$SrTiO_3$ nanocrystalline ceramics. *Scr. Mater.*, 143, 5, 2018.

126. Kumar, R., Kumar, A., Singh, S., Large electrocaloric response and energy storage study in environmentally friendly $(1-x)K_{0.5}Na_{0.5}NbO_3-xLaNbO_3$ nanocrystalline ceramics. *Sustain. Energy Fuels*, 2, 2698, 2018.
127. Rozic, B., Koruza, J., Kutnjak, Z., Malic, B., Kosec, M., Direct measurements of the electrocaloric effect in lead-free $K_{0.5}Na_{0.5}NbO_3$-$SrTiO_3$ ceramics sintered in air. *Proceedings of ISAF-ECAPD-PFM 2012*, IEEE, Aveiro, pp. 1–4, 2012.
128. Wang, X., Wu, J., Dkhil, B., Xu, B., Wang, X., Dong, G., Yang, G., Lou, X., Enhanced electrocaloric effect near polymorphic phase boundary in lead-free potassium sodium niobate ceramics. *Appl. Phys. Lett.*, 110, 063904, 2017.
129. Kumar, R. and Singh, S., Giant electrocaloric and energy storage performance of $[(K0.5Na_{0.5})NbO_3](1-x)$-$[LiSbO_3]_x$ nanocrystalline ceramics. *Sci. Rep.*, 8, 3186, 2018.
130. Kumar, R. and Singh, S., Enhanced electrocaloric response and high energy-storage properties in lead-free (1-x) $(K_{0.5}Na_{0.5})NbO_3$-$xSrZrO_3$ nanocrystalline ceramics. *J. Alloys Compd.*, 764, 289, 2018.
131. Kumar, R., Khurana, D., Kumar, A., Singh, S., Giant negative electrocaloric effect and energy storage response in $0.94(K_{0.5}Na_{0.5})NbO_3$-$0.06SrMnO_3$ nanocrystalline ceramics. *Ceram. Int.*, 44, 20845, 2018.
132. Kumar, S. and Singh, S., Study of electrocaloric effect in lead-free $0.9K_{0.5}Na_{0.5}NbO_3-0.1CaZrO_3$ solid solution ceramics. *J. Mater. Sci.: Mater. Electron.*, 30, 12924, 2019.
133. Yang, J. and Hao, X., Electrocaloric effect and pyroelectric performance in $(K,Na)NbO_3$ based lead-free ceramics. *J. Am. Ceram. Soc.*, 102, 6817, 2019.
134. Böscke, T.S., Müller, J., Bräuhaus, D., Schröder, U., Böttger, U., Ferroelectricity in hafnium oxide thin films. *Appl. Phys. Lett.*, 99, 102903, 2011.
135. Silva, J.P.B., Sekhar, K.C., Negrea, R.F., MacManus-Driscoll, J.L., Pintilie, L., Progress and perspective on different strategies to achieve wake-up-free ferroelectric hafnia and zirconia-based thin films. *Appl. Mater. Today*, 26, 101394, 2022.
136. Park, M.H., Kim, H.J., Kim, Y.J., Moon, T., Kim, K.D., Hwang, C.S., Toward a multifunctional monolithic device based on pyroelectricity and the electrocaloric effect of thin antiferroelectric $Hf_xZr_{1-x}O_2$ films. *Nano Energy*, 12, 131, 2015.
137. Park, M.H., Hoffmann, M., Hwang, C.S., Pyroelectric and electrocaloric effects and their applications, in: *Ferroelectricity in Doped Hafnium Oxide: Materials, Properties and Devices*, pp. 217–244, Elsevier, Woodhead Publishing, Cambridgeshire, United Kingdom, 2019.
138. Allouche, B., Hwang, H.J., Yoo, T.J., Lee, B.H., A negative electrocaloric effect in an antiferroelectric zirconium dioxide thin film. *Nanoscale*, 12, 3894, 2020.
139. Liu, Y.-H., Wang, P.-C., Lin, L.-H., Wang, J.-C., Antiferroelectric titanium-doped zirconia thin films deposited via HiPIMS for highly efficient electrocaloric applications. *J. Eur. Ceram. Soc.*, 41, 3387, 2021.

140. Allouche, B., Hwang, H.J., Yoo, T.J., Lee, B.H., A negative electrocaloric effect in an antiferroelectric zirconium dioxide thin film. *Nanoscale*, 12, 3894, 2020.
141. Hoffmann, M., Schroeder, U., Künneth, C., Kersch, A., Starschich, S., Böttger, U., Mikolajick, T., Ferroelectric phase transitions in nanoscale HfO_2 films enable giant pyroelectric energy conversion and highly efficient supercapacitors. *Nano Energy*, 18, 154, 2015.
142. Park, M.H., Schenk, T., Hoffmann, M., Knebel, S., Gärtner, J., Mikolajick, T., Schroeder, U., Effect of acceptor doping on phase transitions of HfO_2 thin films for energy-related applications. *Nano Energy*, 36, 381, 2017.
143. Park, M.H., Kim, H.J., Kim, Y.J., Moon, T., Kim, K.D., Lee, S.D., Hyun, Y.H., Hwang, C.S., Giant negative electrocaloric effects of $Hf_{0.5}Zr_{0.5}O_2$ thin films. *Adv. Mater.*, 28, 7956, 2016.

5

Ferroelectric/Ferroelastoelectric Materials: Preparation, Improvement, and Characterizations

Serkan Baslayici[1] and Mehmet Bugdayci[1,2*]

[1]*İstanbul Medipol University, Vocational School, Contruction Tech. Department, Istanbul, Turkey*
[2]*Yalova University, Engineering Faculty, Chemical Engineering Department, Yalova, Turkey*

Abstract

Ferroelectric and ferroelastoelectric materials play a critical role in the development of electronic devices, information storage facilities, and sensing applications due to their polarization switching abilities. This book chapter offers an in-depth exploration of how to prepare, improve, and study the characteristics of these materials. The chapter covers a wide range of methods for preparing ferroelectric and ferroelastoelectric materials, including solid-state reaction, sol-gel, and hydrothermal approaches, among others.

The chapter also delves into how to enhance the properties of these materials through techniques like doping and composite structures. The authors discuss how external field influences can be applied to vary the compositions of compound structures and improve their features. Furthermore, the chapter provides insight into inspecting the characteristics of ferroelectric and ferroelastoelectric materials through electrical measurement coupled with X-ray diffraction and scanning electron microscopy.

It is essential to employ specific techniques to understand the peculiarities and tendencies of individual materials, particularly when optimizing their adequacy for specialized uses. This chapter imparts a comprehensive understanding of how to manufacture, enhance, and assess ferroelectric or ferroelastoelectric materials. Therefore, it is highly valuable for engineers and researchers working in these fields.

**Corresponding author*: mehmet.bugdayci@yalova.edu.tr

Inamuddin, Tariq Altalhi and Mohammad Abu Jafar Mazumder (eds.) Ferroic Materials-Based Technologies, (99–124) © 2024 Scrivener Publishing LLC

Keywords: Ferroelectric, ferroelastoelectric, barium titanate, lead zirconate titanate, solid-state, hydrothermal, piezoelectric, sensors

5.1 Introduction

Ferroelectric materials are distinct substances that manifest a remarkable spontaneity for electric polarization that can be conveniently reversed by external electrical fields. This exceptional propensity makes them ideal for diverse applications like nonvolatile memories, switchable capacitors, sensors, microelectromechanical systems (MEMS), and actuators [1, 2].

The most prevalent ferroelectric materials—barium titanate and lead zirconate titanate—are renowned for their heightened responsiveness to electric fields and therefore witness extensive use in electronic devices. As technology continues to advance, scientists delve deeper into the search for newer and more innovative forms of ferroelectric materials with superior conductive qualities or enhanced mechanical attributes cordoning off newer avenues for these exotic substances in electronic device applications. Figure 5.1 presents ferroelectric/ferroelastoelectric material usage areas.

Ferroelectric materials like barium titanate and lead zirconate titanate bear immense importance when it comes to electronics owing to their heightened sensitivity towards electrical fields. However, with technology constantly evolving, researchers are always on the lookout for new ferroelectric crystalline analogs capable of delivering amplified conductivity coupled with improved mechanical features—all aimed at broadening their utility horizons. Relaxor ferroelectrics represent one category of these cutting-edge materials owing to their disordered structure that yields a high dielectric constant along with piezoelectricity and energy density, making them ideal for various energy storage devices such as batteries and capacitors. Additionally, multiferroics exhibit both magnetic (ferromagnetic) and electric (ferroelectric) properties, making them suitable candidates for the next-generation spintronics gadgets that rely on the electron spin for information processing and storage. Finally, ferroelectric thin films provide an excellent way to craft ultra-thin yet high-performance capacitors that find their application in both microelectronics and energy storage realms. The field of ferroelectric materials holds remarkable potential when it comes to their impact on electronics innovations. Researchers are currently pursuing two distinct forms: organic ferroelectrics and hybrid ferroelectric materials. The former uses organic compounds, making them cost-effective yet flexible enough for use in wearable sensor devices whereas the latter combines impressive nanoparticles or polymers with existing ferroelectric

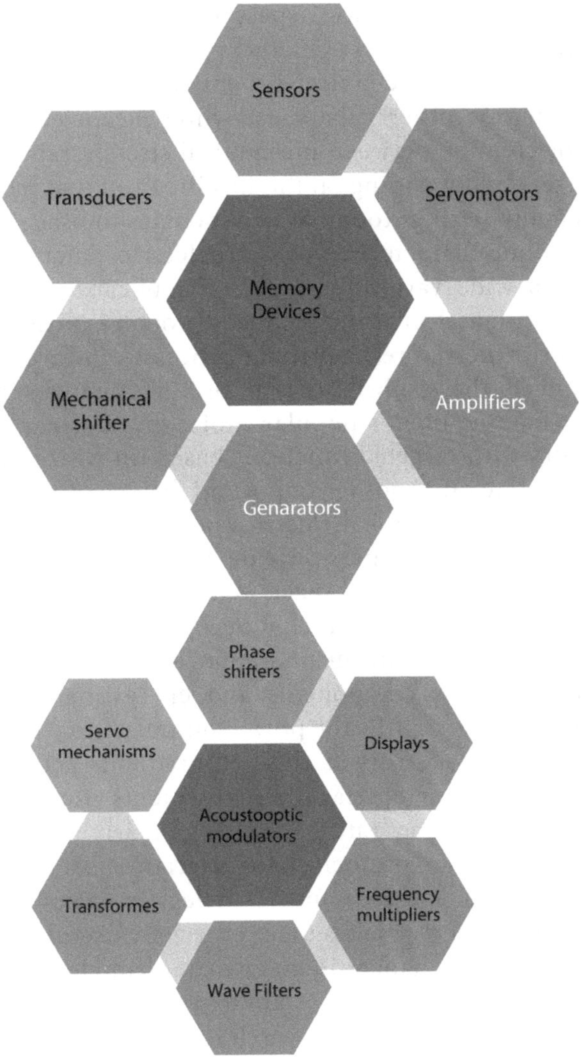

Figure 5.1 Usage areas of ferroelectric materials.

matter, which presents an opportunity to form new composite structures exhibiting singular features not previously seen. As experts dive deeper into this area, they anticipate discovering new varieties of these materials that open even more doors to their utility in various applications. What distinguishes these substances is the singular character they possess thanks to exceptional qualities like spontaneous polarization and strong piezoelectric effects fundamental to all members within this family [3–5].

The spontaneous polarization aspect allows reversal by an external electrical field, making it ideal as memory storage capacity, particularly useful via microelectromechanical system (MEMS) switchable capacitors and numerous other capacities, while piezoelectricity enables destruction or creation between mechanical stress/strain and electricity interchangeably, depending on the specific need that most researchers desire in many settings today. When it comes to usage in electronic devices, the unique characteristics of ferroelectric materials lend themselves well to a wide variety of different applications. Chief among these benefits is their electro-optic effect: through exposure to external electric fields or internal restructuring processes induced by material composition itself, these substances can produce changes in their optical properties making them a popular choice for constructing switches and modulators with targeted functions based on which types of light signals they operate on. Another important feature of these innovative substances is their switchable dielectric constant: this allows engineers and designers to fine-tune microwave devices or sensors' properties by adjusting how the material will react. With such versatility on offer, it is no surprise that they have quickly become an integral component across a range of electronic device applications, including random access memory storage components and energy harvesting devices capable of generating significant power outputs from environmental factors like mechanical stress or thermal gradients actuating mechanisms such as motors or optical tuning elements used for regulating light emissions/absorption rates. One particularly intriguing subset within this class are ferroelastoelectric materials substances exhibiting both ferroelastic (reshapeable) and ferroelectric (polarizable) behaviors. This unique property makes them especially useful as actuators or sensors that can detect changes in an environment electrical condition while also undergoing reversible strain experiences themselves due to applied electric fields. Ongoing research is finding new and improved ferroelastoelectric materials. These materials have properties of both ferroelectric and ferroelastic substances indicating a capacity for spontaneous polarization and strain when exposed to external electric or mechanical fields [6, 7].

Furthermore, they can also show a coupled response to electric and mechanical fields allowing changes in one field to influence the other. This is referred to as the electromechanical coupling coefficient, which plays an essential role in determining the effectiveness of these materials for various applications.

5.2 Structure and Properties of Ferroelectric and Ferroelastoelectric Materials

Ferroelectric materials have a distinct characteristic called spontaneous polarization, which arises from their non-centrosymmetric crystal structure. Figure 5.2 shows a centrosymmetric crystal structure. According to Figure 5.2, a crystal structure is symmetric from octahedral interstices (red point).

In detail, this means that the arrangement of atoms in the crystal lattice lacks an inversion center or center of symmetry. The point within a material's arrangement where its composition remains unaltered despite being inverted through it is known as an inversion center. If such centers are absent, then a non-centrosymmetric characteristic for that material's crystal structure emerges. When applied to ferroelectric polycrystals, this lack of an inversion center gives rise to spontaneous polarization that occurs regardless of whether there is an external electric field present or not. The ions within these materials' structures undergo displacement from their equilibrium positions leading to such polarizations emerging from them. Specific attributes like magnitude and directionality are dependent on aspects like atomic composition and overall crystalline framework for each respective ferroelectric material type identified through its smallest repeating units—otherwise known as its unit cell shape (which can be tetragonal, orthorhombic, or rhombohedral). Such shapes emerge due to the fact that arrangement-specific atoms within its atomic lattice are located along certain existing crystallographic axes. Ferroelectric materials are inherently

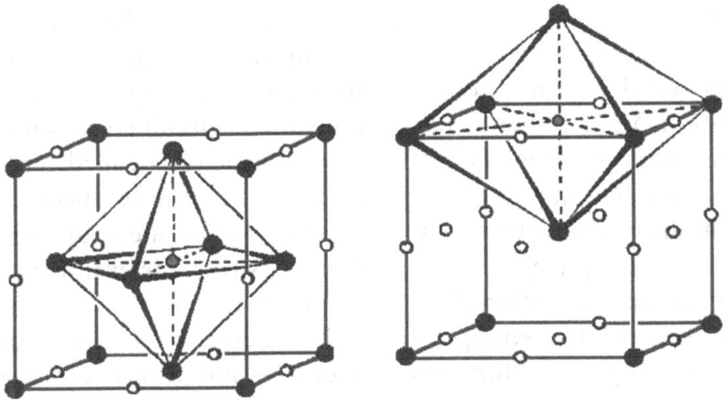

Figure 5.2 Centrosymmetric crystal structure.

symmetrical with crystalline structures, lending themselves well to analysis using group theory mathematical tools. This framework facilitates exploration of symmetry across different systems primarily through studying the way inversion, reflection, and rotation interacts in these contexts. In practice, researchers utilize a mathematical concept known as a point group comprising important components like mirror planes or inversion centers working together to maintain properties of intrinsic symmetry for each ferroelectric material studied. These components vary from material to material, resulting in the establishment of unique point groups for each structure being analyzed [8].

It is essential to understand these materials' intrinsic symmetries since they directly impact key physical properties such as their optical traits or piezoelectric response with spontaneous polarization orientation influencing piezoelectric response based on crystallographic axes. To understand better how certain characteristics are found in ferroelectric components, one must analyze how their symmetrical properties relate to theoretical groups capable of producing spontaneous polarizations within these compounds. They have specific structures that vary from center symmetry due to displacement along coordinate axes resulting from an external electric field that emerged due to many reasons such as ionic distributions or electron density differences [2, 4].

Ferroelectric material separates itself with its distinct quality through reversible spontaneous polarizations caused by a non-centrosymmetric crystal structure. The unique lattice structure can be measured using point groups and group theory analyzing how their ions are displaced from the original position in the lattice when exposed to an electric field resulting in reversed polarization. Ferroelectric substances boast an intriguing behavior referred to as hysteresis when it comes to their polarization pattern. Basically, speaking it means that by altering either magnitude or direction of an applied electric field, one can affect corresponding characteristics of polarization. Yet, another remarkable feature exemplified by some materials possessing both ferroelectricity and elasticity is called ferroelastoelectricity. Its defining trait lies in the ability of such substances' electrical polarizations to interact with mechanical strains by affecting each other's properties and applying enough mechanical stress, for example, and you may induce changes within electrical polarity [5].

This phenomenon emerges from the interaction between polarization–strain coupling within crystal structures found in relevant material samples.

In terms of specifics for either kind (ferroelastic or straight up ferro), descriptions will typically include both magnitude [measured via coulombs per square meter (C/m^2) or dipole moment per unit cell (Debye)] along with directional considerations requiring attention be paid to relevant crystallographic axes and spontaneous polarity orientation relative to those. The domain structure of ferroelectric substances determines their polarizing characteristics through various polarized domains within its composition. Spontaneous polarity is observable—reversible with external electric fields—while ferroelastoelectric elements exhibit an intriguing association between mechanical strain and electrical polarization. Physical attributes are significantly influenced by factors such as directionality, magnitude, and domain structure.

Moving on to dielectric properties, ferroelectric compounds belong to this class through their spontaneous polarity under electric field stimulation. The capacity for storing electrical energy is measured using specific parameters such as permittivity (or relative permittivity) and susceptibility. Ferroelectric material physics introduces fascinating properties into play, one being their orientation-dependent dielectric constants. When put under different applied electrical fields along different crystal axes, the resultant values differ significantly. Interestingly, permittivity also presents similar characteristics for the same matter type. A simple mathematical formula relates both properties: $\varepsilon = \varepsilon_{-r} \times \varepsilon_{-0}$, where free space permittivity is denoted by ε_{-0} and dielectric constant is denoted by ε_{-r} [9–11].

Researchers have found that these properties allow for controlling polarization effects in ferroelectric materials under external electrical fields—material susceptibility changes are measurable for varying field strengths using this approach. Ferroelectric materials' susceptibility exhibits anisotropy due to their inherent spontaneous polarization. An external electric field can alter this property by changing their polarization properties allowing manipulation of various parameters such as permittivity and dielectric constant necessary for storing and manipulating electrical energy under different conditions effectively. Significantly, these depend on directional factors related to crystal axes present within the material structure, thus providing key insights into its electrical behavior under influence by outside stimuli [1].

5.3 Synthesis Methods for Ferroelectric and Ferroelastoelectric Materials

Many different methods can be used for the synthesis of ferroelectric materials. These methods are given in Table 5.1.

Table 5.1 Synthesis methods of ferroelectric materials.

Name of the method
Solid-state reactions
Sol-gel techniques
Hydrothermal synthesis
Chemical vapor deposition (CVD)
Electrochemical deposition
Pulsed laser deposition
Molecular beam epitaxy

5.3.1 Solid-State Reactions

Solid-state reactions represent one robust method utilized widely for creating these desired structures by subjecting different starting solids at high pressure over specific temperature ranges. The solid-state reaction method offers an effective way to synthesize ceramic materials that boast exceptional ferroelectric and ferroelastoelectric qualities. Here is how it works: carefully select your starting ingredients (usually metal oxides or carbonates) for their desired stoichiometry; combine these raw materials with care by thoroughly mixing them via methods like ball milling or mortar and pestle action; perform calcination—heating up the powder mixture in a furnace for several hours to produce a hard solid and grinding/milling process afterwards into smaller sizes; finish off by sintering it once more at elevated temperature for densification. Creating dense and consolidated materials through particle compaction involves a specific process wherein particles come together through bonding. Amidst the creation of ferroelectric products, poling stands as the final stage, which requires an electric field application under high temperatures for electric dipole alignment resulting in net polarization. The synthesis of diverse ferroelectric and ferroelastoelectric elements including lead zirconate titanate (PZT) and barium titanate ($BaTiO_3$) is feasible with solid-state reactions. The reason behind ceramics being manufactured mainly from this method lies in their significant purity levels, dense formation, and controllable microstructure properties along with crystallographic orientation properties [12–15].

5.3.2 Sol-Gel Techniques

Apart from this mechanism, sol-gel techniques encompass hydrolysis followed by polymerization of metal alkoxides to create a sol that undergoes drying as well as calcination for generating strong substances. The sol-gel technique enjoys widespread use for crafting ceramic materials with standout ferroelectric and ferroelastoelectric characteristics. Followed in multiple steps, the process begins by preparing metal alkoxides from alcohol mixing with metal chlorides or nitrates. Hydrolysis next transpires as these alkoxides mingle with water and form metal hydroxides; then, subsequent polymerization produces a gel-like sol substance. After drying away any remaining solvent or water, the material is transformed into a porous solid called xerogel. Finally, high heat exposure envelops this xerogel in a furnace for calcination purposes—resulting in production of a compact solid material. The end goal when dealing with ferroelectric substances is creating net polarization by means of applying higher-temperature electric fields to ensure aligned electric dipoles. Sol-gel reactions provide an effective strategy for synthesizing different types of ferroelectrics and ferroelastoelectric compounds, such as lead zirconate titanate (PZT) or barium titanate ($BaTiO_3$). This method is desirable for creating ceramics due to its ability to produce materials with high purity and density while controlling the microstructure and crystallographic orientation of the final product [16–18].

5.3.3 Hydrothermal Synthesis

The hydrothermal synthesis approach produces ceramics with superb ferroelectric & ferroelastoelectric characteristics under high-temperature and -pressure conditions in an aqueous solution. A precursor solution containing dissolved starting materials is first made as part of this process. When creating a specific material, it is common for metal salts, complexing agents, and surfactants to be included in a solution that will ultimately dictate particle size and morphology. A sealed vessel called an autoclave is then used for heating said solution at high temperatures under pressure usually between 100 and 300°C while reaching several hundred atmospheres, allowing for varied reaction conditions according to the intended outcome of said synthesis. After completion of such reactions, aging further occurs using high temperature as well as pressure, which cause crystals formed through particle growth before cooling down back into room temperature again. To isolate this crystallized product from any impurities, filtration or centrifugation methods followed by thorough washing with water occur next. Lastly speaking only of ferroelectric materials must come poling them after this point, which involves applying an electric field at high temperatures so that dipole alignment occurs, creating a net polarization within the material itself. Generating a wide range of ferroelectric and ferroelastoelectric material choices like lead zirconate titanate (PZT) or barium titanate ($BaTiO_3$) among many others via the hydrothermal synthesis method has become quite commonplace. The key benefits of this practice include easily producing high-purity density ceramics while also holding precise control over particles' morphology and size [19–21].

5.3.4 Chemical Vapor Deposition (CVD)

The chemical vapor deposition (CVD) process differs in that it is utilized to place ultra-thin films onto substrates under closely monitored conditions such as within vacuums or controlled atmospheres. The substance used during CVD must display either ferroelectric or ferroelastic properties to fit with the task at hand. Achieving success within CVD requires mixing precursor gases tailored specifically to the desired outcome beforehand. Creating thin films involves employing volatile organometallic compounds developed through thermal decomposition or plasma-enhanced chemical vapor deposition (PECVD). Precursors are chosen based on their ability to provide particular compositions required for final films. To begin, chambers specific to depositing thin films are utilized by introducing precursors into them so that they may react with substrates consisting typically of silicon or sapphire heated adequately enough to promote precursor breakdown. These reactions produce solid films deposited via several methods such as using plasma for enhanced chemical reactions or pass gas over substrates in flow reactors. After deposition, annealing at high temperatures enhances crystal structure, promoting desirable ferroelectric properties. Thin films require poling to achieve ferroelectric properties. This final step involves applying an electric field at high temperatures to align electric dipoles and establish net polarization in the film typically through electrodes generating high voltages. Chemical vapor deposition (CVD) boasts versatility in depositing thin films of various ferroelastic materials including PZT and $BaTiO_3$. CVD offers precise compositional control on different substrates while also allowing low-temperature processing to maintain substrate compatibility [22–28].

5.3.5 Electrochemical Deposition

When synthesizing thin films of ferroelectrics, researchers can take advantage of electrochemical deposition (ECD). Understanding each facet of the

ECD process requires approaching its components individually. The first step involves preparation of electrolyte; solvent selection plays a crucial role since desired final film composition must be achieved after dissolving specific metal salts within it—organic solvents are sometimes preferred but water may also be used. Deposition secondly ensues, requiring placement of substrate within the said electrolyte solution where it will serve as a cathode while immersion of another electrode constructed from the same or separate material into same solution will act as an anode; voltage applied between these two electrodes causes movement towards the cathode by metallic ions present in the said liquid, which then deposits onto the surface creating a solid film; precision adjustments are utilized here such as variations in voltage plus other deposition parameters like electrolyte concentration and temperature to maintain control over thickness/deposition rate. Lastly, annealing and poling are conducted at higher temperatures post-deposition to improve crystal structure as well as amplify desired ferroelastoelectric or ferroelectric properties. To create ferroelectric materials, the final crucial step involves poling the film by applying electrical fields at high temperatures for aligning electric dipoles towards creating net polarization within the film, thus obtaining a desirable outcome. Electrochemical deposition (ECD) proves to be very versatile in synthesizing various types of ferroelectric and ferroelastoelectric materials (such as PZT and $BaTiO_3$), enabling precise control over thickness and composition as well as easy deposition onto different substrate options for making thin films, unlike the alternative methods requiring higher temperature processing such as sputtering or CVD. While ECD comparatively costs less than other alternatives mentioned above and has considerable potential, its requirement of a conducting substrate could pose limitations with certain applications that utilize non-conductive substrates—an aspect that must be kept in mind while choosing this method! Ferroelectric and ferroelastoelectric materials with thin films require precision synthesis techniques such as pulsed laser deposition (PLD). In this technique, scientists first prepare high-purity ceramic targets for their desired materials. These targets are mounted on rotating stages inside vacuum chambers where they can be irradiated by pulsed lasers. The resulting ablation creates plasma plumes containing atoms and ions, which are then deposited onto properly placed substrates [29–32].

However, it is crucial to optimize factors like electrolyte composition, temperature, and other deposition parameters because they could significantly impact film quality and properties. Scientists must therefore carefully tailor their optimization process to each material and application to ensure success. To improve adhesion and crystallinity of films, heating the substrate at high temperatures is beneficial. To regulate film thickness and deposition rates, adjustment of certain parameters like wavelength, fluence, and repetition rate of lasers becomes necessary. Typically, after depositing films annealing in high temperature creates an ideal environment to refine crystal structures while inducing favorable ferroelastic or ferroelectric properties. Aligning electric dipoles altogether requires an electric field application at high temperatures for polarization in poling for ferroelectric materials.

5.3.6 Pulsed Laser Deposition

Lead zirconate titanate (PZT) and barium titanate ($BaTiO_3$), among many others that are both ferroelastic or/and ferroelectric materials, can be deposited using pulsed laser deposition (PLD). Thin films' synthesis by PLD remains popular due to its precision control over film compositions/thickness deposited on different substrates. PLD is a remarkable technique exhibiting versatility by operating at lower temperatures while offering compatibility with an extensive variety of materials and substrates. Its ability to deposit high-quality crystalline films with impeccable stoichiometry and uniformity along with intricate multi-layered structures featuring outstanding precision and reproducibility underscores its superiority. One limiting factor remains the costly equipment requirements coupled with the need for high-purity target materials. Furthermore, optimizing laser parameters like wavelength, fluence, and repetition rate is essential to determine an ideal films quality for each material type or application. Molecular beam epitaxy (MBE) relies on a highly sophisticated procedure explicitly intended to synthesize thin films comprising ferroelectric or ferroelastoelectric substances that exhibit exceptional crystal quality while maintaining precise control over composition. This process includes several steps that begin by preparing the highest-quality single crystal substrate imaginable before moving forward. Prior to depositing a film on a substrate, it is essential first to clean it up by polishing until all surface contamination and defects are removed. The actual deposition happens inside a vacuum chamber where molecular/atomic species get directed towards the substrate through a beam source that uses various techniques like plasma sources or effusion cells [33–37].

5.3.7 Molecular Beam Epitaxy

High-quality crystalline films result from maintaining temperatures ranging from 400 to 800°C plus manipulation of beam flux and substrate temperature to control the growth rate and composition of the film.

RHEED (Reflection High Energy Electron Diffraction) and STM (Scanning Tunneling Microscopy) are critical in situ characterization techniques that monitor film growth in real time. With them, surface morphology and crystal structures are evaluated, which provides a basis for optimizing growth factors and achieving high-quality films. Following the completion of deposition, an essential procedure called annealing takes place where a high-temperature process helps enhance the crystal structure while simultaneously creating necessary ferroelectric or ferroelastoelectric properties within the film. In case of ferroelectric materials, poling comes up next, which involves applying an electric field at elevated temperatures inducing alignment of electric dipoles that results in the creation of net polarization within the film. The application of MBE offers immense control over composition as well as crystal structure, allowing for synthesis of top-notch quality films that are either ferroelectric or ferroelastoelectric with extremely low defect density. In situ characterization methods assist during deposition, helping keep track of real-time progress, thus ensuring optimum growth conditions leading towards required properties [38–41].

However, expertise accompanied by dedicated equipment requirements makes MBE quite a complex as well as expensive technique.

Ferroelectric materials' diverse applications range from sensors to energy storage devices; however, their functionality hinges on properties like ferroelastic responses and thermal stability, which are subject to improvement using the technique of doping. Doping involves adding impurities that alter the materials' electronic structure leading to improved electrical performance. For example, adding rare earth ions like lanthanum enhances lead zirconate titanate (PZT) ceramics' piezoelectric characteristics; however, availability of high-quality substrates and precise control over beam flux and substrate temperature limit this technique's effectiveness. Thus, improving the quality of substrates alongside precision control methods is essential for maximizing the potential of ferroelectric materials in various fields. An alternate strategy concerns modifying fabrication procedures for these materials by adjusting processing conditions. For instance, microstructure and properties of a specific material can be affected by managing sintering temperature and environment beforehand; this has proven quite promising for upgrading various traits like thermal stability, dielectric constant, or piezoelectric response, among others. Furthermore, nanoscale engineering techniques based on nanoparticle doping or thin-film deposition have shown potential for producing better versions of ferroelectric/ferroelastoelectric materials too: nanoparticle doping improves density along grain boundaries while reducing defects for more ideal dielectric/piezoelectric performance while thin-film deposition using atomic layer deposition (ALD) or chemical vapor deposition (CVD) optimize crystal quality alongside controlling orientation/film thickness to upgrade ferroelectric attributes even further. Advancements in computational materials science have led to newfound success when it comes to anticipating and producing novel ferroelectric/ferroelastoelectric materials with vast amounts of improved functionality. Now, computational methods can facilitate screening for new materials that match desired attributes while simultaneously predicting the effect doping or changing processing conditions will have on them, thereby allowing for more efficient optimization of relevant microstructure and crystal structure. The usefulness of ferroelectric as well as ferroelastoelectric materials for practical applications depends largely on their enhancement characteristics via different strategies such as doping or substitution.

5.4 Improvement of Ferroelectric and Ferroelastoelectric Materials

Doping is an approach used to introduce impurities into a material for modifying their properties—like improving the dielectric behavior and enhancing their piezo- or ferroelectric qualities among others—relevant for these particular materials. For instance, adding elements like lanthanum or niobium to lead zirconate titanate (PZT) ceramics can help heighten their piezoelectric and ferroelectric properties. On the other hand, substitution is an approach that involves replacing one material component element with another to alter its crystal structure or electronic configuration. Substitution proves effective in enhancing the characteristics of ferroelectric and ferroelastoelectric materials. For instance, replacing lead with bismuth in PZT ceramics can lead to a more significant enhancement of their piezoelectric capability due to bismuth's higher atomic mass and polarizability. The procedures for controlling properties of materials vary significantly with factors like grain size playing essential roles. Smaller grain sizes tend to amplify the density of defects and boundaries, leading to enhancements such as fatigue resistance and improved dielectric constants and piezoelectric coefficients, among others observed in ferroelectric and ferroelastoelectric substances. The reduction of grain size within these substances usually requires specialized procedures including high-pressure torsion or severe plastic deformation approaches [42, 43].

Strain engineering is also another commonly used technique for manipulating the features of these substances by exposing them to mechanical or electrical stresses that suit desired alterations uniquely. For example, exposure to compressive strains results in increased efficiency across different ceramic forms like PZT ceramics enhancing their dielectric constants and piezoelectric coefficients. Finally, it is critical also to consider surface modification as another factor facilitating changes within materials' responses under varying environmental conditions besides grain size control. Ferroelectric as well as ferroelastoelectric materials open avenues for innovation through various techniques that can modify their surfaces. Ion implantation is one such technique that can alter the material's surface chemistry as well as its overall morphology. Additionally, applying a coat or etching are other options that could improve different properties of the material such as increasing mechanical strength or enhancing its conductivity. For example, placing graphene on top of these materials could lead to substantial improvements.

Another novel area where progress has been made is through synthesizing new materials that offer better attributes than their predecessors via computational methods using density functional theory that predicts electronic structure or high-throughput experimentation used for screening purposes.

Overall improvement in these types of material could be achieved by playing with different approaches such as doping, substitution, analyzing grain sizes, controlling engineered strain, changing up surface characteristics, or even by developing an entirely new type of material. The customization process of modifying specific applications helps achieve more satisfactory performance levels from ferroelastic and ferroelectric materials through using tailored techniques for each case study. The application mechanisms are different with respect to effectiveness, if we compare them against each other: Doping induces certain unsolicited changes into the lattice of the material by adding impurities; as a result, the material may exhibit better mechanical properties or polarization abilities while its defects increase, which may decrease overall quality. The strain engineering method applies mechanical pressure to alter some properties like inducing ferroelectricity in non-ferroelectric materials, enhancing piezoelectric characteristics, or maximizing polarization levels. However, excessive strain can lead to material failure, which is not convenient. Annealing is a technique where high temperature heats a material up; it leads to the modification of its crystal structure and enhances piezoelectric and ferroelectric properties effectively. Although annealing has some positive effects on materials, sometimes it can also lead to formation of defects or transition between phases, which are not desirable outcomes. Meanwhile, chemical modifications involve changes in materials' chemistry that improve their dielectricity or piezoelectricity; yet, this approach could compromise overall quality due to increased defect introduction risk during processing. Surface modification can also enhance material performance by changing its surface property, but like other methods, it could result in a lower quality due to additional defects. In conclusion, the choice of enhancement method should be based on specific requirements and target properties of the material. Electric field poling is one such enhancement technique that involves application of high electric field for aligning domain polarization in a specified direction leading to higher ferroelectricity or piezoelectricity. To boost materials' ferroelectricity and piezoelectricity properties, several methods exist such as electric field poling, thermal poling, ion beam irradiation, or high-pressure treatment [44–47].

Electric field poling is typically performed at elevated temperatures, reducing coercive fields required for domain switching. Nevertheless, overuse can cause domain wall movements and increase fatigue across the material. Thermal poling strengthens ferroelectricity through heating substances until voltage aligns polarization domains in a specific direction, lowering coercive fields needed for domain switching. Unfortunately, excessive thermal activity can trigger fatigue while creating defects in materials.

Ion beam irradiation promotes structure changes in materials when bombarded with high-energy ions, thereby improving ferroelectric and piezoelectric properties while enhancing their resistance to radiation damage. However, this process could negatively influence material quality through the introduction of defects. High-pressure treatment modifies material properties through exposure to high pressure levels but may result in visible defects that limit performance and affect quality. Applying adequate pressure offers an effective strategy when seeking to improve materials' mechanical strength while enhancing their ferroelectric/piezoelectric features from fabrication processes with distinct advantages over alternative methods; however, excessive pressure must be avoided to prevent material failure. Magnetic field poling represents another approach that has proven successful in aligning polarization domains and improving ferroelectric/piezoelectric properties reducing coercive fields needed for domain switching. However, vigilance is necessary when using this method to prevent domain wall motion and resultant material fatigue. Choosing among these various techniques depends on numerous factors like specific application requirements and desired property configurations. Performance characterization is a crucial aspect of studying these materials that can provide valuable insights into their behaviors and properties under differing external stimuli or influencing parameters. Performance characterization techniques have enabled experts to examine ferroelectric materials effectively. Electrical measurements such as current–voltage (I–V) curves and impedance spectroscopy allow researchers access into variables encompassing dielectric constant polarization or coercive field with respect to their contributions towards identifying electrical properties pivotal towards their study as well known from previous studies [48, 49].

Mechanical evaluations based on stress–strain curves provide valuable insight into elastic moduli or Poisson's ratio, which became a recurring aspect while researching mechanical features concerning some specific types of these materials such as ferroelastoelectric materials found within structural interfaces.

Also, optical measurements have seen practical application in areas like Raman spectroscopy or infrared spectroscopy, which can help determine essential parameters such as crystal structure composition, which is critical for understanding ferroelectric materials better.

Moreover, magnetic measurements through techniques such as magnetization hysteresis loops become crucial when pursuing research endeavors in the domain of magnetic properties affixed to ferroelectric materials.

To optimize the performance of ferroelectric and ferroelastoelectric materials, it is crucial to understand their properties and behavior. Performance characterization techniques provide us with valuable information regarding these materials.

Measurements related to magnetic properties are important for determining key parameters such as magnetic moment and magnetic anisotropy. Other relevant magnetic parameters can also be determined through these measurements.

Thermal measurements such as differential scanning calorimetry (DSC), thermogravimetric analysis (TGA), and thermal conductivity measurements provide a means of understanding how ferroelectric materials respond to changes in temperature. This knowledge is essential for studying thermal properties, including the glass transition temperature and thermal conductivity.

Piezoelectric measurements, particularly those related to the piezoelectric coefficient d33, allow us to examine how well a material can convert mechanical energy into electrical energy and vice versa. These measurements offer insight into the piezoelectric properties of ferroelectric materials.

Overall performance characterization techniques play a critical role in providing valuable information about the behavior of ferroelectric and ferroelastoelectric materials. This information is necessary for optimizing their performance and developing new applications.

5.5 Applications of Ferroelectric and Ferroelastoelectric Materials

Smart materials are designed to respond to external stimuli, such as temperature or electric fields. Ferroelectric and ferroelastoelectric materials have been identified as promising candidates for developing smart materials due to their unique properties that allow them to change in response to external forces or fields. Ferroelectric materials possess the remarkable quality needed to sustain smart material application. The reversible alteration of polarization towards electric fields grants flexibility regarding shape, stiffness, or optical properties. Notably, piezoelectric actuators stand as the most frequent implementation of ferroelectric components by energizing electrical output into mechanical input with various uses such as micro- and nano-positioning, vibration control mechanisms, and robotic systems. Further usage presents itself with the utilization of ferroelectricity by serving as sensors transforming thermal or mechanical energy into an electric signal; pressure sensors or temperature detectors for industry or medicine are among these applications. Moreover, their non-volatile memory potential offers superior storage performance retaining important data regardless of whether power is off. Ferroelectric materials' unique properties render them ideal for various applications like digital cameras, mobile phones, and portable music players due to their versatility and ease of use. These compounds are equally effective within optical devices where they promptly change their refractive index following electrical fields, which effectively regulates light intensity in fiber-optic networks using variable optical attenuators. The immense potential offered by these compounds in developing new smart materials and innovative devices makes them even more valuable [50–53].

Additionally, ferroelectrics properties make them indispensable within microelectronics across several critical areas by enabling retention of polarization even after removal of external electric fields, improving speed rates while consuming less energy during the operation compared to traditional memories via nonvolatile storage systems such as Ferroelectric Random Access Memory (FeRAM). Further ferroelectrics display piezoelectricity when mechanical stress is applied, generating an electrical charge and vice versa. Ferroelectric materials have revolutionized the field of microelectronics by providing workable solutions drawn from their unique properties. Piezoelectricity that manifests within these materials is pivotal in creating functional piezoelectric transducers—a seamless fit for numerous devices such as microphones, sensors, and actuators. Furthermore, ferroelectric substances play an essential role in electro-optic modulators by altering refractive indices under electric fields; this ability allows for efficient light signal modulation necessary for communication infrastructure's smooth operation. Yet, other functions are evident within MEMS technology where thin films function as sensing elements for accelerometers and pressure sensors alongside gyroscope parts—all crucial components driving MEMS devices' functionality. Energy harvesting from mechanical stress-induced voltage among ferroelectric substances remains another vital advantage utilized in powering small electronic gadgets. Indeed, the future holds endless possibilities regarding how best we can use these versatile materials creatively. Sensors and actuators have revolutionized modern technology by making it easier to measure different parameters accurately or transform electrical energy into precise mechanical motion. One material that has emerged as an outstanding choice for designing such systems is ferroelectric material due to its unique electromechanical characteristics. Here are some examples demonstrating how these traits make it highly practical for creating sensors and actuators across various applications:

Pressure Sensors: Ferroelectric materials exhibit unparalleled sensitivity levels, making them an excellent option for developing high-precision pressure sensors. By merely applying external pressures on this type of material, changes in polarization occur, which lead to the creation of an electric field that can be precisely measured to determine the applied pressure.

Accelerometers: Ferroelectric materials are equally effective in creating sensitive accelerometers suitable for use in a variety of applications. These materials respond remarkably well when subjected to acceleration forces by creating mechanical stress on them leading to changes in polarization that can accurately measure acceleration levels.

Piezoelectric Actuators: The strong piezoelectric effect observed in ferroelectric materials enables them to perform with great precision when serving as actuators whereby they convert electrical energy into mechanical motion. The capacity for generating an electric charge due to exposure to external pressure or conversely producing a mechanized response on exposure allows ferroelectric compounds unique advantages over other materials. These capabilities make them ideal for designing piezoelectric actuators known for impeccable control over mechanical movement. Their usefulness extends further to manufacture efficient ultrasonic transducers serving diverse medicinal or industrial purposes by leveraging upon their intrinsic traits. By channeling ferroelectric resources to create ultrasonic waves via mechanical vibrations from electric fields, these transducers offer unparalleled performance levels. Furthermore, the sensitivity of ferroelectric materials towards variations in electric fields allows for constructing sensitive microphones that can detect incoming sound waves with remarkable precision. In summary, this property of ferroelectric substances underlines their suitability as a beneficiary resource for various applications such as sensing and actuating that mandate exceptional sensitivity towards both electrical fields and mechanical stress. Ferroelectric materials possess robust piezoelectric properties due to their spontaneous polarization. As a result, they have found numerous applications in the field of piezoelectricity [54–58]. Examples given in Table 5.2 include the following:

Sensors and Transducers: Piezoelectric sensors and transducers can convert physical forces like pressure, acceleration, or vibration into electric signals commonly used across diverse industries like healthcare, space technology, or automobile engineering.

Ultrasonic Imaging: Ferroelectric materials play an indispensable role in medical imaging equipment such as ultrasound machines because they generate ultrasonic waves that penetrate the body producing images of internal organs and tissues.

Energy Harvesting: The piezoelectric technology allows harvesting energy from mechanical vibrations, converting it into electrical power to charge small electronic devices like sensors without relying on batteries.

Piezoelectric motors: The application of ferroelectric materials in constructing motors with a capacity for converting electrical energy into motion is yet another example of their utility. From precision positioning systems to robotics, ferroelectric-based motors are critical to many

Table 5.2 Applications of ferroelectric materials.

Applications of ferroelectric materials
Electronics
Sensors
Actuators
Energy harvesting
Optical devices
Piezoelectric devices
Data storage

industries that demand speed and accuracy. Additionally, their piezoelectric features allow for the development of acoustic devices like microphones and speakers. Overall, the diverse applications for ferroelectric materials underscore their importance across numerous sectors as key constituents in high-performance machinery.

The one-of-a-kind characteristics inherent to ferroelectrics qualify them as perfect contenders for electricity storage and transformation purposes. By serving as dielectric films in capacitor assemblies, which store much more charge than standard capacitors, ferroelectrics offer a particularly strong promise due to their exceptional dielectric constant and polarization properties.

One notable realm in which this innovation could come in handy is energy storage, including electric car batteries or energy harvesting mechanism. Furthermore, with its ability to convert temperature alterations into electricity, thanks to the pyroelectric phenomenon, ferroelectrics could even prove useful for combatting climate change. Finally, there is the possibility that researchers may someday harness these exceptional materials for piezoelectric energy harvesting by creating an electrical charge based on mechanical stress or deformation—a development that could yield major technological leaps forward. The field of energy harvesting and conversion has seen tremendous development thanks to the potential application opportunities presented by ferroelectric materials' unique properties like high dielectric constant and polarization response. Their impressive capability generates electrical power from mechanical vibrations or strain and positions them perfectly among the next-generation technologies, driving energy efficiency gains; this marks an ongoing active research area with several exciting prospects that should be deployed soon.

In addition to these advancements, substantial progress has also been achieved in electrochemical energy storage devices such as batteries or supercapacitors that require superior charging/discharging rates, featuring another compelling reason behind studies into utilizing ferroelectrics due to their intrinsic features.

Furthermore, nonlinear optical responses have become one prominent attribute that makes ferroelectric materials invaluable within optics/photonics fields, allowing effective light manipulation using devices like frequency doublers, optical switches, and modulators. As an exemplar, periodic poled lithium niobate (PPLN) remains a frequently used ferroelectric material for doubling laser wavelengths. In sum, the future continues to look bright for ferroelectric materials due to their immense potential in various applications besides being a hot topic of research. Ferroelectric materials have shown considerable potential in boosting the performance of solar cells. Their characteristic polarization generates an inherent electric field that separates photogenerated charge carriers mitigating recombination losses and elevating the photovoltaic efficiency. One such material is lead zirconate titanate (PZT), which has been successfully employed to augment silicon solar cell efficiency. Ferroelectric components find extensive utility in electro-optic modulators as they provide means for manipulating light phase and polarization through the application of an electric field that flips their polarization, producing changes in refractive index and birefringence—an attribute highly advantageous in constructing electronic switches/modulators. In addition, ferroelectric materials offer significant prospects in serving as robust photocatalysts for various applications, including water splitting and pollutant degradation. The capacity for built-in electric fields within ferroelectric materials facilitates enhanced separation of photogenerated charge carriers while also bolstering photocatalytic activity. Bismuth ferrite ($BiFeO_3$) stands out as one such material with impressive efficacy in water splitting procedures. Given these desirable attributes, it is of little surprise that researchers have identified the vast potential for applying ferroelectric materials within the optics and solar energy domains.

References

1. Sekine, Y., Akiyoshi, R., Hayami, S., Recent advances in ferroelectric metal complexes. *Coord. Chem. Rev.*, 469, 214663, 2022.
2. Zhu, Y., Sun, K., Wu, S., Zhou, P., Fu, Y., Xia, J., Li, H.F., A comprehensive review on the ferroelectric orthochromates: Synthesis, property, and application. *Coord. Chem. Rev.*, 475, 214873, 2023.
3. Tylczyński, Z. and Wiesner, M., Ferroelastoelectric phase transition in (NH4)2CuCl4·2H2O single crystal. *Mater. Chem. Phys.*, 149, 566, 2015.
4. Silva, J.P.B., Sekhar, K.C., Negrea, R.F., MacManus-Driscoll, J.L., Pintilie, L., Progress and perspective on different strategies to achieve wake-up-free ferroelectric hafnia and zirconia-based thin films. *Appl. Mater. Today*, 26, 101394, 2022.
5. Kong, L.B., Zhang, T.S., Ma, J., Boey, F., Progress in synthesis of ferroelectric ceramic materials via high-energy mechanochemical technique. *Prog. Mater. Sci.*, 53, 207, 2008.
6. Khan, M.A., Nadeem, M.A., Idriss, H., Ferroelectric polarization effect on surface chemistry and photo-catalytic activity: A review. *Surf. Sci. Rep.*, 71, 1, 2016.
7. Jin, X., Zhang, Y.Y., Du, S., Recent progress in the theoretical design of two-dimensional ferroelectric materials. *Fundam. Res.*, 3, 322, 2023.

8. Zhang, J., Hou, X., Zhang, Y., Tang, G., Wang, J., Electrocaloric effect in ferroelectric materials: From phase field to first-principles based effective Hamiltonian modeling. *Mater. Rep. Energy*, 1, 100050, 2021.
9. Ye, F., Zhang, Y., Addiego, C., Xu, M., Huyan, H., Ren, X., Pan, X., Emergent properties at oxide interfaces controlled by ferroelectric polarization. *NPJ Comput. Mater.*, 7, 1–14, 2021.
10. Cui, Y., Guo, P., Dang, P., Sun, H., Jing, P., Tao, X., Improved photodegradation efficiency in Fe3+-doped Bi3TiNbO9 nanosheets through oxygen vacancies introduction and ferroelectric polarization enhancement simultaneously. *Appl. Surf. Sci.*, 575, 151749, 2022.
11. Liu, X., Lv, S., Fan, B., Xing, A., Jia, B., Ferroelectric polarization-enhanced photocatalysis in BaTiO3-TiO2 core-shell heterostructures. *Nanomaterials*, 9, 1116, 2019.
12. Natalia, V., Gustami, A.P., Rahmawati, F., Lestari, W.W., Purwanto, A., Lithium titanate (LTO) synthesis through solid state reaction and its performance for LiFePO4/LTO battery. *J. Math. Fundam. Sci.*, 50, 290, 2018.
13. Yasnó, J.P., Conconi, S., Visintin, A., Suárez, G., Non-isothermal reaction mechanism and kinetic analysis for the synthesis of monoclinic lithium zirconate (m-Li2ZrO3) during solid-state reaction. *J. Anal. Sci. Technol.*, 12, 1–13, 2021.
14. Doppiu, S., Dauvergne, J.L., Del Barrio, E.P., Solid-state reactions for the storage of thermal energy. *Nanomaterials*, 9, 226, 2019.
15. Hidayat, D., Taufik, M., Setianto, S., One-step synthesis of lead zirconate titanate particles using a solid-state reaction method. *Heliyon*, 8, e09125, 2022.
16. Qiao, H., Wang, C., Choi, W.S., Park, M.H., Kim, Y., Ultra-thin ferroelectrics. *Mater. Sci. Eng. R Rep.*, 145, 100622, 2021.
17. Ivanova, Y.Y., Gerganova, T.I., Fernandes, M.H.V., Salvado, I.M.M., Preparation of nanostructured porous SiO2-Al2 O3 oxycarbonitride materials obtained by a new chemical precursor method. *Cent. Eur. J. Chem.*, 7, 42, 2009.
18. Zubair, M.A., Al Mamun, A., McNamara, K., Tofail, S.A.M., Islam, F., Lebedev, V.A., Amorphous interface oxide formed due to high amount of Sm doping (5–20 mol%) stabilizes finer size anatase and lowers indirect band gap. *Appl. Surf. Sci.*, 529, 146967, 2020.
19. Feng, S.H. and Li, G.H., Hydrothermal and solvothermal syntheses, in: *Modern Inorganic Synthetic Chemistry*, Second Edition, R. Xu, and Y.B.T.-M.I.S.C. Xu, (Eds.), pp. 73–104, Elsevier, Amsterdam, 2017.
20. Mazzo, T.M., Pinatti, I.M., Macario, L.R., Avansi, W., Moreira, M.L., Rosa, I.L.V., Mastelaro, V.R., Varela, J.A., Longo, E., Europium-doped calcium titanate: Optical and structural evaluations. *J. Alloys Compd.*, 585, 154, 2014.
21. Yáñez-Vilar, S., Sánchez-Andújar, M., Castro-García, S., Señarís-Rodríguez, M.A., Hydrothermal synthesis of LnMn2 O5 (Ln= Y, Pr, Nd, Sm, Eu, Gd, Tb, Dy, Ho and Er) single crystals. *Bol. Soc. Esp. Ceram. Vidr.*, 47, 219, 2008.

22. Shen, P.C., Lin, C., Wang, H., Teo, K.H., Kong, J., Ferroelectric memory field-effect transistors using CVD monolayer MoS2 as resistive switching channel. *Appl. Phys. Lett.*, 116, 33501, 2020.
23. Io, W.F., Yuan, S., Pang, S.Y., Wong, L.W., Zhao, J., Hao, J., Temperature- and thickness-dependence of robust out-of-plane ferroelectricity in CVD grown ultrathin van der Waals α-In2Se3 layers. *Nano Res.*, 13, 1897, 2020.
24. Ni, G.X., Zheng, Y., Bae, S., Tan, C.Y., Kahya, O., Wu, J., Hong, B.H., Yao, K., Özyilmaz, B., Graphene-ferroelectric hybrid structure for flexible transparent electrodes. *ACS Nano*, 6, 3935, 2012.
25. Pierson, H.O., CVD in optoelectronic and ferroelectric applications, in: *Handbook of Chemical Vapor Deposition (CVD)*, Second E., H.O.B.T.-H. Pierson of C.V.D. (CVD) (Eds.), pp. 384–402, William Andrew Publishing, Norwich, NY, 1999.
26. Fujiwara, Y., Tahara, D., Nishinaka, H., Yoshimoto, M., Noda, M., A preliminary study on mist CVD-derived ferroelectric Hf1-xZr xO2films featuring its possibility of suitable operation for non-volatile analog memory. *Jpn. J. Appl. Phys.*, 59, SPPB09, 2020.
27. Peng, Q., Li, D., Huang, P., Ren, Y., Li, Z., Pi, L., Chen, P., Wu, M., Zhang, X., Zhou, X., Zhai, T., Room-temperature ferroelectricity in 2D metal-tellurium-oxyhalide Cd7Te7Cl8O17via selenium-induced selective-bonding growth. *ACS Nano*, 15, 16525, 2021.
28. Tanaka, S., Fujiwara, Y., Nishinaka, H., Yoshimoto, M., Noda, M., Mist-CVD-derived Hf0.55Zr0.45O2ferroelectric thin films post-annealed by rapid thermal annealing. *AIP Adv.*, 13, 15304, 2023.
29. Kong, L.B., Zhang, T.S., Ma, J., Boey, F., Progress in synthesis of ferroelectric ceramic materials via high-energy mechanochemical technique. *Prog. Mater. Sci.*, 53, 207, 2008.
30. Vasyliev, G., Vorobyova, V., Uschapovskiy, D., Linucheva, O., Local electrochemical deposition of copper from sulfate solution. *J. Electrochem. Sci. Eng.*, 12, 557, 2022.
31. Li, W., Yang, J., Jiang, Q., Li, R., Zhao, L., Electrochemical deposition of PbI2 for perovskite solar cells. *Sol. Energy*, 159, 300, 2018.
32. Tonelli, D., Scavetta, E., Gualandi, I., Electrochemical deposition of nanomaterials for electrochemical sensing. *Sensors (Switzerland)*, 19, 1186, 2019.
33. Zheng, B.J., Lian, J.S., Zhao, L., Jiang, Q., Optical and electrical properties of Sn-doped CdO thin films obtained by pulse laser deposition. *Vacuum*, 85, 861, 2011.
34. Huang, Y., Zhang, L., Wang, J., Chu, X., Zhang, D., Zhao, X., Li, X., Xin, L., Zhao, Y., Zhao, F., Enhanced photoresponse of n-ZnO/p-GaN heterojunction ultraviolet photodetector with high-quality CsPbBr3 films grown by pulse laser deposition. *J. Alloys Compd.*, 802, 70, 2019.
35. Ali, S.B., Alhasan, S.F.H., Salim, E.T., Alsultany, F.H., Dahham, O.S., Pulse laser deposition of HfO2 nanoporous-like structure, physical properties for device fabrication. *J. Renew. Mater.*, 10, 2819, 2022.

36. Sun, Y.T.A., Pan, P.C., Koo, H.S., Lin, N.Y., Growth of low resistivity and high transparency boron-doped zinc oxide film by pulse laser deposition. *Precis. Eng.*, 66, 605, 2020.
37. Pan, P.C., Koo, H.S., Chen, D.X., Chen, C.M., Fabricating high-conduction and high-transparency tungsten-doped zinc oxide films by pulse laser deposition technique. *Crystals*, 12, 1032, 2022.
38. Vuong, T.Q.P., Cassabois, G., Valvin, P., Rousseau, E., Summerfield, A., Mellor, C.J., Cho, Y., Cheng, T.S., Albar, J.D., Eaves, L., Foxon, C.T., Beton, P.H., Novikov, S.V., Gil, B., Deep ultraviolet emission in hexagonal boron nitride grown by high-temperature molecular beam epitaxy. *2D Mater.*, 4, 21023, 2017.
39. Aseev, P., Fursina, A., Boekhout, F., Krizek, F., Sestoft, J.E., Borsoi, F., Heedt, S., Wang, G., Binci, L., Martí-Sánchez, S., Swoboda, T., Koops, R., Uccelli, E., Arbiol, J., Krogstrup, P., Kouwenhoven, L.P., Caroff, P., Selectivity map for molecular beam epitaxy of advanced III-V quantum nanowire networks. *Nano Lett.*, 19, 218, 2019.
40. Mazet, L., Yang, S.M., Kalinin, S.V., Schamm-Chardon, S., Dubourdieu, C., A review of molecular beam epitaxy of ferroelectric BaTiO3 films on Si, Ge and GaAs substrates and their applications. *Sci. Technol. Adv. Mater.*, 16, 36005, 2015.
41. Ginley, T.P., Wang, Y., Law, S., Topological insulator film growth by molecular beam epitaxy: A review. *Crystals*, 6, 154, 2016.
42. Kapustianyk, V., Semak, S., Chornii, Y., Bovgyra, O., Kovalenko, M., Manifestation of the ferroelastoelectric phase transition in the absorption spectra of (NH4)2CuCl4×2H2O crystals. *Phys. B Condens. Matter*, 639, 413929, 2022.
43. Przeslawski, J., Kos, M., Czapla, Z., Calorimetric and optical studies on ferroelastoelectric phase transitions in [(CH 3) 2CHNH 3] 4Cd 3Cl 10 crystal. *Thermochim. Acta*, 546, 49, 2012.
44. Das Mahapatra, S., Mohapatra, P.C., Aria, A.I., Christie, G., Mishra, Y.K., Hofmann, S., Thakur, V.K., Piezoelectric materials for energy harvesting and sensing applications: Roadmap for future smart materials. *Adv. Sci.*, 8, 2100864, 2021.
45. Zhou, X., Shen, B., Lyubartsev, A., Zhai, J., Hedin, N., Semiconducting piezoelectric heterostructures for piezo- and piezophotocatalysis. *Nano Energy*, 96, 107141, 2022.
46. Xu, Q., Gao, X., Zhao, S., Liu, Y., Zhang, D., Zhou, K., Khanbareh, H., Chen, W., Zhang, Y., Bowen, C., Bio-piezoelectric platforms: Construction of bio-piezoelectric platforms: From structures and synthesis to applications *Adv. Mater.*, 33, 2170206, 2021.
47. Covaci, C. and Gontean, A., Piezoelectric energy harvesting solutions: A review. *Sensors (Switzerland)*, 20, 1, 2020.
48. Torsello, D., Casalegno, V., Divitini, G., Ghigo, G., Gerbaldo, R., Fracasso, M., D'Isanto, F., Ching Lai, M., Roux, L., Gutierrez, G., Ducati, C., Cabet, C.,

Ferraris, M., Gozzelino, L., Triple ion beam irradiation of glass-ceramic materials for nuclear fusion technology. *J. Nucl. Mater.*, 567, 153783, 2022.

49. Xiang, X., He, Z., Rao, J., Fan, Z., Wang, X., Chen, Y., Applications of ion beam irradiation in multifunctional oxide thin films: A review. *ACS Appl. Electron. Mater.*, 3, 1031, 2021.
50. Kakimoto, K.I., Ferroelectric materials and their applications. *Jpn. J. Appl. Phys.*, 61, SN0001, 2022.
51. Kaminow, I.P., Principles and applications of ferroelectrics and related materials. *Proc. IEEE*, 66, 1299, 2008.
52. Kim, T.Y., Kim, S.K., Kim, S.W., Application of ferroelectric materials for improving output power of energy harvesters. *Nano Converg.*, 5, 1–16, 2018.
53. Scott, J.F., Applications of modern ferroelectrics. *Science*, 315, 954, 2007.
54. Polking, M.J., Alivisatos, A.P., Ramesh, R., Synthesis, physics, and applications of ferroelectric nanomaterials. *MRS Commun.*, 5, 27, 2015.
55. Liu, X., Geng, X., Liu, H., Shao, M., Zhao, R., Yang, Y., Ren, T.L., Recent progress and applications of HfO2-based ferroelectric memory. *Tsinghua Sci. Technol.*, 28, 221, 2023.
56. Liu, Y., Wang, S., Chen, Z., Xiao, L., Applications of ferroelectrics in photovoltaic devices. *Sci. China Mater.*, 59, 851, 2016.
57. Schroeder, U., Park, M.H., Mikolajick, T., Hwang, C.S., The fundamentals and applications of ferroelectric HfO2. *Nat. Rev. Mater.*, 7, 653, 2022.
58. Blázquez-Castro, A., García-Cabañes, A., Carrascosa, M., Biological applications of ferroelectric materials. *Appl. Phys. Rev.*, 5, 1–19, 2018.

6

Elastocaloric Effect in Ferroelectric Materials

Uzma Hira*, Uswa Ameen and Atfa Ashraf

School of Physical Sciences (SPS), University of the Punjab, New Campus, Lahore, Pakistan

Abstract

With respect to its promising characteristics, the elastocaloric effect (ECE) has been increasing in relevance in the discipline of solid-state cooling. Owing to their distinctive thermodynamic features and phase transition uses, ferroelectric materials are employed for solid-state cooling purposes. ECE states that the temperature of a material changes when it is compressed or expanded without any heat exchange with the surroundings. ECE is further divided into uniaxial and biaxial based on its behavior in microstructure of materials. Less force elasticity and high entropy are favorable in ferroelectric materials for this application. On the basis of their composition, ferroelectric materials are classified into oxides, sulfates, and polymeric materials. They are widely used in mechanical, environmental, and biological fields. These parameters mainly involve temperature change, specific heat capacity, endurance limit, transition temperature, coefficient of performance, entropy change, etc. Due to these advancements, ECE is being fed to develop alternative cooling technologies in the current decade. Devices using this technology are environmentally friendly as they have less potential for greenhouse gas emission and can achieve desirable properties by changing magnetization, voltage, pressure, and temperature according to the requirements. Solid-state refrigeration and elastocaloric refrigerators are the best examples.

***Keywords*:** Elastocaloric effect, ferroelectric materials, oxides, sulfates and polymeric materials

**Corresponding author*: uzma.sps@pu.edu.pk

Inamuddin, Tariq Altalhi and Mohammad Abu Jafar Mazumder (eds.) Ferroic Materials-Based Technologies, (125–156) © 2024 Scrivener Publishing LLC

6.1 Introduction

According to the requirement for highly proficient and eco-friendly refrigeration methods, the topic of solid-state cooling has received a lot of attention from the last decade. One promising approach is based on the elastocaloric effect (ECE), which involves the adiabatic heating or cooling of a material in response to mechanical stress or strain. Ferroelectric materials are now recognized as interesting options for elastocaloric cooling purposes due to their spontaneous mode of polarization that may be reversed by an external electric field effect. The ECE in ferroelectric materials is rooted in their unique thermodynamic properties, such as the presence of a polar-to-nonpolar phase transition that can be induced by mechanical stress. When a ferroelectric material undergoes this transition, it experiences a significant temperature change due to the associated entropy change. By carefully tuning the stress-induced phase transition and manipulating the material's properties, researchers are working to develop highly efficient and sustainable cooling technologies that can compete with conventional vapor-compression refrigeration [1].

6.1.1 Elastocaloric Effect

The elastocaloric effect (ECE), a phenomenon seen in some materials, is where the application of mechanical stress causes a variation in temperature. As presented in Figure 6.1, the notion of ECE is based on the fundamental principle of adiabatic temperature transformation. According to this theory, a substance's temperature changes as it is compressed or extended without any heat being transferred to the surroundings. The elastocaloric effect is a thermodynamic phenomenon that showcases

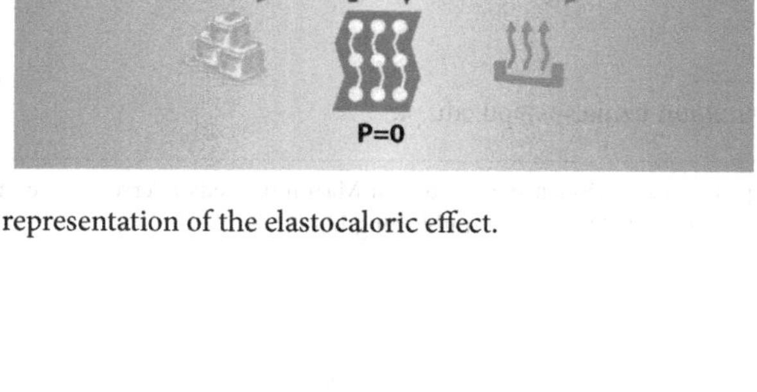

Figure 6.1 A representation of the elastocaloric effect.

remarkable reversibility. This suggests that the substance is capable of experiencing temperature fluctuations through the imposition or removal of mechanical strain, correspondingly [2].

A noteworthy phenomenon known as the elastocaloric effect (ECE) is exhibited in some substances that undergo a phase change whenever subjected to physical stress. The materials that exhibit this intriguing behavior with great frequency are none other than shape memory alloys (SMAs) and ferroic materials, which encompass ferroelectrics, ferromagnetics, and ferroelastics. SMAs are an interesting family of metallic alloys that possess the remarkable property of regaining their initial shape upon being subjected to a particular temperature or mechanical trigger. Ferroic materials are characterized via the manifestation of a spontaneous mode of polarization or magnetization, which can be modulated through imposition of an extrinsic electrical or magnetic fields.

The underlying idea that a substance's entropy changes when it goes through a phase transition is the basis for the elastocaloric effect phenomenon. Entropy can be defined as a quantitative measure of the level of disorder or randomness exhibited by a given system. During a phase transition, the entropy of a substance is altered due to a modification in the organization of its constituent atoms or molecules. The Clausius-Clapeyron equation establishes the correlation between the alteration in entropy and the modification in temperature:

$$\Delta S = \Delta H / T$$

where ΔS is the entropy difference, ΔH is the enthalpy change, and T is the temperature. The enthalpy change refers to the amount of heat that is either absorbed/released via a substance during a phase transition [3].

Using the substance as a refrigerant allows for the use of the elastocaloric effect in solid-state cooling. Upon compression, the material undergoes an endothermic process wherein it absorbs heat from the surrounding environment, resulting in a decrease in temperature. Upon release, the material undergoes an exothermic reaction, causing a liberation of heat and consequent increase in temperature. The aforementioned procedure may be reiterated to establish a refrigeration cycle [4].

6.1.1.1 Types of Elastocaloric Effect

Ferroelectric materials exhibit two distinct types of electrocaloric effect (ECE) that are contingent upon the character of the applied stress or strain.

The elastocaloric effects that can be observed are classified into two categories: uniaxial and biaxial. The uniaxial electromechanical coupling effect (ECE) refers to the phenomenon in which a material experiences stress or strain along a single axis. On the other hand, the biaxial ECE occurs when a material is subjected to stress or strain along two axes simultaneously [5].

6.1.2 Force Elasticity and Entropy Elasticity

6.1.2.1 Force Elasticity

The phenomenon of force elasticity pertains to the intrinsic property of a substance to regain its initial configuration subsequent to experiencing deformation due to an externally applied force. The aforementioned phenomenon can be elucidated by the material's elastic modulus, a parameter that characterizes its rigidity or ability to resist deformation. Substances exhibiting a high elastic modulus possess greater resistance to deformation and exhibit prompt recovery to their initial shape subsequent to deformation [6].

Hooke's law is the predominant theoretical framework for force elasticity relationships. It posits that the extent of deformation exhibited by a given material is directly related to magnitude of the external force exerted upon it. The mathematical expression of Hooke's law is as follows:

$$\sigma = E\varepsilon$$

where, σ, E, and ε are the stress, elastic modulus, and strain, respectively [7].

6.1.2.2 Entropy Elasticity

The concept of entropy elasticity is related to the capacity of a substance to alter its shape or volume in reaction to variations in pressure or temperature. The observed phenomenon can be attributed to the entropy of the material, which is a quantitative measure of its level of disorder or randomness. Substances possessing elevated entropy values exhibit greater flexibility in altering their configuration or magnitude in reaction to variations in temperature or pressure [8].

The Clausius-Clapeyron equation ($\Delta S = \Delta H/T$) is a widely employed model in the realm of entropy elasticity. This equation establishes a correlation between the alteration in entropy and the modifications in temperature and pressure [9].

6.1.2.3 *Relationship between Force Elasticity and Entropy Elasticity*

The complex relationship between force elasticity and entropy elasticity in materials relies on the distinct material and conditions under consideration. In general, substances demonstrating significant strength in elasticity typically manifest reduced elasticity in entropy, and vice versa. This phenomenon can be attributed to the microstructural response of materials to changes in temperature or the imposition of external forces, which can significantly impact the material's deformation properties [10].

6.1.3 Entropy Elastic Stress and Strain Actions for Solid-State Cooling

6.1.3.1 *Basics of Solid-State Cooling*

The growing area of solid-state cooling requires the utilization of solid substances for producing cooling phenomena. The fundamental concept underlying solid-state refrigeration is the Peltier effect, which manifests when an electrical current traverses a junction of two disparate materials. This phenomenon leads to the heat transfer between two substances, resulting in a reduction in temperature. The Peltier effect, albeit a fascinating phenomenon, exhibits suboptimal thermodynamic efficiency and is constrained by the inherent properties of the constituent materials at the junction. The emergence of innovative methodologies, including entropy-elastic cooling, has been observed [11].

6.1.3.2 *Overview of Entropy-Elastic Materials for Cooling*

Entropy-elastic materials are a group of substances that demonstrate a unique capacity to undergo reversible alterations in their elastic characteristics upon exposure to temperature fluctuations. The above-mentioned substances are commonly classified as ferroelectric materials, owing to their inherent ability to reveal a lasting electric dipole moment that can be realigned via influence of an extrinsic electric field. The dipole reorientation phenomenon induces a modification in the crystallographic arrangement of the material, thereby causing a variation in its elastic characteristics. The alteration in the elastic characteristics can be utilized to generate a refrigerating outcome [12].

6.1.3.3 *Entropy and Thermoelectric Performance*

Entropy is a thermodynamic property that quantifies the level of randomness or disorder in a given system. The increase of the Seebeck coefficient, a metric that gauges a material's capability to produce a temperature variance in reaction to an enforced electric field, can be attributed to elevated entropy in thermoelectric substances. The observed expansion in the Seebeck coefficient can be attributed to an increase in the total number of energy states that are at the disposal of charge carriers for occupation. High entropy can also result in a decrease in thermal conductivity, which is preferable in solid-state cooling applications because it reduces heat transmission through the substance [13].

6.1.3.4 *Elastic Stress and Strain Behavior*

The thermoelectric performance of a material is dependent on its elastic stress and strain behavior. Figure 6.2 illustrates how a material's electrical and thermal properties change when it is subjected to an external force, leading to material deformation. The interplay between the exogenous force and the consequent deformation is explicated by the elastic stress–strain characteristics of the material. The Young's modulus of a substance is a fundamental property that describes its elastic behavior, specifically its ability to resist deformation or stiffness [14].

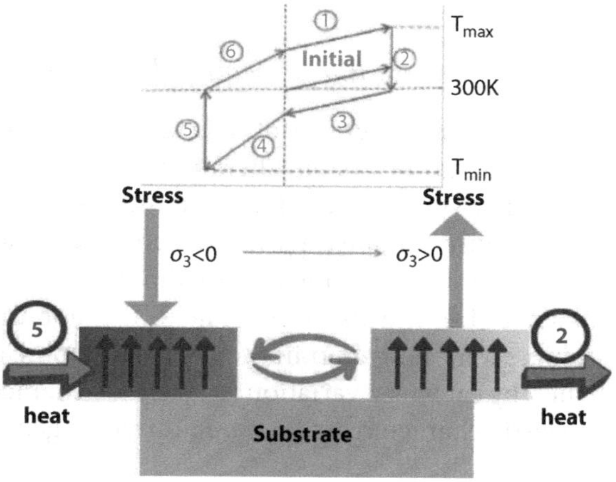

Figure 6.2 Giant room-temperature elastocaloric effect in ferroelectric ultrathin films.

6.1.3.5 *Properties and Characteristics of Entropy-Elastic Materials*

Materials possessing entropy elasticity demonstrate a multitude of attributes and traits that render them apt for employment in the realm of solid-state refrigeration. The reversible alteration in elastic properties is a pivotal characteristic of these substances, enabling a reversible refrigeration phenomenon. Moreover, these substances exhibit elevated thermal conductivity and reduced thermal expansion, making them highly coveted for employment in cooling operations. Furthermore, these materials demonstrate exceptional fatigue resistance and commendable mechanical stability, rendering them highly durable and dependable [15].

6.1.3.6 *Potential Applications of Entropy-Elastic Materials in Cooling Technologies*

Materials exhibiting entropy elasticity possess promising potential for implementation in diverse cooling applications. A plausible implementation lies in the realm of microelectronics, where the preservation of electronic devices' efficiency and dependability necessitates cooling. Entropy-elastic substances possess the potential to fabricate solid-state refrigerators that are both efficient and compact, rendering them a perfect fit for deployment in microelectronic gadgets. Other potential applications include cooling of infrared sensors, medical equipment, and space-based applications [16].

In this chapter, we will discuss in detail the use of entropy-elastic materials for solid-state cooling as a promising approach that offers several advantages over traditional refrigeration techniques. These materials exhibit reversible changes in their elastic properties when subjected to a change in temperature, which can be harnessed to create a cooling effect. Additionally, they possess several desirable characteristics like high thermal conductivities values, lower thermal expansion, and better mechanical stability. With further research and development, entropy-elastic materials have the potential to modernize the area of solid-state cooling.

6.2 Ferroelectric Materials

6.2.1 Introduction to Ferroelectric Materials

6.2.1.1 *Definition and Characteristics of Ferroelectric Materials*

Ferroelectric materials are classified as a dielectric class of materials whose polarization persists even after the applied electric field has been removed.

Also, by adding an electrical field, the orientation of the dipole moment changed [17].

It is presented as a property of a material with two characteristics [18], namely

- The presence of spontaneous polarization
- The ability to reverse when exposed to external electric fields

The phenomenon initially appeared in Rochelle salt and acquired that name due to its resemblance to ferromagnetism, a magnetic property of a material with an ongoing moment of magnetization. Domains, the hysteresis loop, and the Curie temperature (TC) are similarly analogous concepts. The prefix ferro, which means iron (Fe), was employed since magnetic substances included iron (Fe) at that point in time. Nevertheless, there exists no link connecting Fe and ferroelectricity. While certain ferroelectric substances incorporate Fe, this element was not originally liable for the material's creation. When a compound is polarized through external electrical field effect, the imposed polarization, or dielectric polarization, is usually inversely correlated through size of the imposed external electrical field (E). Ferroelectric substances are paraelectric above the Curie temperature, TC. An outside electric field generates nonlinear polarization in this scenario. As a consequence, the slant of the curve of polarization indicates that electric permittivity is not consistent. Since the applied field (E) was a value of zero along with the nonlinearity, there was spontaneous non-zero polarization at the ferroelectric condition. A hysteresis loop forms as spontaneous polarization is reliant on both the current electrical field and its past behavior, which may be reverted via an adequately strong electrical field. Changes in the crystal lattice may have an impact on the dipoles' strength or the amount of spontaneous mode of polarization because of the interaction between the electric dipoles and the ferroelectric material's crystal lattice. Without applying an external field across a ferroelectric material used to make a capacitor, the change in spontaneous polarization causes a variation in the surface charges, which result in current flow. A ferroelectric material can change its lattice structure in response to two stimuli:

- Temperature
- Mechanical force

Piezoelectricity is the process of creating surface charge in reaction to an external stress, whereas pyroelectricity is the process of changing the spontaneous mode of polarization in reaction to temperature gradient.

6.2.2 Historical Overview

In 1920, Joseph Valasek made the discoveries of hysteretic switching and persistent polarization. Rochelle salt, a form of potassium sodium tartrate tetrahydrate, reported to exhibit piezoelectric action, was the subject of an experimental research by Valasek. The earlier discoveries that Rochelle salt displays peculiar behavior in response to exerted pressure and electrical fields—for example, piezoelectric charge production that departs from typical linear correlation with the force being applied or dielectric characteristics that alter with employed voltage—were the inspiration for Valasek's study. Ferroelectrics' distinguishing characteristics, persistent polarization and the impact of hysteresis on changing, were established in 1920 at a session of the "American Physical Society".

In December, an article explaining how ferroelectrics are similar to ferromagnets and highlighting the unique properties of Rochelle salt was submitted to the *Physical Review* as a result. Others immediately became motivated and began looking for substitute ferroelectrics as a result. Thirteen years passed until Paul Scherrer and Georg Busch discovered potassium di-hydrogen phosphate, or KDP, and they discovered that it possessed a ferroelectric nature below the temperature of 122 K. Nevertheless, Rochelle salt and KDP's first uses have been restricted since they are both delicate and water-soluble. With the discovery of long-lasting ferroelectricity in an impermeable substance, specifically $BaTiO_3$, which had been assumed to have a dielectric constant that was high, it was determined that ferroelectrics were limited to use for industrial and military purposes, including ultrasonic instruments as well as purposes like ferroelectric recollections. The discovery of new compounds like $PbZrO_3$-$PbTiO_3$ mixes that are currently employed as piezoelectrics/ $BaTiO_3$ renditions that are employed in capacitors came about as a result of investigations into other perovskite oxides that were sparked by the finding of ferroelectricity in an inorganic material. A. F. Devonshire created the phenomenological theory of ferroelectricity in the 1950s after doing extensive investigations of the modifications to the structure of the crystal that occur throughout the ferroelectric phase transition using the $BaTiO_3$ crystal structure. According to this theory, the barrier of energy separating the two polarization states in a double-well potential may be used to represent a ferroelectric material's unlimited energy. This approach has shown to be effective in illuminating ferroelectrics. However, it does not mean that because of that time, ferroelectric research has stopped. Research on ferroelectrics is still active. In one instance, Rolf Landauer noted that since ferroelectric shifting has a free-energy barrier, the local curves must be negative in some area, suggesting that capacitance may also be negative there on the free-energy map. It has only recently been recognized that negative capacitance, which is present in thin ferroelectric sheets, can be used to amplify voltage. $Hf_{0.5}Zr_{0.5}O_2$ was subsequently employed for examining the double-well prospective, which is critical to ferroelectric activity. When this material is formed as part of a ferroelectric thin film, it is said to be ferroelectric. It exhibits ferroelectricity; however, when it is in its bulk form, it does not. This suggests the potential of improved growth strategies. In this regard, a ferroelectric instability is created in $(SrTiO_3)_{n-m}$ $(BaTiO_3)_m SrO$ superlattices using controlled chemical pressure at the atomic length scale in an article by Darrell Schlom and colleagues in this issue. As a result, at frequencies in the range of 100 GHz, there is extremely little loss of dielectric and tunability that may be significant for microwave electronics and wireless uses within 5G—probably considerably more than Valasek could have imagined. Despite the fact that the notion of ferroelectricity is often dated to 1920, Valasek did not mention it in his piece of writing. Though he did not coin the term ferroelectricity until a long time afterwards, Erwin Schrödinger first referred to substances having persistent electric polarization as "ferroelektrisch" in his habilitation of 1912. Since that first illustration and "Valasek's" revolutionary revelation of hysteresis, studies into the development of analogous systems have advanced quickly. In accordance to our predictions, the following era of ferroelectricity will be just as lucrative as the previous one [19].

6.2.3 Structure and Properties of Ferroelectric Materials

They are non-centrosymmetric, or without a center of symmetry, in accordance with the relationship between their static dielectric constant and temperature (see Figure 6.3a). If $T > Tc$, then ($\varepsilon = C/T - Tc$), where C is a constant that is unaffected by temperature. Curie-Weiss law is the name of this relationship. Characteristics of a dielectric are the C and T; C is commonly referred to as the Curie constant parameter, and Tc is Curie temperature.

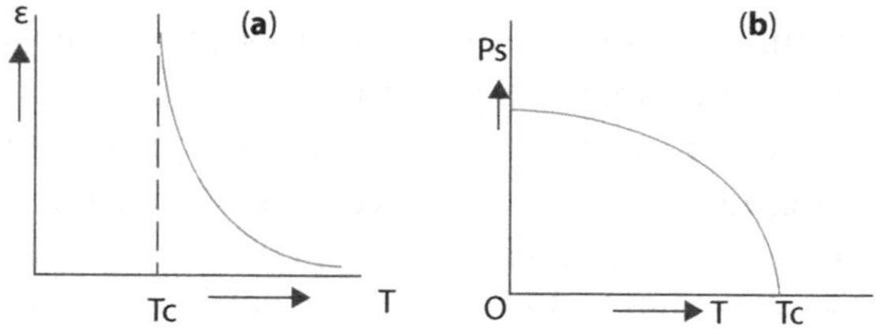

Figure 6.3 (a) Dielectric constant vs. temperature curve and (b) spontaneous polarization vs. temperature curve.

They are spontaneously polarized electrically; that is, they are polarized without the aid of an outside field. Nevertheless, this polarization only manifests itself in the *Tc* temperature range. The relationship is plotted and can be seen in Figure 6.3b.

Figure 6.4 displays a hysteresis loop when an alternating voltage is applied. A ferroelectric material can have a ferroelectric hysteresis loop parameter, or a polarization vs. electrical field curve. The B–H curve of a ferromagnetic material can be found to be quite similar to the curve. Ferroelectric materials behave similarly to ferromagnetic materials with respect to magnetic fields in terms of how they respond to electric fields.

Nonsymmetric crystals and crystals with a center of symmetry are two different types of crystals. One or more crystallographically distinct directional axes can be found in a crystal that lacks a center of symmetry. All non-centrosymmetric point groups exhibit a piezoelectric effect along specific directional axes. Ten of the twenty point groups that exhibit the piezoelectric effect have just one distinct direction axis. These crystals are known as polar crystals because they exhibit spontaneous polarization. Ferroelectric materials can have several different crystal structures, including perovskite, tungsten bronze, and pyrochlore. Ferroelectric materials have a complex domain structure, which is the arrangement of regions of polarized material within the crystal lattice. These domains can have different sizes, shapes, and orientations, and they are separated by domain walls. The domain structure of ferroelectric-type materials can be manipulated via the course of an applied electrical field, which can cause domain switching.

Above *Tc*, these materials exhibit dielectric behavior. Ionic polarizability is related to ferroelectricity. The ions in the crystal are thought to experience an eventual asymmetrical shift in their locations as a result of the local fields created by the polarization itself growing faster than the elastic

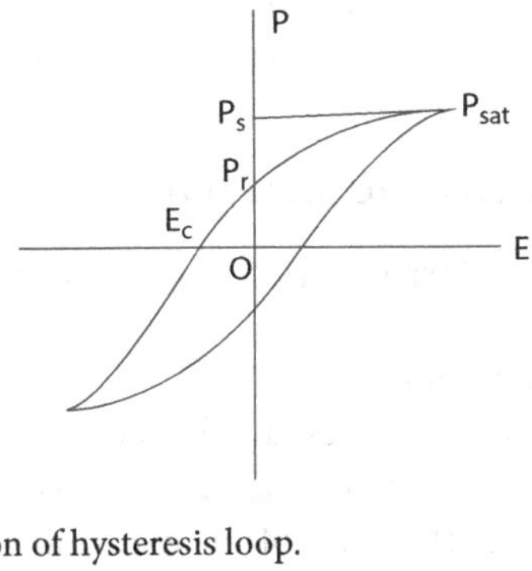

Figure 6.4 A representation of hysteresis loop.

restoring forces. Thus, using the Clausius-Mossotti connection, one can determine the dielectric constant *e* of a ferroelectric substance

$$\varepsilon - 1/\varepsilon + 2 = 4\pi/3.\ N\alpha$$

$$4\pi/3.\ N\ \alpha = \beta N$$

Assuming that the polarizability of an ion is independent of temperature, *N* here stands for the number of ions per unit volume. Because of thermal expansion, the quantity *N* changes with temperature, and tiny changes in *N* can result in substantial changes in *E*. This is how the quantity is temperature dependent.

The polarization that occurs naturally is determined by the quantity of the dipole moment in a given volume. Typically, spontaneous polarization originates across a single crystal plane. An increase or decrease in the temperature of the crystal causes an alteration in its polarization, which may be seen. This means that the value of spontaneous polarization relies on temperature. The pyroelectric effect was first seen in tourmaline by Teophrast in 314 B.C., and Brewster assigned it the title in 1824. The pyroelectric coefficient can be used to describe the pyroelectric phenomenon. The spontaneous polarization vector *Ps*, which is determined by, $\Delta Ps = \lambda\Delta T$, changes when a crystal's temperature, *T*, changes slightly and gradually. A researcher explained the relationship between temperature and *Ps* (spontaneous polarization) of a $BaTiO_3$ ferroelectric crystal. A rise in temperature results in a fall in spontaneous polarization. The pyroelectric coefficient of $BaTiO_3$ is negative. On heating the crystal over the Curie temperature, this polarization abruptly drops to zero. By using an external field that is strong enough to reverse the polarization direction of a pyroelectric crystal, the crystal is said to be ferroelectric. It should be mentioned that a crystal's atomic structure or crystal structure alone is solely responsible for a crystal's intrinsic properties of piezoelectricity and pyroelectricity. On the other hand, a pyroelectric crystal experiences ferroelectricity when an external electric field is applied [20].

6.2.4 Types of Ferroelectric Materials

Some important classes of ferroelectric materials are discussed as follows:

- Class 1—Ferroelectric Oxides

Oxygen with a metal cation makes up the family of materials known as ferroelectric oxides, which show ferroelectricity, or spontaneous electrical mode of polarization with the lack of an external electrical field. Due to their special characteristics, these materials have a wide range of uses in electronics, data storage, sensing, and energy harvesting [21].

There are several types of ferroelectric oxides:

1. Perovskite Ferroelectric Oxides
   Perovskite ferroelectric oxides have a perovskite crystal structure, which is a cubic lattice of oxygen ions with metal cations located at the corners and center of the lattice. Strontium bismuth tantalate, $Pb[Zr_{1-x}Ti_x]O_3$, and $BaTiO_3$ are a few examples of perovskite ferroelectric oxides. Perovskite ferroelectric oxides have a high dielectric constant, which makes them useful for capacitors and energy storage applications.
2. Aurivillius Ferroelectric Oxides
   Aurivillius ferroelectric oxides have a layered crystal structure composed of alternating metal oxide layers and oxygen layers. Examples of Aurivillius ferroelectric oxides include bismuth-layered ferroelectrics such as bismuth titanate ($Bi_4Ti_3O_{12}$) and bismuth ferrite ($BiFeO_3$). Excellent ferroelectric characteristics, such as a significant remnant polarization, a small coercive field, and a high Curie temperature, are present in Aurivillius ferroelectric oxides.
3. Tungsten Bronze Ferroelectric Oxides
   The intricate crystal structure of tungsten bronze ferroelectric oxides is composed of metal cations situated within oxygen octahedral cages. Some examples involve tungsten bronze ferroelectric oxides that include $MgNb_2O_9Pb_3$ and $Ba_{1-x}Sr_xTiO_3$. Ferroelectric oxides of tungsten bronze exhibit a remarkable dielectric constant and are extensively employed in various applications, including electro-optic modulators and tunable microwave devices.
4. Pyrochlore Ferroelectric Oxides
   The ferroelectric oxides of pyrochlore exhibit a crystal structure that is cubic in nature, consisting of oxygen ions that are arranged in a lattice resembling that of a diamond. The metal cations are positioned at the corners of the lattice. Pyrochlore ferroelectric oxides, such as lead zirconate pyrochlore (PZP) and bismuth pyrochlore ($Bi_2Ti_2O_7$), are among the known examples. Pyrochlore ferroelectric oxides exhibit a remarkable property of possessing a high Curie temperature and low dielectric loss, rendering them highly advantageous for deployment in high-temperature applications.

- Class 2—Triglycine Sulfates

Triglycine sulfate (TGS) is a prominent ferroelectric substance that has been extensively employed in the realm of ferroelectricity since the 1950s. TGS is a crystalline substance that manifests a remarkable dielectric constant, low dielectric loss, and robust ferroelectric reaction at ambient temperature, rendering it a valuable material for diverse applications. The tetragonal crystal lattice of TGS is composed of two distinct assemblages of glycine molecules that are interconnected via hydrogen bonding, in addition to the presence of sulfate ions. The non-centrosymmetric crystal lattice of TGS is responsible for its ferroelectric characteristics. The crystal structure's lack of centrosymmetry induced charge separation between positive and negative charges, leading to the TGS's spontaneous polarization.

TGS exhibits unique characteristics that render it highly efficient in diverse applications. TGS finds significant utility in the realm of piezoelectricity, serving as a transducer to facilitate the transformation of mechanical form of energy to electrical energy or vice versa. Triglycine sulfate (TGS) has been employed as a viable candidate for infrared detection applications, owing to its unique material properties. Additionally, TGS has been utilized as a substrate for the growth of various ferroelectric materials.

The Curie temperature of TGS is comparatively lower in relation to other ferroelectric materials, thereby restricting its utility in applications requiring high-temperature endurance. The advantageous low-temperature characteristics exhibited by TGS render it a valuable substance for implementation in cryogenic detectors and other low-temperature sensors.

- Class 3—Polymeric Ferroelectrics

Polymeric ferroelectrics are a class of ferroelectric substances that consist of polymers, which are extended chains of recurring molecular units. Owing to their facile processing, versatility, and cost-effectiveness, these materials have garnered significant interest for deployment in diverse applications such as actuators, sensors, high-density data storage, and energy harvesting devices.

The manifestation of ferroelectricity in polymeric materials can be attributed to the existence of polar functional moieties, including carbonyl, nitro, or cyano groups, either within the polymer backbone or as pendant groups. TGS exhibits unique characteristics that render it highly efficient in various applications. TGS finds significant utility in the domain of piezoelectricity, serving as a transducer for the transformation of mechanical form of energy to electrical energy and vice versa. Triglycine sulfate (TGS) has exhibited potential as a viable candidate for employment in infrared detectors, in addition to serving as a substrate for the growth of other ferroelectric materials.

These polar groups create an asymmetric charge distribution, leading to a net dipole moment within the polymer chain. The degree of polarity in a polymeric ferroelectric can be enhanced through the incorporation of bulky side groups or by introducing electron-rich or electron-poor aromatic rings into the polymer backbone.

One of the most well-known polymeric ferroelectrics is polyvinylidene fluoride (PVDF), which is a semicrystalline polymer with a high dipole moment that can be easily aligned by mechanical stretching or through application of an external electrical field. PVDF can be processed into a variety of forms, such as films, fibers, and coatings, and has been extensively studied for its use in sensors and energy harvesting devices. Other polymeric ferroelectrics include poly(vinyl acetate), polyacrylonitrile, and polyimides. These materials can exhibit higher ferroelectric properties than PVDF, with larger remnant polarization values and lower coercive fields, making them attractive for use in high-performance applications.

6.2.5 Applications of Ferroelectric Materials

Ferroelectric substances offer a wide range of uses owing to their distinctive features. Here are some of the most important applications of ferroelectric materials:

- Mechanical Applications
- Ferroelectric RAM

Non-volatile memory called Ferroelectric RAM (FeRAM) uses ferroelectric materials as the storing medium. Ferroelectric materials can maintain their polarization state even after the power supply is turned off, making them ideal for data storage applications. FeRAM is used in applications that need high-speed and low-power consumption, like smart cards and electronic tags [22].

- Piezoelectric Transducers

The piezoelectric effect's existence is demonstrated by ferroelectric substances, which can transform mechanical force into electrical energy and the other way around. This property makes them ideal for use in piezoelectric transducers, which are devices that can convert mechanical vibrations into electrical signals. Sensors, ultrasound imaging, actuators, and acoustic devices are just a few of the many applications for piezoelectric transducers.

- Capacitors

Ferroelectric materials have a high dielectric constant, which makes them suitable for use in high-capacitance capacitors. Computers, cell phones, and televisions are just a few of the electrical gadgets that use ferroelectric capacitors.

- Electro-optic Modulators

Ferroelectric materials exhibit the electro-optic effect, which means that they can change the refractive index of light when an electric field is applied. This property makes them ideal for use in electro-optic modulators, which are devices that can control the intensity, phase, or polarization of light. Electro-optic modulators are applied in a wide range of usages like laser technology, telecommunication, and data storage.

- Energy Harvesting

Ferroelectric materials can be used to harvest energy from mechanical vibrations or heat. They are appropriate for use in energy harvesting systems because of this feature, which allows waste energy to be converted into valuable electrical energy. Applications for energy-harvesting devices include wireless communication, sensors, and environmental monitoring.

- Nonlinear Optics

Ferroelectric materials exhibit nonlinear optical properties, which means that they can generate new frequencies of light when an intense laser beam is applied. This property makes them suitable for use in nonlinear optics applications, including frequency conversion, optical switching, and optical parametric amplification.

- Biological Applications

Due to its distinctive qualities, including a high dielectric constant, piezoelectricity, and electro-optic effects, ferroelectric materials have demonstrated considerable potential for a variety of biological applications. Ferroelectric materials have some of the most significant biological uses in [22].

- Tissue Engineering

Ferroelectric materials can be used to stimulate the growth and differentiation of cells in tissue engineering applications. For example, polycrystalline ferroelectric materials can be used to enhance the differentiation of stem cells into bone cells. This property makes ferroelectric materials suitable for use in bone regeneration and other tissue engineering applications.

- Biosensors

Ferroelectric materials can be used in biosensors for the detection of various biomolecules, including proteins, nucleic acids, and bacteria. The high dielectric constant of ferroelectric materials enhances the sensitivity of biosensors, making them suitable for use in medical diagnostics, environmental monitoring, and food safety applications.

- Drug Delivery

Ferroelectric materials can be used to enhance drug delivery by using them as carriers for drug molecules. When an external electrical field is used, the ferroelectric materials' piezoelectric action can be exploited to release drug molecules. Ferroelectric materials are useful for targeted drug administration because of this feature, which increases the efficiency and decreases the negative effects of medications.

- Neural Stimulation

Ferroelectric materials can be used to stimulate neural activity by applying an external electrical field. The electro-optic effect of ferroelectric materials can be used to control intensity, phase, or polarization of light, making them suitable for use in optogenetics, a technique for controlling neural activity with light.

- Bioimaging

Ferroelectric materials can be used in bioimaging applications, such as fluorescence microscopy, due to their nonlinear optical properties. The ability of ferroelectric materials to generate new frequencies of light can be used to enhance the contrast and resolution of bioimaging techniques [23].

6.3 Performance Indicators

The choice of elastocaloric materials for such a solid-state coolant is important to compare their performance. The performance can be evaluated by considering the following parameters:

6.3.1 Elastocaloric Effect (ΔT)

A reversible change in temperature is associated to the caloric effect via external fields that are imposed adiabatically. They might alternatively be understood to be an isothermal change in entropy due to the application of the fields. Additionally, only when the system's degree of order increases (temperature rises) is the system entropy reduced. Entropy rises as the disordered state of the substance has returned after stretching load is relaxed (the temperature decreases). A supporting molecular chain can therefore be thought of as an entropical spring [24]. Let us now have a look at a thermodynamic system represented by the variables [25, 26], where X is a general displacement, Y is the conjugation field, T is the temperature, and the caloric effect occurs when displacement varies and it can be represented by the following equation:

$$\Delta S = \int \Delta y \left(\frac{\partial Y}{\partial T} \right) Y dy$$

where consideration for the generalized Maxwell's relation $\left(\frac{\partial S}{\partial Y} \right) \boldsymbol{T} = \left(\frac{\partial X}{\partial T} \right) Y$ has been made. The induced isothermal entropic variation characterizing elastocaloric effect is as follows:

$$\Delta \boldsymbol{S}(0 \rightarrow \sigma) = \int_0^{\sigma} \left(\frac{\partial X}{\partial T} \right) Y \, dy$$

The parameter determines the entropy variation associated with an isothermal change of strain from zero to σ.

$$\Delta \boldsymbol{S}(0 \rightarrow \varepsilon) = \int_0^{\varepsilon} \left(\frac{\partial \sigma}{\partial T} \right) \varepsilon \, d\varepsilon$$

where, in this case, the Maxwell relation $\left(\frac{\partial S}{\partial E} \right) T = -\left(\frac{\partial \sigma}{\partial T} \right) \varepsilon$ is directly employed. Clearly, if E strain and stress are in exact harmony. $\Delta S(0 \rightarrow \sigma) = \Delta S(0 \rightarrow \varepsilon)$.

As a data element for refrigeration systems, the refrigeration capacity (RC) is defined as:

$$\boldsymbol{RC} = \int_{\Delta T}^{T} \Delta Siso(\Delta T) dT$$

The usual first-order martensitic transformation (MT) is associated with a large entropy shift in the SMAs that show a significant elastocaloric effect, $\left(\frac{\partial \sigma}{\partial T} \right) \sigma$, which is extremely close to the transformation temperature. The compacted Fe-Pd single crystal's strain curves are shown in Figure 6.5a with stress up to 100 MPa in the direction. Figure 6.5b shows the curve for the compacted Fe-Pd single crystal with stress up to 100 MPa in any direction. In all of the temperature values looked at, the strains were bidirectional or stress hysteresis was minor [25]. The strain at 100 MPa rose as the temperature reduced. Strain at 300 K is 1.6%, but at 240 K, it is 3.7% as

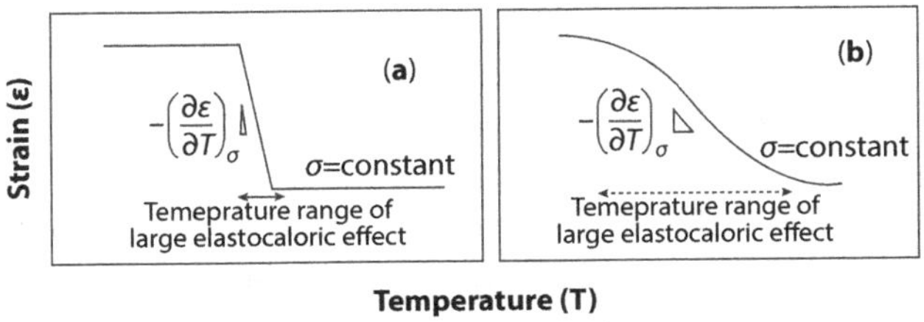

Figure 6.5 The temperature range of a meaningful elastocaloric effect is shown diagrammatically in (a) a normal stress-induced 1st-order MT and (b) a sustained expansion under a steady yield stress [25].

the temperature dropped. For Cu-Zn-Al single crystals, the elastocaloric effect replaces the role of the magneto-caloric effect, with higher ΔS and ΔT values.

In reality, research into the elastocaloric effect is becoming more and more popular. The researchers have reported significant temperature changes for Ti-Ni wires and Ti-Ni films have demonstrated that the effect is stable even after repeated cycling. Uniaxial stress causes the cubic-to-martensitic phase transition in Ti-Ni films, Fe-Pd, and Cu-Zn-Al alloys, despite negligible volume change and shear strain (along the axis) [26]. Here, we look into the elastocaloric effect in a solid $Ba_{0{\cdot}5}Sr_{0{\cdot}5}TiO_3$ ferroelectric alloy. The class of $Ba_xSr_{1-x}TiO_3$ ferroelectric alloys exhibits the most promising properties. The Curie temperature varies greatly depending on the concentration x, from 0 K in $SrTiO_3$ to as high as 405 K in $BaTiO_3$. Since the majority of the caloric effects are amplified close to the Curie threshold, such tunability is particularly appealing for caloric applications. $Ba_{0{\cdot}5}Sr_{0{\cdot}5}TiO_3$ material, which falls in the middle of the compositional spectrum, is a good example of the entire family of $Ba_xSr_{1-x}TiO_3$ alloys. Three ferroelectric transitions occur in the $Ba_{0{\cdot}5}Sr_{0{\cdot}5}TiO_3$ compound. In the heating simulations, the starting temperature is $T_0 = 260$ K above the Curie point. The temperature rises together with the tensile stress and, as a result, asserts that this material has an elastocaloric effect [27–29].

6.3.2 Specific Heat Capacity

When selecting a material for solid-state refrigeration, it is crucial to take its specific heat capacity (C_p) into account since it enables a high temperature flow rate during each cycle, which lowers the number of cycles needed to achieve the required chilling. This is especially useful for elastomers like natural rubber (NR), where a high specific heat capacity is sought because it determines right once how much heat the material can absorb. As the optimum ΔTSMA for phase transition materials like SMAs is ascertained by L/C_p, a smaller C_p is preferred for a given value of L. The absence of perceptible heat means that this has no impact on the SMAs' capacity to cool. An example of this is a study characterizing a substantial cooling capacity in a (Fe-31.2Pd atomic %) monocrystalline SMA. The martensitic transition exhibited a low measured ΔTSMA of about 2K despite having a higher cooling capacity of 2 MJ m^{-3} and a substantial latent heat under a 100-MPa applied stress [30]. Similar developments are anticipated for ferroelectric materials, where an increased specific heat capacity may lower the adiabatic T but has no impact on heat extraction capability or overall performance. A substance with a smaller specific heat capacity or capacity, corresponding to the literary works, will ultimately benefit from a larger caloric impact, enhancing its ability to modify itself to different working and temp parameters.

6.3.3 Endurance Limit

Fatigue is a declined performance of a material caused by an increase in working cycles, which can lead to material failure and property degradation. The endurance limit of a material is the highest amount of load it can support through a considerable number of cycles without significantly degrading its performance. Materials that experience numerous loading cycles in order to accomplish long-term cooling require a high endurance limit. Structure-related fatigue and functional fatigue are the two main forms of elastocaloric material fatigue. Thermal stresses that can be reversed cause functional fatigue while cyclic loads cause structural fatigue. Two methods (fracture propagation and crack initiation) were considered in a literature review on rubber fatigue analysis procedures for evaluating the fatigue behavior of rubber elements. The impacts of alternating loads as well as external factors including frequency, stress intensity, environment, and temperature have been explored in relation to essential material properties such as polymer structure, molecular weight, and crosslinking. The study also covered methods for prolonging polymer fatigue life, including the use of fillers such as short glass fibers that stop fatigue fracture initiation and tiny gaps made during kneading that stop crack propagation. Reducing the quantity of plasticizer or solvents used during polymerization can help delay the rate of fatigue crack growth. A different study looked at thermo-mechanical treatments to increase the endurance limit and thermo-mechanical fatigue of SMA materials. A research study into the operational and structural exhaustion of NiTi smart memory alloy material realized that exceptional productivity can be maintained for up to ~106 cycles of loading with only minor variations in execution, stipulating that the stress needed to induce pseudo elastic phase transition is less than the stress at yield of the austenite phase. According to Xu *et al.*, [30], dual-phase alloys will have a larger durability capacity than single-phase compositions. Their research showed that, for single-phase samples of Ni-Fe-Ga alloys, mechanical fatigue can start after just 10 cycles, but dual-phase samples can continue to be resistant to fatigue beyond 100 cycles. This discovery makes it possible to take a material's phase into account when assessing a material's viability for elastocaloric applications. NiTi wires maintain their functional integrity for more than 400 cycles, according to a study by Tušek *et al.* [26], and Schmidt *et al.* [31] examined the effects of heat training on functional and mechanical fatigue of magnetic SMA ribbons of NiTiCuV. The findings demonstrated that different temperature profiles were formed depending on the type of training, with dispersion patterns and local temperature peaks that were heavily affected by the amount of cycles performed. The NiTiCuV ribbons showed good tolerance to functional fatigue, enabling them to be appropriate for use in real-world applications. Even though SMA single crystals have a high endurance limit (>1000 cycles), their high cost prevents them from being widely used. A high strength limit is also preferred because mechanical failure is a possibility and the load needed for cooling can be large (up to 700 MPa). Frequent use can cause the material's shape memory and elastocaloric properties to degrade, which can also lead to fatigue in the form of thermal cycles. A larger endurance limit would improve the economy and reliability of SMAs and NR even though their mean time between failures is high enough. Numerous studies analyzing their fatigue behavior have revealed ferroelectric materials to be outstanding candidates for stress-driven solid-state refrigeration. These materials can withstand prolonged operation times due to their strong mechanical and functional endurance. They are also appropriate for use where neither polymers nor SMAs are efficient because they only need a small strain (0–0.4%) and stress (0–200 MPa). Contrary to conventional ferroelastic materials, which see a decline in performance as they are downscaled to micro- or nano-levels, ferroelectrics perform significantly better. As a result, there are numerous potential uses for ferroelectrics, and further studies will probably be conducted in this area.

6.3.4 Inversion Temperature

The phase transition between martensite (M) and austenite (A) in SMA materials causes the entropy change. The wide temperature range in which this transformation occurs leaves no clear definition of it. Two temperatures, start (s) and finish (f), which, respectively, denote the beginning and conclusion of the phase formation, are associated with each phase. Figure 6.6 shows that the transition temperature is the temperature at which there is a 50% concentration of both the martensite and austenite phase.

The right level of a material's transition temperature is necessary for efficient heat extraction. If somehow the transition temperature is too elevated in relation to the working temperature, heat extraction will be limited. The temperature at which transition occurs should be close to the temperature of operation to improve heat extraction. If the heat being supplied has no ability to initiate the transition of phase to the austenite, no heat will be evacuated. The substance will continue to exist in the austenite phase if the temperature required for the transition towards the martensitic phase is low enough. The temperature at which the gradient of the linear thermal expansion coefficient shifts is known as the inversion temperature in the context of elastomeric materials. When there are small stresses (up to 10%), thermal expansion coefficient is inversely related to the temperature. On the other hand, thermo-elastic inversion occurs when the temperature–strain graph approaches the transition temperature, and a negative linear thermal expansion coefficient is described as:

$$\alpha^3_0 = (1+2\lambda_0 T/1-\lambda_0 T_i) \exp 3\lambda_0 (T_i - T_0)$$

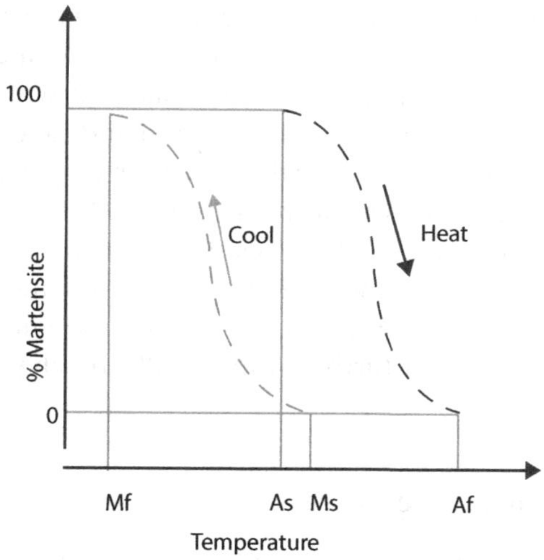

Figure 6.6 Shape memory alloys' phase transition diagram.

In this context, α denotes the strain of an elastic material at a reference temperature, or L/L_0 (T_0), whereas λ_0 denotes the elastic material's coefficient of linear expansion under unloaded conditions. Ti is a representation of the inversion temperature, and T_0 depicts the reference temperature. The above equation demonstrates that if $T_0 = 298.15$ K $= T_i$ and $\lambda_0 = 2.2104$ K^{-1}, then the required value of λ_0 is 1.067. (i.e., isothermal stretching). This shows that a lower strain of 6.7% can cause it to happen at room temperature. To cool to the necessary inversion temperature, which is the bottom limit, at least this strain is required. As a result, it is crucial to carefully consider both the corresponding strain and inversion temperature. The Curie or Néel temperatures, respectively, are used to represent the inversion temperatures for ferroelectric, ferromagnetic, and antiferromagnetic materials in ferroic materials. Above this temperature, paramagnetic properties appear and the ferroelectric/ferromagnetic characteristic completely disappears. Because it is impossible to achieve an entropy-ordered state above this temperature, it serves as the maximum limit for proper functioning of ferroelectric materials. This Curie temperature is influenced by external force and also by its structure and composition. The existence of nano-domains has been reported to provide ferroelectric materials with a wide transition temperature range. Additionally, recent advancements in the ferroelectric elastocaloric phenomena have shown a correlation between ferroelectric material temperature and stress-driven adiabatic temperature change. As per reported studies on PMN-32PT monocrystals and $Ba_{0.85}Ca_{0.15}Zr_{0.1}Ti_{0.9}O_3$ mass ceramics, a larger caloric effect is seen close to the temperature of the phase transition, whether it be ferroelectric–ferroelectric or ferroelectric–paraelectric [32].

6.3.5 Coefficient of Performance (COP)

In a refrigerating system, coefficient of performance is described as:

$$\mathbf{COP_{cooling} = Q_C/W = T_{cold} / T_{hot} - T_{cold}}$$

The variable Q_c in this case stands for the heat drawn from the reservoir, whereas the variable W stands for the amount of labor put into the system. The second component of the equation is notably utilized when a Carnot cycle is under consideration. Because it specifies how much heat is extracted for each unit of work, this expression is a useful performance metric. Recent research investigating the elastocaloric properties of SMAs found the maximum value of COP = 3.07 and of COP relative = 21.6%, which is suggestive of a more efficient mechanism according to the following equation:

$$\mathbf{COP_{Relative} = COP_{cycle}/COP_{Carnot}}$$

The material is dynamically loaded to the point where it reaches the limit of its elasticity in this situation. If mechanical work could be produced during the cycle's unloading phase, such as pumping, the COP might rise to a value of 11.8, or a relative COP of 83.7%. In order to achieve high efficiency, mechanical effort should be carefully extracted during unloading. The likelihood of a temperature rise from the lesser level obtained during unloading is the only disadvantage of this strategy. This is because there is not much time for the materials to complete mechanical tasks, which could cause unloading curves to veer off the adiabatic route and result in very little heat transfer to the environment. Moya *et al.* [33] go into great lengths regarding the basic idea of the requisite and recovering work for caloric processes in ferroic compounds in a published article. This article details the kind and amount of the real work prerequisite to enter the compound phase change. The material efficiency equal to Q/W was used to compare the efficacy of various caloric effects and within a single caloric effect. Comparisons were done with a scenario that featured comparable heat flow and extraction rates (Q). This assertion was made despite the finding that the comparative productivity of stress-derived caloric effect is low, since high-grade electric work can be recovered and used in electrocaloric materials. However, elastocaloric effects can still achieve efficiencies that are greater than electrocaloric effects even when this premise is ignored. Furthermore, this information does not take into account recently published values for elastocaloric effects in ferroelectric materials and only applies to conventional shape-memory alloys of ferroic materials, which demand high strains. Furthermore, while high-intensity magnetic and electric fields must be produced with specialized equipment, mechanical strains can occasionally be derived physically. This idea has been supported by Tušek *et al.* [34], who provide a thorough description of the operation and performance assessment of a real solid refrigerant device that makes use of elastocaloric effects. A unique regenerative cycle was used to examine the cooling capabilities of two distinct SMAs, NiTi and Cu-Zn-Al. A number of operational factors, including operation frequency, loading circumstances, and operating temperatures, were taken into consideration when evaluating the performance. The effectiveness was also contrasted with a magneto-caloric cooler based on Gd. Along with providing the capability for gradual temperature change needed for ambient temperature refrigeration systems, it was shown that SMAs can even surpass magnetic refrigerators and the electrocaloric effect at 30 K or higher temperature. As a result, mechanical refrigeration is a versatile option for application in far-off places [48].

6.3.6 Other Important Parameters

When evaluating a material's actual performance related to its molecular effectiveness and volumetric efficiency, three additional considerations should be made: An extensive operating range near room temperature (temperature invariant performance), the isothermal entropy variation, and field normalized caloric impact, commonly referred to as material efficiency, round out the list of desirable characteristics. The isothermal entropy change measures the material's ability to extract heat per unit volume; a higher number indicates a material's ability to extract heat more effectively in each cycle. Performance that operates throughout a wider temperature range guarantees consistent performance under fluctuating heat loads. A material's performance can be assessed using an adiabatic temperature change curve, which has a parabolic shape, with a steeper slope denoting worse performance under varied thermal loads. The chemical also needs to work well in field-normalized conditions, which includes caloric effect for less stress. The system would function more effectively and efficiently overall if the material had a higher T for a lower input. Therefore, when selecting the appropriate material for elastocaloric solid-state refrigeration, these factors should be taken into account.

6.4 Challenges and Future Potential

Alongside the previously stated material parameters, the exploration of substances for elastocaloric purposes presents substantial prospects for progress towards attaining triumphant integration into cutting-edge technologies. The primary obstacle in this field of study pertains to the creation of appropriate substances that can exhibit linked caloric phenomena. The simultaneous coupling of multiple caloric effects in a material, specifically the co-existence of "electric," "elastic", and "magnetic" entropy transitions, can lead to noteworthy heat transfers. The quest for an appropriate material chemistry that can effectively link multi-ferroic ordered phase transitions is presently in progress. The thin film of the PZT ($PbTiO_3$-$PbZrO_3$) ferroelectric compound has been found to exhibit a significant electrocaloric effect [35]. PZT is a composite material consisting of distinct ferroelectric phases of lead-titanate and -zirconate that are independent of each other. By modulating the concentration of constituent components and introducing dopants like lanthanum (La), one can conveniently customize its characteristics to transition between the two distinct phases [36, 37]. Furthermore, contemporary research has revealed the presence of an elastocaloric phenomenon in the $PbTiO_3$ utilizing a fundamental principle methodology. Through precisely manipulating the chemical nature of the substance, it is possible to achieve a simultaneous occurrence of the elastocaloric and electrocaloric effects at a specific temperature. The aforementioned phenomenon will bring about a significant alteration between states characterized by elevated and reduced entropy, ultimately causing a substantial linked caloric effect.

In the scenario of magnetic shape memory alloys (SMAs), a comparable event has been documented. Multiple investigations have demonstrated that significant amounts of Ni-Mn alloy and other magnetic shape memory alloys exhibit a barocaloric/elastocaloric phenomena in addition to magnetically ordered phase transitions [38, 39]. The elastocaloric phenomenon, which is triggered solely by mechanical strain, is observable in SMAs. This effect can be synchronized with the magnetic ordered phase transition, resulting in the generation of significant entropy fluctuations. By modifying the proportion of nickel and manganese alongside other elemental components, one can effectively manipulate magnetic phase transition order [40]. It is anticipated that the highest entropy alteration in ferroic substances will manifest in close proximity to their ferroelectric–paraelectric transition. Upon surpassing the Curie or Néel temperature of a ferroic substance, a phase transition occurs, leading to the emergence of a paraelectric/paramagnetic phase and the consequent loss of all functional characteristics. Therefore, it is desirable for the Curie/Néel temperature to closely align with the designated operating temperature in order to attain optimal material efficacy and COP in any real-world implementation. It is critical to recognize that performing any activity in a substance exceeding the Curie/Néel point entails the use of exterior magnetic or electric fields to rearrange the domains following the material's return to its original temperature values. The new electro-thermomechanical process that combines the elasto- and electrocaloric phenomena in ferroelectric individual crystal systems leads to increased efficiency [41].

It is anticipated that comparable occurrences will manifest in alternative ferroelectric compositions, and ongoing investigations are currently underway in this area. Now, accessibility of multi-ferroic compounds with pseudoelastic phase shift is inadequate. The development in complex multi-ferroic substances has the ability to provide considerable advantages. These composite materials are designed to possess a minimum of two distinct phases that exhibit entropy transition in close proximity to the desired operating temperature. Although direct coupling is not anticipated in such materials, the resulting caloric effect can surpass that of its constituent parts [42].

Aside from the linked caloric implications, any cooling process based on the elastocaloric effect has practical implementation challenges. The primary goal is to lower the amplitude of the actuating field and the quantity of energy spent by the system components as a result of an outside stimulus within the realm of caloric occurrences. Whereas the electrocaloric as well as magneto-caloric effects' impact is non-mechanical in origin, the elastocaloric action needs the use of an appropriate loaded cell capable of administering required forces and accommodating the connected stresses. This phenomenon is attributed to the high elasticity of elastomers, which allows for a significant degree of elastic deformation, often exceeding several hundred percent. Furthermore, the advancement of thermal pumps and flow apparatuses for these methodologies is yet to be perfected. Several methodologies have been suggested to attain this aim, including thermal switches, diodes, Peltier modules, and convection via liquid crystals. However, additional research is necessary to fully explore these options. Nevertheless, endeavors are being undertaken to surmount the concomitant constraints and prototyping is currently in progress. The noteworthy contributions of Schmidt and Tušek *et al.* cannot be overlooked, as they have successfully devised and evaluated the efficacy of a functional refrigeration mechanism that utilizes the elastocaloric phenomenon [26, 31]. It is anticipated that the forthcoming decade will witness the development of dependable and effective systems that are primed for widespread implementation.

6.5 Sustainability and Environmental Impact

With an increase of about 0.6 K during the previous century, there is now a noticeable rise in world temperatures. The UN Intergovernmental Panel on Climate Change's most current study has indicated that this trend is set to continue, with projections estimating a further increase of 1.4–4.5 K in the average global temperature by the year 2100. The magnitude of this increase holds the potential to influence the seasonal patterns of rainfall, alter the migratory habits of diverse organisms, and intensify the occurrence of severe weather phenomena, among other effects. The imperative to mitigate climate change has prompted endeavors to curtail activities and discharges that contribute to this phenomenon. The primary endeavor towards this objective is exemplified by the Kyoto Protocol [43–47]. One of the primary objectives of the protocol is to impose legal obligations on developed nations to curtail their combined discharge of greenhouse gases. Additionally, the protocol has prohibited the utilization of diverse chlorofluorocarbons (CFCs) in the refrigeration sector. The process of refrigeration is a significant consumer of energy in modern society and constitutes a substantial portion of the electrical power usage in Europe. As a result of its significant carbon footprint, this industry is also recognized as one of the primary contributors to greenhouse gas emissions. The utilization of halofluorocarbons (HFCs) in vapor compression systems, coupled with low operational efficiencies, presents a potential issue [48]. The greatest degree of efficiency attainable by a vapor compression system is limited to a maximum of 60% of the Carnot efficiency, according to theoretical considerations. However, it is noteworthy that a significant number of refrigeration systems utilize compressors that exhibit considerably diminished performances, thereby resulting in an overall efficiency that ranges between 40% and 55%, which is realistically achievable. Furthermore, despite the discontinuation of the hazardous CFCs, prevalent refrigeration systems in the market still utilize HFCs. The aforementioned gases exhibit lower toxicity levels and do not participate in the depletion of the ozone layer. Nevertheless, these compounds are categorized as super-greenhouse gases and make a significant contribution to the phenomenon of global warming. An alternative approach that can be considered is the utilization of organic refrigerants with low boiling points, such as pentane or ammonia. However, the restricted scope of operational temperatures poses a constraint on the extensive implementation [49–51].

Given these circumstances, solid-state refrigeration presents a viable option for achieving high reduction potentials in refrigeration and cryogenic applications on a medium to small scale. The phase transitions of ferroic materials of higher order are known to exhibit significant changes in entropy, leading to remarkable caloric effects. The low thermal inertia exhibited by such systems restricts their usage to lower loading systems. However, solid-state refrigerating techniques offer a multitude of advantages such as high efficiency and COP. Polycrystalline Ni-Ti SMA has reportedly shown predicted COP levels that are nearly 83.7% of Carnot [52]. This advancement signifies a significant enhancement compared to traditional vapor compression methodologies. Additionally, solid-state conditioning works by using the substance being cooled as a means of heat exchange, negating the requirement for refrigerants like HFCs. In addition to their eco-friendly nature, the reduction of moving components can result in prolonged functionality, decreased malfunctions, and improved safety measures. A key step for the creation of environmentally friendly solid-state technology is the suggestion of an ongoing life cycle evaluation for ferroelectric substances.

The initial stride towards this objective encompasses the advancement of ferroelectric materials that are devoid of lead. Elastocaloric materials based on alloys and polymers have the advantage of being easily reusable and recyclable, which further simplifies the matter. Hence, it is imperative to conduct additional investigations in the realm of solid-state cooling substances and methodologies. These particular arrangements not only exhibit potential for economic advantages but also demonstrate a reduced ecological footprint in comparison to currently utilized conventional apparatus [53, 54].

6.6 Conclusions

The chapter looks into the intrinsic characteristics that govern the elastocaloric effect in diverse materials, such as shape memory alloys, polymeric compounds, and ferroelectric substances, from a chemical standpoint. The subject matter centers around the two varieties of elasticity—entropy elasticity and force elasticity—as well as the associated fundamental mechanisms. Thermodynamic equations have been employed to deduce expressions for the elastocaloric phenomenon, while also examining the factors that impact the efficacy of substances utilized in elastocaloric refrigeration. The aforementioned parameters encompass the specific heat capacity, elastocaloric magnitude, endurance limit, coefficient of performance (COP), and transition temperature. The detailed discussion pertains to the potential practical implementation of mechanical refrigeration in multi-ferroic materials, with a particular focus on the coupled caloric effects. Furthermore, this section gives a concise summary of the present technological constraints that impede the practical use of the elastocaloric phenomenon in applications in industry. The practicality and ecological ramifications of solid-state cooling systems are also being debated. In a nutshell, this study has the potential to spark more research into how to make use of the elastocaloric phenomenon in the area of chemistry to develop environmentally friendly and effective solid-state refrigeration.

References

1. Chauhan, A., Patel, S., Vaish, R., Bowen, C.R., A review and analysis of the elasto-caloric effect for solidstate refrigeration devices: Challenges and opportunities. *MRS Energy Sustain.*, 2, 126, 2015.
2. Chauhan, A., Patel, S., Vaish, R., Elastocaloric effect in ferroelectric ceramics. *Appl. Phys. Lett.*, 106, 127, 2015.

3. Yoshida, Y., Yuse, K., Guyomar, D., Capsal, J.F., Sebald, G., Elastocaloric effect in poly(vinylidene fluoride-trifluoroethylene-chlorotrifluoroethylene) terpolymer. *Appl. Phys. Lett.*, 108, 242904, 2016.
4. Bai, G., Liu, D., Gao, C., Phenomenological analysis of elastocaloric effect in ferroelectric poly(vinylidene fluoride-trifluoroethylene) copolymers. *J. Appl. Phys.*, 126, 127, 2019.
5. Hou, H., Cui, J., Qian, S., Catalini, D., Hwang, Y., Radermacher, R., Takeuchi, I., Overcoming fatigue through compression for advanced elastocaloric cooling. *MRS Bull.*, 43, 285, 2018.
6. Chen, J., Lei, L., Fang, G., Elastocaloric cooling of shape memory alloys: A review. *Mater. Today Commun.*, 28, 102706, 2021.
7. Qian, S., Geng, Y., Wang, Y., Ling, J., Hwang, Y., Radermacher, R., Takeuchi, I., Cui, J., A review of elastocaloric cooling: Materials, cycles and system integrations. *Int. J. Refrig.*, 64, 1, 2016.
8. Seekell, R.P., Sarangapani, P.S., Zhang, Z., Zhu, Y., Relationship between particle elasticity, glass fragility, and structural relaxation in dense microgel suspensions. *Soft Matter.*, 11, 5485, 2015.
9. Buehler, M.J. and Wong, S.Y., Entropic elasticity controls nanomechanics of single tropocollagen molecules. *Biophys. J.*, 93, 37, 2007.
10. Koga, T., Seo, Y.S., Shin, K., Zhang, Y., Rafailovich, M.H., Sokolov, J.C., Chu, B., Satija, S.K., The role of elasticity in the anomalous swelling of polymer thin films in density fluctuating supercritical fluids. *Macromolecules*, 36, 5236, 2003.
11. Chen, Z., Li, M., Guo, R., Wang, Y., Zhou, D., Chen, Z., Ang, R., Performance evolution of thermoelectric modules under constant heat flux. *Mater. Today Phys.*, 35, 101136, 2023.
12. Snyder, G.J. and Toberer, E.S., Complex thermoelectric materials. *Nat. Mater.*, 7, 105, 2008.
13. Baker, I., Interstitials in f.c.c. high entropy alloys. *Metals (Basel)*, 10, 130, 2020.
14. Li, S., Lou, X., Li, X., Zhang, J., Li, D., Deng, H., Liu, J., Tang, G., Realization of high thermoelectric performance in polycrystalline tin selenide through Schottky vacancies and endotaxial nanostructuring. *Chem. Mater.*, 32, 9761, 2020.
15. Fan, Z., Wang, H., Wu, Y., Liu, X.J., Lu, Z.P., Thermoelectric high-entropy alloys with low lattice thermal conductivity. *RSC Adv.*, 6, 52164, 2016.
16. Kim, T.Y., Kim, S.K., Kim, S.W., Application of ferroelectric materials for improving output power of energy harvesters. *Nano Converg.*, 5, 131, 2018.
17. Kong, L.B., Huang, H., Li, S., Fundamentals of ferroelectric materials, in: *Ferroelectric Materials for Energy Applications*, pp. 1–31, 2018.
18. A century of ferroelectricity. *Nat. Mater.*, 19, 129, 2020.
19. Cui, C., Xue, F., Hu, W.J., Li, L.J., Two-dimensional materials with piezoelectric and ferroelectric functionalities. *NPJ 2D Mater. Appl.*, 2, 18, 2018.
20. Whatmore, R., Ferroelectric materials, in: *Springer Handbooks*, S. Kasap and P. Capper (Eds.), p. 1, Springer International Publishing, Cham, 2017.

21. Bowen, C.R., Kim, H.A., Weaver, P.M., Dunn, S., Piezoelectric and ferroelectric materials and structures for energy harvesting applications. *Energy Environ. Sci.*, 7, 25, 2014.
22. Blázquez-Castro, A., García-Cabañes, A., Carrascosa, M., Biological applications of ferroelectric materials. *Appl. Phys. Rev.*, 5, 139, 2018.
23. Guyomar, D., Li, Y., Sebald, G., Cottinet, P.J., Ducharne, B., Capsal, J.F., Elastocaloric modeling of natural rubber. *Appl. Therm. Eng.*, 57, 33, 2013.
24. Xiao, F., Fukuda, T., Kakeshita, T., Significant elastocaloric effect in a Fe-31.2Pd (at.) single crystal. *Appl. Phys. Lett.*, 102, 141, 2013.
25. Binek, C. and Burobina, V., Near-room-temperature refrigeration through voltage-controlled entropy change in multiferroics. *Appl. Phys. Lett.*, 102, 141–143, 2013.
26. Tušek, J., Engelbrecht, K., Mikkelsen, L.P., Pryds, N., Elastocaloric effect of Ni-Ti wire for application in a cooling device. *J. Appl. Phys.*, 117, 141, 143, 145, 151, 2015.
27. Mañosa, L., Jarque-Farnos, S., Vives, E., Planes, A., Large temperature span and giant refrigerant capacity in elastocaloric Cu-Zn-Al shape memory alloys. *Appl. Phys. Lett.*, 103, 143, 2013.
28. Lisenkov, S. and Ponomareva, I., Intrinsic electrocaloric effect in ferroelectric alloys from atomistic simulations. *Phys. Rev. B - Condens. Matter Mater. Phys.*, 80, 149, 2009.
29. Xiao, F., Fukuda, T., Kakeshita, T., Significant elastocaloric effect in a Fe-31.2Pd (at.) single crystal. *Appl. Phys. Lett.*, 102, 143, 2013.
30. Xu, Y., Lu, B., Sun, W., Yan, A., Liu, J., Large and reversible elastocaloric effect in dual-phase Ni54Fe19Ga27 superelastic alloys. *Appl. Phys. Lett.*, 106, 144, 145, 2015.
31. Schmidt, M., Ullrich, J., Wieczorek, A., Frenzel, J., Schütze, A., Eggeler, G., Seelecke, S., Thermal stabilization of NiTiCuV shape memory alloys: Observations during elastocaloric training. *Shape Mem. Superelasticity*, 1, 132, 2015.
32. Patel, S., Chauhan, A., Vaish, R., Multiple caloric effects in (Ba0.865Ca0.135Zr0.1089Ti0.8811Fe0.01)O3 ferroelectric ceramic. *Appl. Phys. Lett.*, 107, 147, 2015.
33. Moya, X., Defay, E., Heine, V., Mathur, N.D., Too cool to work. *Nat. Phys.*, 11, 202, 2015.
34. Tušek, J., Engelbrecht, K., Millán-Solsona, R., Mañosa, L., Vives, E., Mikkelsen, L.P., Pryds, N., The elastocaloric effect: A way to cool efficiently. *Adv. Energy Mater.*, 5, 1500361, 2015.
35. Mischenko, A.S., Zhang, Q., Scott, J.F., Whatmore, R.W., Mathur, N.D., Giant electrocaloric effect in thin-film PbZr0.95Ti 0.05O3. *Sci. (80-.)*, 311, 1270, 2006.
36. Lisenkov, S., Mani, B.K., Chang, C.M., Almand, J., Ponomareva, I., Multicaloric effect in ferroelectric PbTiO3 from first principles. *Phys. Rev. B - Condens. Matter Mater. Phys.*, 87, 150, 2013.
37. Barr, J.A., Beckman, S.P., Nishimatsu, T., Elastocaloric response of $PbTiO_3$ Predicted from a first-principles effective Hamiltonian. *J. Phys. Soc. Japan*, 84, 24716, 2015.

38. Mañosa, L., Planes, A., Acet, M., Advanced materials for solid-state refrigeration. *J. Mater. Chem. A*, 1, 4925, 2013.
39. Krenke, T., Duman, E., Acet, M., Wassermann, E.F., Moya, X., Manosa, L., Planes, A., Inverse magnetocaloric effect in ferromagnetic Ni-Mn-Sn alloys. *Nat. Mater.*, 4, 450, 2005.
40. Castillo-Villa, P.O., Mañosa, L., Planes, A., Soto-Parra, D.E., Sánchez-Llamazares, J.L., Flores-Zúñiga, H., Frontera, C., Elastocaloric and magnetocaloric effects in Ni-Mn-Sn(Cu) shape-memory alloy. *J. Appl. Phys.*, 113, 150, 2013.
41. Chauhan, A., Patel, S., Vaish, R., Multicaloric effect in Pb(Mn1/3Nb2/3)O3-32PbTiO3 single crystals. *Acta Mater.*, 89, 384, 2015.
42. Vopson, M.M., Theory of giant-caloric effects in multiferroic materials. *J. Phys. D. Appl. Phys.*, 46, 345304, 2013.
43. Root, T.L., Price, J.T., Hall, K.R., Schneider, S.H., Rosenzweig, C., Pounds, J.A., Fingerprints of global warming on wild animals and plants. *Nature*, 421, 57, 2003.
44. Schlumberger, The impact of global warming. *Am. Econ. Rev.*, 2009, 1309, 2008.
45. Collins, M., An, S.I., Cai, W., Ganachaud, A., Guilyardi, E., Jin, F.F., Jochum, M., Lengaigne, M., Power, S., Timmermann, A., Vecchi, G., Wittenberg, A., The impact of global warming on the tropical Pacific Ocean and El Nĩo. *Nat. Geosci.*, 3, 391, 2010.
46. Miller, J.R. and Russell, G.L., The impact of global warming on river runoff. *J. Geophys. Res.*, 97, 2757, 1992.
47. French, D., Kyoto protocol to the United Nations framework convention on climate change. *J. Environ. Law*, 10, 215, 1998.
48. Tully, S., UNEP: Montreal protocol on substances that deplete the ozone layer, 1987, in: *International Documents on Corporate Responsibility*, Edward Elgar Publishing, Cheltenham, UK, 2013.
49. 96/00806 1995, ASHRAE Handbook, HVAC applications, Si edition. *Fuel Energy Abstr.*, 37, 49, 1996.
50. Chakravarthy, V., Weber, J., Rashad, A.A., Acharya, A., Bonaquist, D., Oxygen liquefier using a mixed gas refrigeration cycle. *Am. Soc. Mech. Eng. Heat Transf. Div. HTD*, 374, 491, 2003.
51. Nikitin, S.A., Myalikgulyev, G., Annaorazov, M.P., Tyurin, A.L., Myndyev, R.W., Akopyan, S.A., Giant elastocaloric effect in FeRh alloy. *Phys. Lett. A*, 171, 234, 1992.
52. Dingquan, X., Environmentally conscious ferroelectrics research - Present and prospect. *Ferroelectrics*, 231, 133, 1999.
53. Saito, Y., Takao, H., Tani, T., Nonoyama, T., Takatori, K., Homma, T., Nagaya, T., Nakamura, M., Lead-free piezoceramics. *Nature*, 432, 84, 2004.
54. Chauhan, A., Patel, S., Vaish, R., Bowen, C.R., A review and analysis of the elasto-caloric effect for solidstate refrigeration devices: Challenges and opportunities. *MRS Energy Sustain.*, 2, 153, 2015.

7

Effective Flexomagnetic/Flexoelectric Sensitivity in Ferroics/Nanosized Ferroic Materials

Uzma Hira* and Asifa Safdar

School of Physical Sciences (SPS), University of the Punjab, New Campus, Lahore, Pakistan

Abstract

Flexoelectricity is the term for the mechano-electrical connection between electric polarization and strain gradient, as well as the mechano-electrical response of coupling among gradients of the electrical field and the mechanical stress. An interesting size phenomenon, which is typically substantial when the material size is reduced, is demonstrated by this exceptional outcome. In this chapter, we explain the flexoeffect (flexomagnetic and flexoelectric) alterations in a ferroic nanosystem and their attributes that result from the inhomogeneity of order parameters. This effect exists spontaneously. The equation framework of FCT (flexoelectric coupling tensor) for various point groups using direct matrix approach, including the quantity of nonzero independent components, has been studied. The efficient and effective flexoelectric/flexomagnetic behavior has been investigated for a number of point symmetries that are crucial for different applications. Additionally, it has been seen in many other types of materials, including insulators, biological materials, liquid crystals, and semiconductors.

Keywords: Flexoresponse, flexomagnetic, flexoelectricity, ferroic nanoparticle, direct matric approach, point symmetry

**Corresponding author*: uzma.sps@pu.edu.pk

Inamuddin, Tariq Altalhi and Mohammad Abu Jafar Mazumder (eds.) Ferroic Materials-Based Technologies, (157–172) © 2024 Scrivener Publishing LLC

7.1 Introduction

When an object is subjected to an inhomogeneous impact of elastic force or strain gradient, a polarization or magnetization reaction manifests itself. When mechanical stress or strain appears in reaction to a change in the order of parameter gradient, this is the opposite of the flexoeffect. Let us emphasize how flexocoupling modifies the intrinsic gradient for order parameters as well as the system's reaction to external impact [1]. The flexoelectric effect is a typical illustration of this phenomenon, as it results from the coupling of a polarization barrier within an elastic strain. Like well-studied flexoelectric phenomena, the flexomagnetic response exists in different materials such as in some antiferromagnetic [2] exhibit symmetry without time inversion.

Mashkevich and Tolpygo [3] anticipated the flexoelectric phenomenon; later, Tagantsev [4] investigated the phenomenon theoretically in the bulk crystals, while Cross and Ma [5–7] and Zubko *et al.* [8] measured the flexoelectric tensor components experimentally. Catalan *et al.*'s article [9] serves as a starting point for updating the theoretical detail of the flexoelectric response of different nanomaterials, whereas Kalinin, Majdoub *et al.* [10] and Meunier's papers detail more recent accomplishments. However, the link between intrinsic parameters, such as the spontaneous polarization gradient and strain, was not considered in these publications; instead, flexoelectric response was viewed as a relationship between intrinsic and extrinsic characteristics [11]. The ferroic nanomaterials containing strong-order variables show spontaneous flexocoupling response [12]. Bulk materials barely exhibit this phenomenon due to the role of surfaces in different properties of nanosystems.

It is important to emphasize the fact that estimations of dimensions when the phenomenological method is valid are required by the phenomenological approach, which has been widely used as the global scientific literature in description of nanosystems [13, 14] and by the bulk of the aforementioned theoretical studies. The smallest feasible shift in the order parameter of a ferroelectric can be determined by its correlation radii; if these radii are larger than the lattice constant, then the phenomenological technique can be used. The experimentally observed ferromagnetic size range from 7.0 to 30.0 nm or the ferroelectric critical length range from 1.0 to 10.0 nm [15–17] agrees reasonably with the critical sizes derived from microscopic or phenomenological calculations [18] for the presence of long-range order in ferroics. Therefore, the range of critical size is of several nanometers, making the zone of validity of the phenomenological method sufficiently large. As an example, Majdoub *et al.* [14] investigated dielectric characteristics of thin film considering the polarization gradient enabling the flexoelectric action or the reaction of a cantilever-shaped beam to an external stress gradient. In order to compare the results with microscopic modeling, they employ phenomenological macroscopic models on nanosized systems.

The phenomenological model that incorporates flexoelectricity and polarization gradient was found by Majdoub *et al.* [14] to qualitatively or in some cases quantitatively duplicate results from the microscopic modeling. The flexoeffect occurs when the order parameter is not uniformly distributed in space, and how it modifies the properties of ferroic nanostructures. The flexoelectric connection influences different properties and, in particular, reduces the polarization gradient, renormalizes the radius of correlation, and stabilizes the ordered phase using concrete examples of mechanically free nano ferroelectric pills, rods, and wires.

7.2 Basic Mathematical Form for Flexoeffect Contribution in Ferroic Nanomaterials

Regarding gradient energy, depolarization, removal fields, surface energy, flexoeffect, and mechanical stress, we will employ the classic Landau–Ginzburg–Devonshire (LGD) phenomenological technique [19, 20] to explain flexocoupling response in ferroic nanomaterials. Minimizing Helmholtz energy F is important because flexoeffects in nanoparticles produce the body propelling via different order-parameter shadings, which carry out the actual work. The Helmholtz free energy value F [17] of second-order ferroics can be decomposed into a bulk (F_V) and a surface (F_S) component, and their LGD expansions on the order parameters or strain tensor component uij have the form:

$$F_V \int_V d3r \begin{bmatrix} aij * T/2\, ninj + a_{ijkl}/4 n_i n_j n_k n_l - n_i(E_{0i} + Ei^d/2) + g_{ijkl} * 1/2(\partial n_i \partial x_j \partial n_k \partial x_l) \\ -f_{ijkl} * 1/2\, n_k \partial u_{ij}/\partial x_l - u_{ij} \partial n_k/\partial x_l - q_{ijkl} u_{ij} n_k n_l + c_{ijkl}/2\, u_{ij} u_{kl} + \\ v_{ijklmn}/2(\partial u_{ij}/\partial x_m \partial u_{kl}/\partial x_n \end{bmatrix}$$

$$F_s = \int_s d2\, r[a^s ij/2\, n_i n_j + a^2{}_{ijkl}/4\, n_i n_j n_k n_l + \mu^s{}_{\alpha\beta} u_{\alpha\beta} + d^s{}_{ijk} u_{ij} n_k + w^2{}_{ijkl}/2\, u_{ij} u_{kl} + \ldots) \tag{7.1}$$

Temperature T has an influence on the coefficients a_{ij}. Coefficients a^s_{ij}, a_{ijkl}, and a^s_{ijkl} are temperature-independent variables. The terms regarding different tensors are defined. g_{ijkl} and v_{ijklmn} determine the energy gradient. w^s_{jklm} is the surplus elastic moduli of surface, $\mu^s_{\alpha\beta}$ is the stress of surface [21, 22], and d^s_{ijk} is the surface piezoelectric [19]. q_{ijkl} is the coefficient of bulk striction and c_{ijkl} are elastic stiffness components [20].

Practically, only Lifshitz invariant $f_{ijkl}/2\ (n_k\ \partial u_{ij}\ \partial_{xl} - u_{ij}\partial n_k\ \partial_{xl})$ is relevant to materials in bulk. The gradient terms like $v_{ijklmn}\ (\partial u_{ij}/\partial x_k)(\partial u_{lm}/\partial x_n)$ are responsible for smooth scattering at nonzero strain gradients [21]. While considering tensorial case, the terms $v_{ijklmn}\ (\partial u_{ij}/\partial x_k)\ (\partial u_{lm}/\partial x_n)$ can be ignored following condition $f^2_{\ klmn}/2 < g_{ijkl}c_{ijmn}$.

E_0 is an extrinsic field joined with the η (order parameter) in Equation 7.1. E^d is demagnetization or depolarization due to divergence of nonzero order η in a confined system. [23, 24]

The equation of state $\delta F_V/\delta n_i = 0$ and $\delta F_V/\delta u_{ij} = \sigma_{ij}$, where σ_{jk} is the tensor of stress and δ is a variation derivative that is attained through changes in the bulk free energy. Equation 7.1 should be solved with the help of mechanical equilibrium equation $\delta\ \sigma_{ij}\ (x)/\delta xi = 0$ and compatibility equations of the mechanical displacement [25]. The changes in the bulk and surface free energy on η_i gives Euler–Lagrange equations with different boundary conditions.

$$(g_{kjim}n_k\ \partial\eta_m/\partial x_j + a^s_{ij}\ n_j + f_{jkim}/2\ u_{jk}n_m\ )_s = 0 \qquad (7.2)$$

Ferroic surfaces have exterior normal components denoted by n_k. The heterogeneous boundary conditions are the most noticeable effect of the flexocoupling. In the following, we ignore the contributions to the exterior energy (Equation 7.1) from the excessive elastic modulus of surface, the stress tensor of surface, and the piezoelectric effect on surface in order to show that the spontaneous flexoeffect is responsible for the nanoferroic characteristics. Here, we take into account mechanically unrestricted nanoparticles devoid of misfit dislocations that showed result only in an "external" flexoeffect. Misfit dislocations were carefully investigated by Catalan *et al.*, as a potential contributor from the flexoelectric impact in thin films of material.

First, let us calculate the elastic strains uij using the free energy Equation (7.2). After an elastic solution has been found, it can be used to replace the n_j components of the order parameter in the Euler–Lagrange equations.

7.3 Symmetry and Definition of the Flexoelectric Coupling

Researchers are always interested in exploring new elements of flexoelectricity in solids, whether they be theoretical or practical. The direct effect of electric polarization gradient *Pk/xl* on elastic strain *uij* constitutes the static flexoelectric response, while the inverse strain effect gradient *uij/xl* on polarization *Pi* constitutes the static flexoelectric effect [26–30]. The Lifshitz invariant static flexible tricks coupling (for convenience "flexocoupling") is part of a ferroic's free energy in the continuum-medium approximation [31–34].

$$F_{FL} = \int_{v} d^3r \ f_{ijkl} / 2(P_k \partial u_{ij} / \partial x_l u_{ij} \partial P / \partial x_l) \tag{7.3}$$

where the tensor parts of strain [$u_{ij} = (\partial U_i/\partial x_j + \partial U_j/\partial x_i)/2$] represent different types of the displacement vector parameters U_i (symmetrized). In general, the anti-symmetric structure of the relocation gradients, ($\partial U_i/\partial x_j$-$\partial U_j/\partial x_i$), did not influence physical characteristics. Particularly, it affects the flexoelectric response of the entire material. Due to the obvious index-permutation symmetrical ($u_{ij} \equiv u_{ji}$), each of the elements of the rigid flexocoupling tensor f_{ijkl} is balanced with a pair of permutation.

$$f_{ijkl} \equiv f_{jikl} \tag{7.4a}$$

All 32 crystalline groupings of points provide the static flexoelectric response. The number of the flexoelectric tensors for independent components is determined for various symmetries by Shu, [29] Quang, and He [35]. Notably, f_{ijkl} (flexocoupling tensor) permutation invariance (Equation 7.4a) with regard to the first pair of indices was utilized by Quang, Shu *et al.*, and He [29, 35].

In continuous media approximation, surface integrals of an unbounded bulk ferroic are ignored while integration by parts is used to solve Equation 7.3. As a result of this procedure, additional, "hidden" index-permutation symmetry features of infinite ferroic are made apparent, including:

$$f_{ijkl} = f_{ilkj} \tag{7.4b}$$

Note that Equation 7.4b can only hold true for an infinite material, and is thus not applicable to finite or nanoscale structures.

Kvasov and Tagantsev [36] made a valuable prediction about the presence of the static flexoelectric response on a cubic lattice using the lattice's dynamics theory. They proposed that minimizing the Lagrange operator $L= F - T$ yields dynamical equations of form for a compact medium.

In nanoparticles and curved domain structures like polar vortices and labyrinthine domains seen in ferroelectric thin films, flexoelectric coupling plays a crucial role at the nanoscale [37].

7.4 Symmetry and Definition of the Flexomagnetic Coupling

Flexomagnetic coupling can be initiated by inhomogeneous strains attached to magnetic forces that are expected to exist in environments that are continuous to homogeneously scattered moments of magnetism (for example, close to domain walls, surfaces, and interfaces) [27, 28].

It is possible to define the flexomagnetic response like the stationary flexoelectric response. Similar to how electric magnetization gradients M_k/x_l cause elastic strain u_{ij} to develop, the flexomagnetic effect can be defined as the emergence of magnetization Mi in response to strain gradients u_{ij}/x_l. In the continuum-medium approximation, the flexomagnetic coupling of a ferroic can be defined as a Lifshitz invariant in the LGD free energy.

$$F_{FL} = \int_v d^3r \; m_{ijkl} / 2(M_k \partial u_{ij} / \partial x_l - u_{ij} \partial M_k / \partial x_l) \tag{7.5a}$$

Because the strain tensor, u_{ij} is obviously index-permutation symmetric, the flexomagnetic tensor, m_{ijkl}, is also index-permutation symmetric with regard to each permutation of the first pairing of indices, i.e., $u_{ij}=u_{ji}$.

$$m_{ijkl} \equiv m_{jikl} \tag{7.5b}$$

Integrating Equation 7.5a by parts allows us to ignore surface integrals for an infinite bulk ferroic, as allowed by the continuous medium approximation. This process, identical to Equation 7.4b, reveals index-permutation symmetry features for infinite ferroic:

$$m_{ijkl} = m_{ilk} \tag{7.5c}$$

Equation 7.5c is similar to Equation 7.5b in that it is only true for the infinite material and fails for finite and nanoscale systems.

The flexomagnetic effect, like the flexoelectric coupling, is predicted to play an important role in the thermodynamics and kinetics of curved domain structures such as antiferromagnetics, ferromagnetic small films, nanoparticles, skyrmions, and others at the micro- and nano-scale. In fact, the flexomagnetic response has been explored by Chen *et al.* [29] in relation to the emergence of magnetism at the ferroelectric domain boundaries. While considering the response of inhomogeneous strain (tension) above the assembled graphene structure, Naumov and Bratkovsky [30] adopted a flexomagnetic coupling idea. In their experimental investigation, they considered a heterostructure made up of magnetic and ferroelectric layers. Spurgeon *et al.* [31] found evidence of inhomogeneous tension across their entire system due to flexomagnetic coupling. Considering the flexomagnetic coupling, Ray modeled the magnetic effect of a bent nanomaterial. Ghobadi *et al.* [32] investigated the incorporation of flexomagnetic response in phenomenological analysis of a nanoplate subjected to a magnetic field.

Although the flexoelectric coupling constant has been extensively investigated, the research on flexomagnetic response is less. By defining the flexomagnetic coefficient as the coupling of gradient strain tension and magnetism dipole momentum divided by the atom of Mn, Lukashev *et al.* determined an approximate value of $1.95\mu_B$ A° for the antiperovskite Mn_3GaN.

It seems interesting that the flexomagnetic effect cannot exist if a symmetry group only has time as well as distance inversion operations. These two processes need to be connected in the symmetry group of the material in order to show a flexomagnetic effect. Using a symmetry theory and LGD technique, Eliseev *et al.* determined its framework of the flexomagnetic influence tensor. In the past, researchers were able to use a unified method to determine the fundamentals of piezomagnetic, magnetoelectric, inhomogeneous magneto electric, and flexomagneto-electric phenomena [33, 34, 38].

7.5 The Chapter Structure and Motivation

By performing a symmetry analysis, it is possible to determine how symmetry affects the effective flexible tricks and flexomagnetic response (henceforth "flexoresponse") of various ferroics. Only the effective flexocoupling constant can be derived directly from bending experiments, which is a quite direct and easy method. Raman and inelastic neutron scattering are two examples of other indirect experimental approaches.

To the greatest of our ability, accurate flexoresponse calculations and alignment analysis are not applicable to the arbitrary alignment of the sample. In this section, we analyze the anisotropy and angular dependencies of the calculated effective flexoresponse of the bending plate for every given sample orientation and then visualize it. The calculations are useful because it is possible to avoid testing the flexoresponse by continually rotating the crystal if the angular dependency is known. Instead, measurements along various crystallographic orientations are sufficient enough at this stage.

7.6 Flexocoupling Response in Ferroics

7.6.1 Response of Flexoelectric Coupling in Different Ferroics Having Lower and Cubic Symmetry

The system of linear equations for algebra that results from applying the index sequence symmetry (Equation 7.4a) directly towards the flexoelectric matrices f_{ijkl} (shortly, the "direct method") is as follows:

$$f_{ijkl} = C_{ii'}C_{jj'}C_{kk'}C_{ll'}f_{i'j'k'l'}$$

$$f_{ijkl} = f_{jikl} \tag{7.6a}$$

Both index-permutation symmetry (Equation 7.4a) and point-group (shown by the first line) symmetry are reflected in the second line of Equation 7.6a. Therefore, $C_{ii'}$ represent concrete point-group symmetry. The significance of the findings is revealed by comparing Tables 7.1 and 7.2 for various point symmetries, which are provided below. In particular, an index-permutation symmetry (Equation 7.4b) lessens the overall number of f_{ijkl} elements that are independent from three to two for a cubic asymmetry and from eight to five for a rectangular point equilibrium. The index-permutation symmetrical (Equation 7.4b) minimizes different numbers of autonomous elements *fijkl* from 9 to 15. This is the case for the orthorhombic geometry point's symmetry referred to as the most common form.

Table 7.1 Structure of the static flexoelectric effect tensor calculated from Equation 7.6a for several point symmetry groups.

Point symmetry	Number of nonzero elements	Number of independent elements	Relation between nonzero elements
Tetragonal			
4'2m,422, 4mm, 4/mmm	**21**	8	f_{3333},f_{1111}=f_{2222},f_{1122}=f_{2211}, f_{1133}= f_{2233},f_{3311}=f_{3322},f_{1212}=f_{2112}=f_{1221} =f_{2121}, f_{1313}=f_{3113}=f_{2323}=f_{3223}, f_{1331}=f_{3131}=f_{2332}=f_{3232}
Orthorhombic			
222, mm2, mmm	21	15	f_{1111}, f_{1122}, f_{1133}, f_{2211} f_{2222}, f_{2233}, f_{3311}, f_{3322}, f_{3333},f_{1212}=f_{2112},f_{1221}=f_{2121}, f_{1313}=f_{3113},f_{1331}=f_{3131}, f_{2323}=f_{3223}, f_{2332}=f_{3232}
Cubic			
m3m, 432, 4'3m	21	3	f_{1111}=f_{2222}=f_{3333}, f_{1122}=f_{2211}= f_{1133}=f_{3311}=f_{2233}=f_{3322},f_{1221}= f_{2112}=f_{1331}=f_{3113}=f_{2332}= f_{3223}=f_{1212}= f_{2121}=f_{1313}=f_{3131}= f_{2323}=f_{3232}

7.7 Flexomagnetic Behavior of Coupling in Ferroics Having Cubic Symmetry

The linear equation system is obtained by applying the point-group symmetry calculations Equations 7.5a, 7.5b, and 7.5c directly on the flexomagnetic tensor *mijkl.*

$$\{m_{ijkl} = (-1)^R det(C^{\wedge})C_{ii'}C_{jj'}C_{kk'}C_{ll'}f_{i'j'k'l'},$$
$$\{m_{ijkl} = m_{jikl},$$
$$\{m_{ijkl} = m_{ilkj}\}$$

Table 7.2 Structure of the static flexoelectric effect tensor calculated from Equation 7.6b for several point symmetry groups.

Point symmetry	Number of nonzero elements	Number of independent elements	Relation between nonzero elements
Tetragonal			
4'2m, 422, 4mm, 4/mmm	**21**	5	f_{3333}, $f_{1111}=f_{2222}$, $f_{1122}=f_{1221}=f_{2121}=f_{2211}=f_{2112}=f_{1212}$, $f_{1133}=f_{1331}=f_{3131}=f_{2233}=f_{3232}=f_{2332}$, $f_{1313}=f_{3311}=f_{3113}=f_{2323}=f_{3322}=f_{3223}$
Orthorhombic			
222, mm2, mmm	21	9	f_{1111}, f_{2222}, f_{3333} $f_{1122}=f_{1221}=f_{2121}$, $f_{2211}=f_{2112}=f_{1212}$, $f_{2233}=f_{2332}=f_{3232}$, $f_{3311}=f_{1313}=f_{3113}$, $f_{3322}=f_{3223}=f_{2323}$, $f_{1133}=f_{1331}=f_{3131}$
Cubic			
m3m, 432, 4'3m	21	3	$f_{1111}=f_{2222}=f_{3333}$, $f_{1122}=f_{2211}=f_{1133}=f_{3311}=f_{2233}=f_{3322}=f_{1221}=f_{2112}=f_{1331}=f_{3113}=f_{2332}=f_{3223}=f_{1212}=f_{2121}=f_{1313}=f_{3131}=f_{2323}=f_{3232}$

The number of nonzero elements is 21 for all considered examples, but it can be different for other groups.

Therefore, we introduced the transformation framework or determinant, *det of C = 1 det of A = 1*; the factor *R* indicates whether or not the spatial transformation C_{ij} is linked with a time-reversal operation, with *R*=1 indicating the presence and *R*=0 indicating the absence of time reversal. $m_{ijkl}=m_{jikl}$ *and* $m_{ijkl}=m_{ilkj}$ are index-permutation symmetries. $C_{ii'}$ (matrix elements) are referred to as concrete points that can be obtained using the matrix form. For all 90 magneto stability groups, m_{ijkl} forms include the variety of independent, nonzero elements that are given by the solution of Equation 7.5c.

Analyses in Table 7.2 for two different cubic symmetries (m'3m' and m'3m') provide insight into the significance of the findings. These groups have fewer significance than three m_{ijkl}-independent components due to the index-permutation symmetry $m_{ijkl}=4_{milkj}$, which decreases the number to two and one, respectively. Both cubic groups under consideration have a total of 21 nonzero components.

7.8 Effective Flexoresponse

The f_{ijkl} of a paraelectric $SrTiO_3$ precursor with $m3m$ symmetry was experimentally determined. The flexoelectric constants have also been determined from basic principles for the metallic group $Ba_xSr_{1-x}TiO_3$ with a value of x=0.50, which contains $m3m$ as a parent geometry or symmetry. The three-stage knife setup for experiments and accompanying sample orientation typical for measurements of successful flexoresponse is depicted in Figure 7.1 [34, 38–40].

7.9 Flexoelectricity in Different Materials

Despite the fact that all insulators should theoretically exhibit flexoelectric coupling, research on the flexoelectric units in different materials is still limited as a flexoelectric coefficient is inherent. A few dielectrics and polymers have been used to explore their flexoelectric units (coefficients) using those direct and indirect techniques. As demonstrated by TiO_2 ceramic

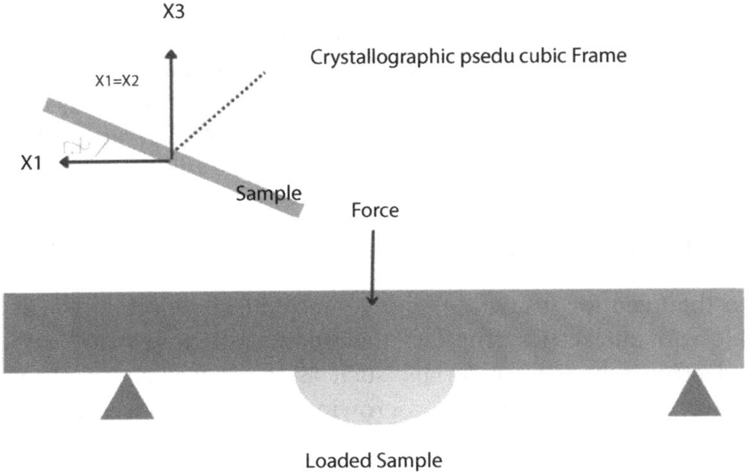

Figure 7.1 Flexoresponse measuring apparatus.

and polyvinylidene fluoride (PVDF) [40–43], the flexoelectric units in typical polymers and dielectrics fall in the range of 1.0–10.0 nCm^{-1}, which is in good relation with the intrinsic properties of flexoelectric units.

However, the outcomes are considerably different in some types of single crystals and ferroelectric ceramics. A number of high-permittivity ferroelectrics, including $BaTiO_3$, were shown to have extremely increased flexoelectricity [44, 45]. In comparison to typical dielectrics like TiO_2, their reported flexoelectric units are three to five times greater in magnitude. It is interesting to note that such amplification is not a common occurrence in ferroelectrics.

7.9.1 Flexoelectricity in Biological Materials

Numerous biological materials including bones, hair, and bio-membranes may exhibit exceptional flexoelectric response, according to recent investigations [46–50]. Williams and Breger's initial research on the flexoelectricity of biological materials dates back to 1975 [51].

This original study suggested that "gradient polarization" may have contributed to a variety of electromechanical characteristics of bones but the exact process is still unclear. Vasquez-Sancho *et al.* [52] used the conventional cantilever technique to compare the flexoelectricity effect of pure hydroxyapatite (a common mineral found in bone) and bone in order to explain the genesis of this electromechanical feature in bones.

7.9.2 Flexoelectricity in Liquid Crystal

One of the key characteristics of these kinds of materials has been thought to be the liquid crystal flexoelectricity, which dates back to the 1970s. In liquid crystals, the element e_1 and the element e_3, which are splay and bend deformation, respectively, are primarily responsible for representing the flexoelectric coefficients [53].

7.9.3 Flexoelectricity in Semiconductors

Due to the need for insulation, electromechanical coupling is often only possible with dielectrics. The fact that semiconductors may be thought of as electromechanical materials, such as the piezoelectric ZnO semiconductor, has come to light in a growing number of research, however [54]. In 2016, researchers saw a major development in flexoelectricity, and Narvaez *et al.* [55] worked considerably to expand the idea of flexoelectricity into semiconductors. The observed flexoelectricity in certain types of semiconductors may reach a level of up to 1000.0 Cm^{-1}, and a dead layer mechanism has been put forth.

7.10 Conclusion

Because of surface and size effects, ferroic nanosystems have an inherent inhomogeneity of different order parameters, which is the reason for spontaneous flexocoupling exhibiting mechanical tension or strain. This particular electromechanical coupling mechanism enables the electrical monitoring of strain gradient data, which is considered as a sensitive indication of the structural health state, in mechanical systems. The flexomagnetic effect, a unique type of magneto-elastic effect that involves the connection of magnetism and strain gradients, has drawn significant attention in the interdisciplinary areas of materials science, electronics, and mechanical engineering. It has the potential to be applied to the creation of innovative devices with low power consumption and miniaturization, such as energy harvesters and actuators.

References

1. Narvaez, J., Vasquez-Sancho, F., Catalan, G., Enhanced flexoelectric-like response in oxide semiconductors. *Nature*, 538, 219, 2016.
2. Yudin, P.V. and Tagantsev, A.K., Basic theoretical description of flexoelectricity in solids, in: *Flexoelectricity in Solids: From Theory to Applications*, pp. 1–29, World Scientific, Singapore, 2016.
3. Ouyang, J., Zhang, W., Alpay, S.P., Roytburd, A.L., Effect of elastic domains on electromechanical response of epitaxial ferroelectric films with a three-domain architecture. *J. Adv. Ceram.*, 2, 1, 2013.
4. Ma, W. and Cross, L.E., Strain-gradient-induced electric polarization in lead zirconate titanate ceramics. *Appl. Phys. Lett.*, 82, 3293, 2003.
5. Eliseev, E.A., Fomichv, Y.M., Kalinin, S.V., Vysochanskii, Y.M., Maksymovich, P., Morozovska, A.N., Labyrinthine domains in ferroelectric nanoparticles: Manifestation of a gradient-induced morphological transition. *Phys. Rev. B*, 98, 054101, 2018.
6. Zhang, S., Liang, X., Xu, M., Feng, B., Shen, S., Shear flexoelectric response along 3121 direction in polyvinylidene fluoride. *Appl. Phys. Lett.*, 107, 2015.
7. Eliseev, E.A., Kalinin, S.V., Morozovska, A.N., Finite size effects in ferroelectric-semiconductor thin films under open-circuit electric boundary conditions. *J. Appl. Phys.*, 117, 2015.
8. Ma, W. and Cross, L.E., Flexoelectricity of barium titanate. *Appl. Phys. Lett.*, 88, 232902, 2006.

9. Kretschmer, R. and Binder, K., Surface effects on phase transitions in ferroelectrics and dipolar magnets. *Phys. Rev. B*, 20, 1065, 1979.
10. Majdoub, M.S., Sharma, P., Cagin, T., Enhanced size-dependent piezoelectricity and elasticity in nanostructures due to the flexoelectric effect. *Phys. Rev. B - Condens. Matter Mater. Phys.*, 77, 119904, 2008.
11. Tagantsev, A.K., Electric polarization in crystals and its response to thermal and elastic perturbations. *Phase Transitions*, 35, 119, 1991.
12. Popov, A.I., Plokhov, D.I., Zvezdin, A.K., Symmetry and magnetoelectric effects in garnet crystals and films. *Phys. Rev. B - Condens. Matter Mater. Phys.*, 90, 214427, 2014.
13. Glinchuk, M.D., Morozovska, A.N., Eliseev, E.A., Blinc, R., Misfit strain induced magnetoelectric coupling in thin ferroic films. *J. Appl. Phys.*, 105, 84108, 2009.
14. Ghobadi, A., Beni, Y.T., Golestanian, H., Size dependent thermo-electro-mechanical nonlinear bending analysis of flexoelectric nano-plate in the presence of magnetic field. *Int. J. Mech. Sci.*, 152, 118, 2019.
15. Lines, M.E. and Glass, A.M., Study of soft modes using infrared and optic techniques, in: *Principles and Applications of Ferroelectrics and Related Materials*, M.E. Lines, and A.M. Glass, (Eds.), pp. 209–240, Oxford University Press, Walton Street, Jericho, 2010.
16. Tagantsev, A.K., Piezoelectricity and flexoelectricity in crystalline dielectrics. *Phys. Rev. B*, 34, 5883, 1986.
17. Majdoub, M.S., Maranganti, R., Sharma, P., Understanding the origins of the intrinsic dead layer effect in nanocapacitors. *Phys. Rev. B - Condens. Matter Mater. Phys.*, 79, 115412, 2009.
18. Chen, Y.C., He, Q., Chu, F.N., Huang, Y.C., Chen, J.W., Liang, W.I., Vasudevan, R.K., Nagarajan, V., Arenholz, E., Kalinin, S.V., Chu, Y.H., Electrical control of multiferroic orderings in mixed-phase BiFeO 3 films. *Adv. Mater.*, 24, 3070, 2012.
19. Zhang, S., Liu, K., Xu, M., Shen, H., Chen, K., Feng, B., Shen, S., Investigation of the 2312 flexoelectric coefficient component of polyvinylidene fluoride: Deduction, simulation, and mensuration. *Sci. Rep.*, 7, 3134, 2017.
20. Eliseev, E.A., Morozovska, A.N., Glinchuk, M.D., Blinc, R., Spontaneous flexoelectric/flexomagnetic effect in nanoferroics. *Phys. Rev. B - Condens. Matter Mater. Phys.*, 79, 165433, 2009.
21. Morozovska, A.N. and Eliseev, E.A., Surface and finite size effect on fluctuations dynamics in nanoparticles with long-range order. *J. Appl. Phys.*, 107, 44101, 2010.
22. Smirnova, E.P., Sotnikov, A.V., Smirnov, S.I., Weihnacht, M., Ferroelastic transition and the relaxor state in SrTiO3- PbMg1/3Nb2/3O3. *Phys. Solid State*, 50, 2145, 2008.
23. Lukashev, P. and Sabirianov, R.F., Spin density in frustrated magnets under mechanical stress: Mn-based antiperovskites. *J. Appl. Phys.*, 107, 09E115, 2010.
24. Zhang, S., Liu, K., Wu, T., Xu, M., Shen, S., Experimental approach for measuring cylindrical flexoelectric coefficients. *J. Appl. Phys.*, 122, 144103, 2017.

25. Trabi, C.L., Brown, C.V., Smith, A.A.T., Mottram, N.J., Interferometric method for determining the sum of the flexoelectric coefficients (e1+ e3) in an ionic nematic material. *Appl. Phys. Lett.*, 92, 223509, 2008.
26. Lukashev, P. and Sabirianov, R.F., Flexomagnetic effect in frustrated triangular magnetic structures. *Phys. Rev. B - Condens. Matter Mater. Phys.*, 82, 094417, 2010.
27. Rychetský, I., Deformation of crystal surfaces in ferroelastic materials caused by antiphase domain boundaries. *J. Phys. Condens. Matter*, 9, 4583, 1997.
28. Zubko, P., Catalan, G., Buckley, A., Welche, P.R.L., Scott, J.F., Erratum: Strain-gradient-induced polarization in srTiO3 single crystals (Physical Review Letters (2007) 99 (167601)). *Phys. Rev. Lett.*, 100, 167601, 2008.
29. Pyatakov, A.P. and Zvezdin, A.K., Flexomagnetoelectric interaction in multiferroics. *Eur. Phys. J. B*, 71, 419, 2009.
30. Narvaez, J., Saremi, S., Hong, J., Stengel, M., Catalan, G., Large flexoelectric anisotropy in paraelectric barium titanate. *Phys. Rev. Lett.*, 115, 037601, 2015.
31. Naumov, I.I. and Bratkovsky, A.M., Gap opening in graphene by simple periodic inhomogeneous strain. *Phys. Rev. B - Condens. Matter Mater. Phys.*, 84, 245444, 2011.
32. Migozzi, J.B., Progress in diffractive optics for. H.U.D. *J. Phys. C Solid State Phys.*, 18, 40, 1985.
33. Raphael, R.M., Popel, A.S., Brownell, W.E., A membrane bending model of outer hair cell electromotility. *Biophys. J.*, 78, 2844, 2000.
34. Tanygin, B.M., Symmetry theory of the flexomagnetoelectric effect in the magnetic domain walls. *J. Magn. Magn. Mater.*, 323, 616, 2011.
35. Williams, W.S. and Breger, L., Piezoelectricity in tendon and bone. *J. Biomech.*, 8, 407, 1975.
36. Morozovska, A.N., Fomichov, Y.M., Maksymovych, P., Vysochanskii, Y.M., Eliseev, E.A., Analytical description of domain morphology and phase diagrams of ferroelectric nanoparticles. *Acta Mater.*, 160, 109, 2018.
37. Ma, W. and Cross, L.E., Flexoelectric polarization of barium strontium titanate in the paraelectric state. *Appl. Phys. Lett.*, 81, 3440, 2002.
38. Shchukin, V.A., Ledentsov, N.N., Bimberg, D., Self-organization phenomena at crystal surfaces, in: *Epitaxy of Nanostructures*, V.A. Shchukin, N.N. Ledentsov, D. Bimberg, (Eds.), Springer Berlin Heidelberg, Berlin, Heidelberg, 2004.
39. Catalan, G., Sinnamon, L.J., Gregg, J.M., The effect of flexoelectricity on the dielectric properties of inhomogeneously strained ferroelectric thin films. *J. Phys. Condens. Matter*, 16, 2253, 2004.
40. Bao, N., Shen, L., Gupta, A., Tatarenko, A., Srinivasan, G., Yanagisawa, K., Size-controlled one-dimensional monocrystalline BaTiO3 nanostructures. *Appl. Phys. Lett.*, 94, 253109, 2009.
41. Kvasov, A. and Tagantsev, A.K., Dynamic flexoelectric effect in perovskites from first-principles calculations. *Phys. Rev. B - Condens. Matter Mater. Phys.*, 92, 054104, 2015.

42. Erdem, E., Semmelhack, H.C., Böttcher, R., Rumpf, H., Banys, J., Matthes, A., Gläsel, H.J., Hirsch, D., Hartmann, E., Study of the tetragonal-to-cubic phase transition in PbTiO3 nanopowders. *J. Phys. Condens. Matter.*, 18, 3861, 2006.
43. Kang, H., Hou, Z., Qin, Q.H., Experimental study of time response of bending deformation of bone cantilevers in an electric field. *J. Mech. Behav. Biomed. Mater.*, 77, 192, 2018.
44. Alpay, S.P., Misirlioglu, I.B., Sharma, A., Ban, Z.G., Structural characteristics of ferroelectric phase transformations in single-domain epitaxial films. *J. Appl. Phys.*, 95, 8118, 2004.
45. Glinchuk, M.D., Eliseev, E.A., Morozovska, A.N., Blinc, R., Giant magnetoelectric effect induced by intrinsic surface stress in ferroic nanorods. *Phys. Rev. B - Condens. Matter Mater. Phys.*, 77, 024106, 2008.
46. Jiang, X., Huang, W., Zhang, S., Flexoelectric nano-generator: Materials, structures and devices. *Nano Energy*, 2, 1079, 2013.
47. Chu, B. and Salem, D.R., Flexoelectricity in several thermoplastic and thermosetting polymers. *Appl. Phys. Lett.*, 101, 103905, 2012.
48. Petrov, A.G., Flexoelectricity in lyotropics and in living liquid crystals, in: *Flexoelectricity in Liquid Crystals: Theory, Experiments and Applications*, pp. 177–210, Imperial College Press, Covent Garden, London, UK, 2012.
49. Glinchuk, M.D., Morozovska, A.N., Eliseev, E.A., Ferroelectric thin films phase diagrams with self-polarized phase and electret state. *J. Appl. Phys.*, 99, 114102, 2006.
50. Shu, L., Wei, X., Pang, T., Yao, X., Wang, C., Symmetry of flexoelectric coefficients in crystalline medium. *J. Appl. Phys.*, 110, 104106, 2011.
51. Abou-Dakka, M., Herrera-Valencia, E.E., Rey, A.D., Linear oscillatory dynamics of flexoelectric membranes embedded in viscoelastic media with applications to outer hair cells. *J. Nonnewton. Fluid Mech.*, 185–186, 1, 2012.
52. Bates, L., Electrodynamics of continuous media By L. D. Landau and E. M. Lifshitz (translated from the Russian by J. B. Sykes and J. S. Bell). Pp. x + 417. Pergamon Press, London. 1960. 84s. net. *Endeavour*, 20, 167, 1961.
53. Theory of elasticity. Timoshenko and Goodier . McGraw-Hill. New York 1951. 493 pp. 270 diagrams. 81s. net. (New Edition.). *J. R. Aeronaut. Soc.*, 56, 308, 1952.
54. Spurgeon, S.R., Sloppy, J.D., Winkler, C.R., Jablonski, M., Kepaptsoglou, D., Balachandran, P., Nejati, S., Karthik, J., Damodaran, A.R., Johnson, C.L., Ambaye, H., Goyette, R., Lauter, V., Ramasse, Q., Idrobo, J.C., Lau, K.K., Lofland, S.E., Rondinelli, J., Martin, L.W., Taheri, M.L., Evidence for a thickness-dependent crossover from charge- to strain-mediated magnetoelectric coupling in La0.7Sr0.3MnO3 / PbZr0.2Ti0.8O3 thin film oxide heterostructures. *Microsc. Microanal.*, 19, 1964, 2013.
55. Vasquez-Sancho, F., Abdollahi, A., Damjanovic, D., Catalan, G., Correction to: Flexoelectricity in bones. *Adv. Mater.*, 30, 1801413, 2018.

8

Advancements in Ferroic Thin Films, Multilayers, and Heterostructures

Abdur Rehman Mushtaq Ahmad, Maria Wasim*, Aneela Sabir, Muhammad Shafiq, Rafi Ullah Khan and Farhan Asghar

Institute of Polymer and Textile Engineering, University of the Punjab, Lahore, Pakistan

Abstract
Ferroic thin films, multilayers, and heterostructures have received a great deal of attention recently because of the unique properties that make them ideal for a variety of applications. The background, properties, and characterization methods of ferroic materials and their thin film analogs will be covered in this chapter. The relevance of atomically controlled thin films and interfaces will be highlighted by recent developments in the growth and characterization of these materials. We will also look at the numerous uses of ferroic multilayers, heterostructures, and thin films in microelectronics, data storage, sensing, and energy conversion. Finally, we explore the possibilities for these materials in the future, including ongoing research aimed at enhancing their characteristics and broadening the scope of their uses.

***Keywords*:** Ferroic thin films, multilayers, heterostructures

8.1 Ferroic Materials

Ferroic materials are those that spontaneously order their physical characteristics, such as ferromagnetic, ferroelectric, ferroelastic, and ferrotoroidal ordering. These materials have been researched for more than a century because of their special characteristics and applications in several domains, including sensing, actuation, and data storage. The creation of ferroic thin

**Corresponding author*: maria-be24@hotmail.co.uk; mariawasim.ipte@pu.edu.pk

Inamuddin, Tariq Altalhi and Mohammad Abu Jafar Mazumder (eds.) Ferroic Materials-Based Technologies, (173–206) © 2024 Scrivener Publishing LLC

films, multilayers, and heterostructures has attracted increasing attention recently since they provide fresh opportunities for tailoring and enhancing these materials' characteristics for various applications.

Ferroic materials have been researched for more than a century and employed in a variety of industries. Thin films, multilayers, and heterostructures formed of ferroic materials are now more prevalent than ever, and there are more options than ever to customize and enhance them for particular applications. Because of their unique qualities, these buildings are used a lot in microelectronic devices, spintronic, data storage, sensing, and actuation. More progress in these fields is likely to come from the continued development of ferroic materials and their use in devices [1].

8.1.1 Ferroic Thin Films

Ferroic thin films are just a few-micron-thick layers of ferroic materials. Thin films have unique properties that cannot be found in the bulk material they come from because they are so small. This changes how thin ferroic films behave physically and chemically. This process is called "dimensionality reduction."

Ferroic thin films can be produced by a variety of techniques, including atomic layer deposition (ALD), pulsed laser deposition (PLD), physical vapor deposition (PVD), and chemical vapor deposition (CVD). Each of these approaches employs a somewhat different set of processes to achieve the desired film thickness and composition. As a result, designers have a lot of creative latitude while using films. Which deposition technique to employ depends on the required coating's thickness, content, and crystal structure. Thin-film geometries are advantageous for use in microelectronic devices due to their low power consumption, compact size, and quick response times [2].

8.1.1.1 Historical Developments of Ferromagnetic Thin Films

Ferromagnetic thin films have been the subject of much research over the past few decades, which has greatly improved both their production process and material composition. Thanks to this research, scientists now have a better understanding of the characteristics and behavior of these materials, which has led to new applications for them in a variety of sectors. Vacuum evaporation techniques were employed in the 1960s to investigate thin ferromagnetic films. The magnetic characteristics of these thin films have been shown to be strongly tied to the deposition process, and it has been found that even modest changes in temperature and pressure can have a significant impact on the final properties of the films. Although these films have potential, they frequently suffer from repeatability and quality issues, which make it challenging to regularly produce high-quality videos [3].

Researchers began developing novel techniques for producing ferromagnetic thin films in the 1970s, including molecular beam epitaxy (MBE) and sputtering. Sputtering involves hitting a target material with high-energy ions in order to deposit atoms on a substrate. By applying atoms or molecules to a substrate one at a time using MBE, it was feasible to precisely regulate the film's characteristics. Researchers discovered that MBE was better for producing thin films with high structural perfection and low defect density whereas sputtering was better for producing thin films with a high magnetic moment and good uniformity [4].

The definition and potential applications of ferromagnetic thin films have become clearer as a result of research. In numerous domains, including spintronics, magnetic data storage, and magnetic sensing, they have made them more versatile.

8.1.1.2 Historical Developments of Ferroelectric Thin Films

Ferroelectric thin films can have their naturally occurring electric polarization reversed by the application of an electric field. "Ferroelectricity" is the term for this. Since the discovery of ferroelectricity in bulk materials in the early 1900s, this area of study has had a lengthy history. Thin ferroelectric films have made tremendous strides since that time and have a wide range of applications. Researchers began utilizing methods like vacuum evaporation and sputtering in the 1960s and 1970s to investigate ferroelectric thin films. However, it was discovered that the properties of these films greatly depend on the method of deposition, making it challenging to obtain accurate and reproducible results.

Thin films could be produced on a greater variety of substrates and their properties could be changed more easily due to chemical vapor deposition (CVD) technology in the 1980s. A thin film forms as a result of the chemical interaction that is started between the gases above a heated substrate in CVD. With the help of this technique, thin ferroelectric films with controlled thickness and composition may be produced.

Scientists discovered new ferroelectric materials in the 1990s that outperformed more traditional ferroelectric materials in terms of fatigue resistance and Curie temperatures. For instance, the 1990s saw the discovery of the ferroelectric materials lead scandium tantalate (PST) and bismuth titanate (BIT). High polarization values are one of the fascinating characteristics of ferroelectric thin films, making them useful in a variety of applications, including energy storage and non-volatile memory.

It was quite popular to grow epitaxial thin films in the late 1990s and early 2000s. The electrical characteristics of an epitaxial layer improve and defects are eliminated when the crystal structure is optimized for the substrate. Numerous techniques, including molecular beam epitaxy (MBE) and pulsed laser deposition (PLD), can be used to create epitaxial layers.

In recent years, researchers have worked very hard to create ferroelectric thin films at the nanoscale. These films have special qualities that make them helpful in a variety of contexts, including energy storage and object detection. Few nanometer-thick lead zirconate titanate (PZT) thin films have demonstrated increased piezoelectric characteristics.

New materials and deposition techniques have been discovered during the history of ferroelectric thin films, and we now know more about the relationship between film properties and deposition conditions. Due to these developments, it is now possible to produce thin films of exceptional quality that have special features and can be applied in a variety of contexts [5].

8.1.1.3 Importance of Thin Films in Ferroic Materials

Thin films are an essential part of research into ferroic materials because of their particular properties and various potential uses. Ferroic materials are particularly advantageous in a range of technologies, including sensors, actuators, and memory components, due to their capacity to spontaneously polarize in the electric, magnetic, or elastic directions. This lecture will focus on new research findings from academic journals as it examines the significance of thin films in ferroic materials.

The ability to tweak and perfect the characteristics of thin films is one of the finest features of ferroic materials. This increases the variety of applications for materials by enabling the production of materials with flexible properties. For instance, recent studies have demonstrated that the addition of thin films can enhance ferroelectric materials' piezoelectric and pyroelectric capabilities, which are crucial for applications in sensing and actuation.

Thin films can also improve the dependability and durability of ferroic materials, which is another advantage. Because they have a higher surface-to-volume ratio than bulk materials, thin films are more mechanically and thermally stable. They are therefore ideal for usage in challenging circumstances where variations in temperature and pressure may have an impact on how things operate. For instance, recent studies have

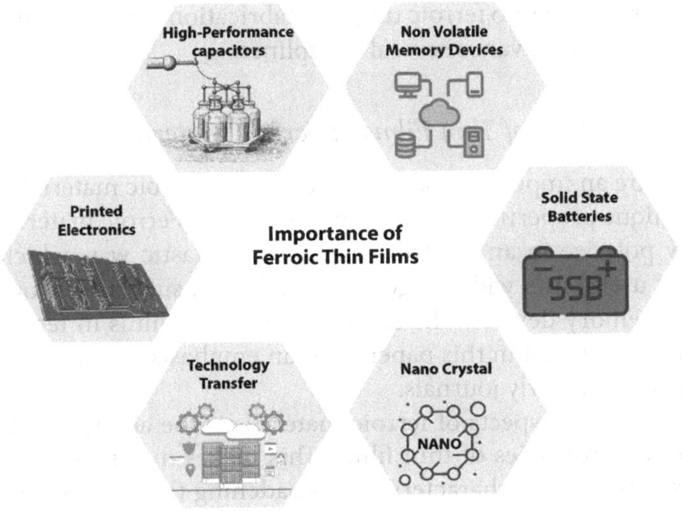

Figure 8.1 Importance of ferroic thin films.

demonstrated that the incorporation of thin films into ferromagnetic materials can enhance both their magnetic and thermal stability [4]. The Figure 8.1 describes the importance of ferroic thin films.

Furthermore, it has been shown that thin films have distinctive properties from those of bulk materials. For example, the material's transition temperature could change, and new phases might form that can be used for a number of things. Recent investigations have shown that new phases in ferroelectric materials, such as multiferroic phases, which have both magnetic and electric polarization, may be produced using thin films. These multiferroic materials could be used in spintronics and high-density data storage.

Thin films have also been used for devices that use ferroic materials. For instance, ferroelectric thin films have helped non-volatile memory devices and high-performance capacitors function better. Magnetic sensors and spin valves were also made possible by the use of thin films in ferromagnetic materials. A few sectors that have used these technologies are communications, space travel, and electronics [6].

Due to their special characteristics and many applications, thin films are frequently utilized with ferroic materials. Because of their ability to be adjusted and controlled, thin films are employed in a wide range of industries. This enables the creation of materials with unique features. Additionally, the utilization of thin films has improved the stability and dependability of ferroic materials and sparked the creation of novel device phases and functionalities. These findings highlight the significance of continuing research into ferroic thin film fabrication, which has the potential to significantly advance several disciplines.

8.1.1.4 Properties of Thin Films in Ferroic Materials

Thin films are an important part of research into ferroic materials because of their unique properties and many possible uses. Ferroic materials spontaneously polarize in an electric, magnetic, or elastic way, which makes them very useful in a wide range of technologies, such as sensors, actuators, and memory devices. The significance of thin films in ferroic materials will be discussed in this paper, with an emphasis on current research findings from scholarly journals.

One of the best aspects of ferroic materials is the ability to adjust and fine-tune the properties of thin films. This makes it possible to produce materials with flexible characteristics, broadening the range of their uses. For instance, recent research has shown that the addition of thin films can improve the piezoelectric and pyroelectric properties of ferroelectric materials, which are essential for applications in sensing and actuation.

Another benefit of thin films is that they can increase the reliability and stability of ferroic materials. Thin films are more mechanically and thermally stable than bulk materials because they have a larger surface-to-volume ratio. As a result, they are perfect for use in difficult situations where changes in temperature and pressure may have an influence on how things function. For instance, recent research has shown that the addition of thin films to ferromagnetic materials can strengthen their magnetic properties and increase their thermal stability [4].

Additionally, it has been discovered that thin films possess peculiar qualities that are absent from bulk materials. The material's transition temperature, for instance, might alter and new phases can arise, which can be employed for a variety of purposes. According to recent studies, thin films can be used to create novel phases in ferroelectric materials, such as multiferroic phases, which exhibit both magnetic and electric polarization. These multiferroic materials could be used in spintronics and high-density data storage.

Thin films have also been used to improve the way devices work with ferroic materials. For example, thin films made of ferroelectric materials have helped improve the performance of high-performance capacitors and non-volatile memory devices. In the same way, magnetic sensors and spin valves were made possible by using thin films in ferromagnetic materials. These devices have been used in many fields, such as electronics, communications, and space travel [6].

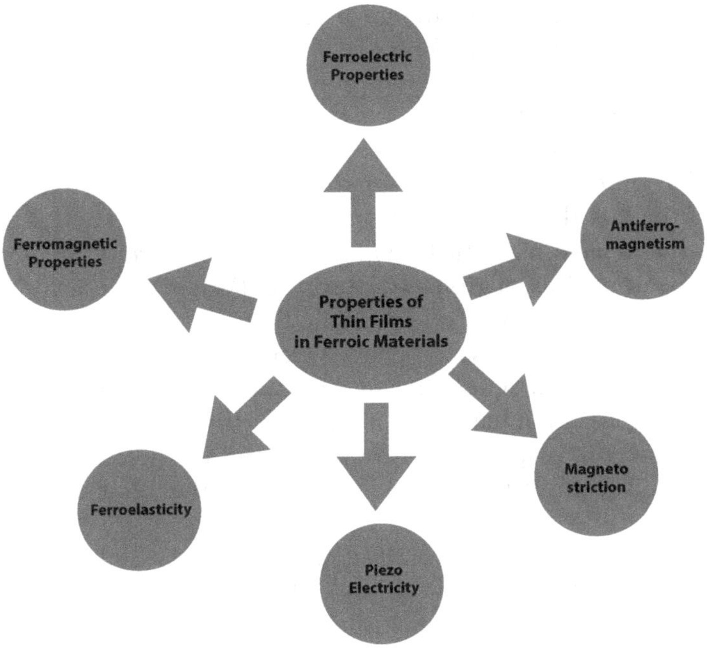

Figure 8.2 Properties of thin films.

Thin films are used a lot with ferroic materials because of their unique properties and wide range of uses. Thin films are used in many different fields because their properties can be tuned and controlled. This makes it possible to make materials with custom properties. Also, the use of thin films has made ferroic materials more stable and reliable and has led to the development of new phases and functions in devices. These results show how important it is to keep researching how to make thin films out of ferroic materials, which could lead to big improvements in a lot of different fields. The Figure 8.2 shows the properties of thin films.

Ferroelectric Properties

Due to their small size and the fact that they have substrate-induced stresses, thin films of ferroelectric materials have unique properties. Srinivasan *et al.* looked at the ferroelectric properties of thin films made of lead zirconate titanate (PZT) (2018). Sol-gel was used to put thin layers of PZT on silicon substrates. The films were looked at with scanning electron microscopy (SEM) and X-ray diffraction (XRD). A well-defined ferroelectric hysteresis loop was found in thin PZT films, which showed that spontaneous polarization was happening. It was found that the thickness of the film and the strain caused by the substrate affect the coercive field and the remnant polarization of thin films [7].

Ferromagnetic Properties

Ferromagnetic materials have their own magnetism, which can be changed by putting an outside magnetic field on them. Due to their small size and the fact that they have magnetic anisotropy at their interfaces, materials with ferromagnetic thin films have unique properties. On MgO substrates, thin films of iron-platinum (FePt) that were ferromagnetic were made by sputtering. To look at the films, X-ray diffraction (XRD) and magnetic force microscopy (MFM) were used. In the FePt thin films, a well-defined ferromagnetic hysteresis loop was seen, which suggests that the magnetization happened on its own. It was found that the coercive field and the magnetization of thin films are affected by both the magnetic anisotropy caused by the interface and the thickness of the film [8].

Thin films of cobalt iron boron (CoFeB) were also looked at to see if they were ferromagnetic. Thin coatings of CoFeB were made on silicon substrates by sputtering with a magnetron. The results showed that the CoFeB thin films had a strong remnant magnetization and a well-defined ferromagnetic hysteresis loop. It was found that the coercive field of thin films was affected by both the thickness of the film and the magnetic anisotropy caused by the interface [9].

Antiferromagnetism

Antiferromagnetic materials have no net magnetic moment because there are no magnetic forces between the atoms. They have recently gotten a lot of attention because they could be used in spintronic devices, magnetic memories, and the storage of information. Due to interfaces and flaws, magnetic order is often broken in thin antiferromagnetic films.

Ferroelasticity

Ferroelasticity is the ability of a material to change shape when a force is applied and then go back to its original shape and size without breaking. Thin films that are ferroelastic could be used in sensors, actuators, and microelectromechanical systems (MEMS). But making high-quality ferroelastic thin films is hard because it is hard to make these materials and figure out what they are like at the nanoscale.

Recently, pulsed laser deposition (PLD) technology was used to make thin films of ferroelastic $BaTiO_3$. At 280 MPa, the films showed a reversible deformation of 0.03%, which is a lot more than bulk $BaTiO_3$. With a stress of 140 MPa, the films deformed by 2%, which was more than the bulk NiTi. These results show that thin ferroelastic films could be very useful in high-performance MEMS devices [10].

Piezoelectricity

When put under mechanical stress, piezoelectric materials create an electric field, and when put in an electric field, they can also create mechanical stress. Because of this, they can be used as sensors, energy-harvesting devices, and actuators, among other things. Researchers have focused on thin films of piezoelectric materials because they could be used in microelectromechanical systems and other small devices.

Magnetostriction

Magnetostriction is a property of materials that change shape or size when they are in a magnetic field. These changes can be reversed. Thin films of magnetostrictive materials are getting a lot of attention right now because they could be used in sensors, actuators, and energy harvesting devices.

Thin films of magnetostrictive materials have been studied a lot because they could be used in microelectromechanical systems and other nanoscale devices. Several studies have shown that high-quality thin films of magnetostrictive materials with good magnetostrictive properties can be made successfully.

These results show that thin sheets of magnetostrictive materials could be very useful in nanoscale devices like magnetic field sensors, actuators, and energy harvesting devices. Also, high-quality thin films of magnetostrictive materials need to be made in order to make modern microelectromechanical systems that work better and are more reliable.

8.1.1.5 Recent Research on New Ferroic Thin Film Materials

Materials with ferroic characteristics have either ferromagnetic, ferroelectric, or both. By applying a mechanical force or an external electric or magnetic field, these materials can change between two or more stable states, such as polarization, magnetization, or strain. A range of sectors, including electronics, sensors, and energy storage, may benefit from ferroic materials because of their unique properties.

New multiferroic, magnetoelectric, and piezoelectric ferroic thin-film materials with enhanced or distinctive features have been the focus of study in recent years. A subclass of ferroelectric materials known as multiferroic materials also exhibit ferromagnetism and other ferroic characteristics. A magnetoelectric substance has the same magnetic and electric properties. It is well known that mechanical stress has the power to induce spontaneous electric polarization in piezoelectric materials.

$EuTiO_3$-$BiFeO_3$ is a new multiferroic material that was found as a result of recent multiferroic material research. Due to its combination of ferromagnetic, ferroelectric, and ferroelastic properties, this substance is an excellent option for creating multiferroic heterostructures with improved magnetoelectric coupling. The $EuTiO_3$-$BiFeO_3$ heterostructure had significantly stronger magnetoelectric coupling than did the individual layers, the researchers demonstrated. This implies that it might be applied to sensors and magnetic memory systems [3].

8.1.1.6 Characterization Methods for Ferroic Thin Films

Ferroic thin films are often looked at with X-ray diffraction (XRD), scanning electron microscopy (SEM), and atomic force microscopy (AFM) [11].

X-ray Diffraction (XRD)

Finding out the crystal structure, orientation, and crystallinity of thin films of ferroic materials may be done extremely well using X-ray diffraction (XRD). In a study, XRD was utilized to determine the multiferroic $BiFeO_3$ thin films' crystal structure and how much they had grown epitaxially. The films were discovered to have excellent ferroelectric and magnetic characteristics. This was partially due to the rhombohedral crystal structure of their preferred (111) orientation.

Scanning Electron Microscopy (SEM)

Ferroic thin films' surface morphology and appearance may be studied using SEM. In a research, the form and roughness of thin lead-free piezoelectric sheets were investigated using SEM. The surface roughness of the films changed as they thickened, affecting how the films responded to compression, as seen by the researchers.

Atomic Force Microscopy (AFM)

Ferroic thin films may have their surface shape and topography studied and characterized using the high-resolution technique known as atomic force microscopy (AFM). AFM was used in a study to examine the polarization switching and domain structure in thin ferroelectric films (AFM). It was discovered that the films' polarization, which had a distinct domain structure, depended on their composition and thickness.

X-ray Photoelectron Spectroscopy (XPS)

A thin ferroic film's surface composition and wiring may be identified using the surface-sensitive technique known as XPS. In a work on multiferroic $BiFeO_3$ thin films, XPS was used to determine the valence states and the kind of bonding between the Fe and Bi ions. Researchers discovered that the mixed valence state of the Fe ions was partly responsible for the films' magnetic and ferroelectric characteristics.

Vibrating Sample Magnetometry (VSM)

VSM is used to evaluate the magnetic properties of thin ferroic films. VSM was used to determine the magnetization of thin magnetoelectric films by monitoring how the films responded to a magnetic field. The films were discovered to have a high degree of magnetoelectric coupling, making them ideal for use in magnetoelectric devices.

8.1.2 Ferroic Multilayers

Alternating layers of various ferroic materials, such as ferroelectric, ferromagnetic, or ferroelastic materials, are used to create composite structures called ferroic multilayers. The distinctive and superior aspects of these structures that go beyond their component pieces have received a lot of attention in recent years. Distinct coupling effects between different ferroic orders, such as magneto-electric and strain-mediated magnetization switching, can be seen in ferroic multilayers. They are a fantastic option for creating new devices with new functions because of this.

Ferroic multilayers have special characteristics that result from interactions between the various ferroic orders at the layer boundaries. These interactions have the power to introduce new properties or enhance those that already exist. The layers' thickness, makeup, and arrangement can all be precisely regulated. This implies that the composite structure's characteristics can be tailored to meet particular requirements.

Sensors, data storage, and electronic devices are only a few applications for ferroic multilayers. For instance, new forms of memory devices that incorporate the advantages of both ferroic orders can be created using ferroelectric/ferromagnetic multilayers. The magnetic characteristics of multilayers of ferroelastic and ferromagnetic materials can be exploited to create sensors that can detect mechanical stress based on changes in those properties.

Ferroic multilayers are now being studied to learn more about them and how they function as well as to develop new technologies and applications for them. You need to have a thorough knowledge of the interactions between various ferroic orders in order to accomplish these goals. Thus, research into ferroic multilayers can aid in our understanding of the physics underlying them and pave the way for the creation of new materials with particular features and applications [12].

8.1.2.1 *History of Ferroic Multilayers*

Ferroic multilayers have existed since the middle of the 20th century, when the first ferromagnetic/ferroelectric multilayers were created by depositing thin layers of various materials on a substrate.

The first experimental proof of magneto-electric coupling in multiferroic materials was documented in a publication from the 1970s. When magnetic and electric fields interact, this is known as magneto-electric coupling. An electric field can become polarized by a magnetic field, and vice versa. Due to this discovery, interest in ferroic multilayers was revived, and new multilayer kinds constructed of various ferroic materials were developed.

Materials that are both magnetic and ferroelectric are referred to as "multiferroics" as of the 1980s. As a result, new varieties of multilayers constructed of various ferroic materials were developed, including ferroelectric/ferromagnetic and ferroelastic/ferromagnetic multilayers. Molecular beam epitaxy and pulsed laser deposition are two recent advances in thin-film deposition techniques that have made it simpler to create multilayer structures with increasing complexity.

Multilayer structures were becoming more complex during the 1990s, but advances in thin-film deposition techniques and characterization technology made it viable to investigate them. This resulted in the identification of novel phenomena like strain-mediated magnetization switching and magnetoelectric coupling. The interaction of the magnetic and electric fields is known as magnetoelectric coupling. An electric polarization can result from a magnetic field, and vice versa. By applying mechanical stress, a mechanism known as "strain-mediated magnetization switching" can change magnetic properties like magnetization. Mechanical stress has the potential to alter magnetic properties, such as magnetization.

An intriguing type of material that has numerous potential applications is ferroic multilayers. These materials are anticipated to exhibit more intriguing features and potential applications as a result of contemporary research, which could result in the development of novel devices and technologies. It is crucial to comprehend how various ferroic orders interact with one another in order to accomplish these objectives and create novel materials with particular qualities and applications. The Figure 8.3 represents the importance of ferroic materials.

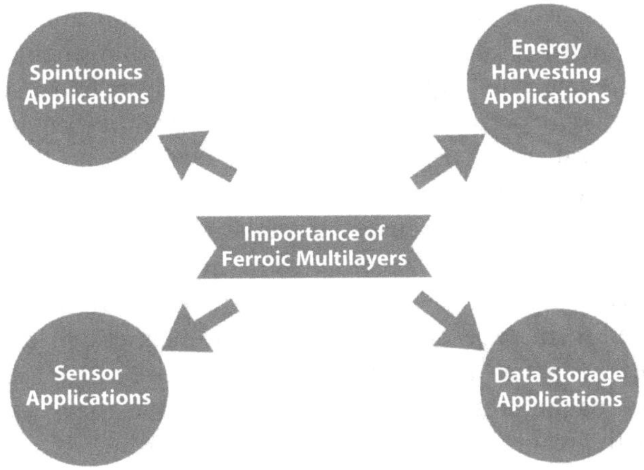

Figure 8.3 Importance of ferroic materials.

8.1.2.2 Importance of Ferroic Multilayers

Due to their distinctive characteristics and extensive variety of potential applications, ferroic multilayers are a crucial area of research. Alternating layers of various ferroic materials, such as ferroelectric, ferromagnetic, or ferroelastic materials, are used to create these composite structures. The interactions between the various ferroic orders at the layer interfaces are what give these formations their characteristics. This article will discuss the significance of ferroic multilayers in many sectors based on research publications.

Ferroic multilayers have received a lot of attention in the realm of data storage because of their capacity to have both ferroelectric and ferromagnetic ordering. They are therefore very promising for developing future data storage devices. In a 2017 study, Lee *et al.* demonstrated the potential application of ferroelectric/ferromagnetic multilayers in magnetic random access memory (MRAM). A strong tunneling magnetoresistance effect, which gauges how resistance changes when a magnetic field is introduced from the outside, was discovered to exist in the multilayers. Additionally, the study revealed that the multilayers exhibited good magnetic stability, indicating that they would be suitable for high-density data storage.

Sensor Applications

Because ferromagnetic multilayers can display both ferroelastic and ferromagnetic ordering, they may potentially be used in sensors. In a recent work, flexible and magnetic layers were used to create magnetic field sensors.

The research discovered that the multilayers had a significant magnetostriction effect, which is a measurement of how much they changed shape in response to an external magnetic field. The multilayers had a considerable influence on magnetoresistance, which suggests that they might be used in sensors with great magnetic field sensitivity, according to the study [13]. Considering that ferromagnetic multilayers can exhibit both ferroelastic and ferromagnetic ordering, sensors may take advantage of them. Magnetic field sensors were developed recently using flexible and magnetic layers. The magnetostriction effect, which is a measurement of how much the multilayers altered form in reaction to an external magnetic field, was shown to be important in the investigation. The multilayers had a considerable influence on magnetoresistance, which suggests that they might be used in sensors with great magnetic field sensitivity, according to the study [13].

Energy Harvesting Applications

Because of their remarkable capacity to transform mechanical energy into electrical energy, ferroic multilayers have received a lot of attention, opening up new opportunities for energy harvesting. Multilayers of ferroelectric and ferromagnetic materials may be utilized in energy-harvesting devices, according to a recent study. The study discovered a high magnetoelectric coefficient, demonstrating a potent magnetoelectric coupling effect. Making very effective energy collecting systems may now be viable thanks to this finding [14].

Spintronics Applications

Ferroic multilayers could be useful in spintronics because of their spin-dependent transport characteristics. The enormous spin Hall effect of the multilayers, which serves as a gauge of their spin-dependent transport characteristics and demonstrates their potential for usage in highly effective spintronic devices, was discovered [14].

8.1.2.3 Properties of Ferroic Multilayers

Multiferroic behavior, giant magnetoresistance (GMR), anisotropic behavior, layer thickness dependency, interfacial effects, switchable behavior, and the possibility for new applications are only a few of the intriguing characteristics of ferroic multilayers. They are becoming common for many various applications due to their adaptability, including energy-collection devices, high-frequency transformers, and magnetic storage systems.

Ferroic multilayers exhibit behaviors that are not present in single-phase materials because of the significant influence the borders between the various layers have on their physical characteristics. Overall, ferroic multilayers may contribute to the development of new technologies with increased functionality.

Multiferroic Behavior

Multiferroic behavior, which is uncommon in materials, is shown by ferroic multilayers. They also possess one or more additional ferroic features, such as ferroelectricity, ferroelasticity, or ferromagnetism. This is because the ferroic materials in the thin films interact with each other, giving rise to new functions that are not found in the individual layers. Ferroic multilayers are useful for many things, such as high-frequency transformers and magnetic memory devices, because they behave in a way that makes them multiferroic.

Giant Magnetoresistance

When put in a magnetic field, ferroic multilayers show a property called enormous magnetoresistance (GMR). This effect causes a big difference in resistance, which led to the creation of magnetic sensors and read heads for hard drives. The GMR is caused by how the magnetic moments of the different layers in a multilayer structure interact with each other. This changes how electrons move through the material and how resistant it is.

Anisotropic Behavior

Ferroic multilayers have properties that depend on which way you measure them. This means that their properties change depending on which way you measure them. The anisotropic behavior of ferroic multilayers comes from the fact that each layer in the structure has its own anisotropy and that the layers interact with each other. This can be useful, for example in magnetic memory devices, where the direction of the magnetization needs to be controlled very precisely.

Layer Thickness Dependence

Ferroic multilayers are very easy to change because their properties depend closely on the thickness of each layer. Materials that are naturally magnetic can be turned into something else by putting a magnetic field near them. Therefore, it is possible to precisely manage the intensity of the contacts between the layers, resulting in a variety of features, including multiferroic behavior and enormous magnetoresistance (GMR). In a way, the layer thickness in ferroic multilayers can be compared to a dial that can be turned to adjust the material's characteristics. This approach to creating materials with specialized purposes shows promise.

Interfacial Effects

Ferroic multilayers exhibit behaviors that are not present in single-phase materials because of the significant influence the borders between the various layers have on their physical characteristics. Strain, charge transfer, and other phenomena that alter the material's characteristics can occur where the layers converge. To improve the performance of things like magnetoelectric memory devices, interfacial effects can be employed.

Switchable Behavior

Ferroic multilayers may exhibit switchable behavior, in which case an external force, such as an electric or magnetic field, can turn one or more of their characteristics on or off. This is so that the properties of the material as a whole can vary as a result of an external force altering how the layers of a multilayer construction interact with one another.

Potential for Novel Applications

Ferroic multilayers could be utilized in products like magnetic memory chips, high-frequency transformers, and energy harvesting systems due to their special features. It is feasible to develop new technologies with improved performance and usefulness by utilizing the multiferroic behavior, GMR, anisotropic behavior, layer thickness dependence, interfacial effects, and switchable behavior of ferroic multilayers.

Interface Effects

The properties of a ferroic multilayer are significantly influenced by the interfaces between the various layers. The construction of the multilayers, how they conduct electricity and magnetism, and how they appear can all be altered by these linkages. How strong the magnetoelectric interaction is depends significantly on the interface between the ferromagnetic and ferroelectric layers. The strength of the magnetoelectric connection can also be improved by altering the interface between the ferroelectric and ferromagnetic layers.

Magnetoelectric Coupling

It has been demonstrated that ferroic multilayers exhibit high magnetoelectric coupling, the connection between the electric and magnetic orders. Due to this relationship, magnetic characteristics can be controlled by electric fields and vice versa. The magnetoelectric interaction in multilayers that are both ferroelectric and ferromagnetic was studied by Chen *et al.* in 2019. According to the research, the magnetoelectric coefficient, which is a gauge of how strong the coupling is, was high for the multilayers.

The study also demonstrated that by modifying the thickness and makeup of the multilayers, the magnetoelectric coupling can be enhanced [14].

Strain Effects

Ferroic multilayers can exhibit considerable strain effects because the various ferroic materials have varying lattice characteristics and thermal expansion coefficients. The significant strain effects that are evident in these multilayers are brought on by this. The multilayers' structure and electrical conductivity are both susceptible to strain. In a paper published in 2020, Wang and his coworkers examined the impact of strain on ferroelectric and ferromagnetic multilayers. The multilayers had a significant piezoelectric response, which is the capacity to produce an electric field when mechanical strain is applied, according to the research. The study additionally revealed that the multilayers had a substantial magnetostriction effect, or the propensity of a material to change shape when exposed to an external magnetic field [15].

8.1.2.4 Advances in Ferroic Multilayers

New materials have been created with higher performance and attributes as a result of advancements in ferroic multilayers. These materials were created by modifying the interface's layer thickness, chemical makeup, and quality. These qualities can now be precisely controlled thanks to thin-film deposition methods and improved measurement tools. Multilayers with new capabilities and potential applications have resulted as a result of this. Sensors, actuators, and memory devices all now use ferroic multilayers, and as the area develops, more and more innovative applications are anticipated to appear [16]. The Figure 8.4 reflects the advances in ferroic materials.

Growth Techniques

Ferroic multilayers' characteristics and functionality are greatly influenced by the process of growth. Ferroic multilayers have been created recently using several novel methods of material growth, including atomic layer deposition (ALD), molecular beam epitaxy, and pulsed laser deposition (PLD). In 2019, Pradhan *et al.* investigated the feasibility of fabricating ferroelectric/ferromagnetic multilayers using PLD. According to the study, the PLD approach produced high-quality multilayers with significant magnetoelectric coupling. Kim *et al.* (2020) investigated the application of MBE to the fabrication of ferroelectric/ferromagnetic multilayers. The study discovered that the MBE approach produced multilayers with superior quality and magnetic characteristics [17].

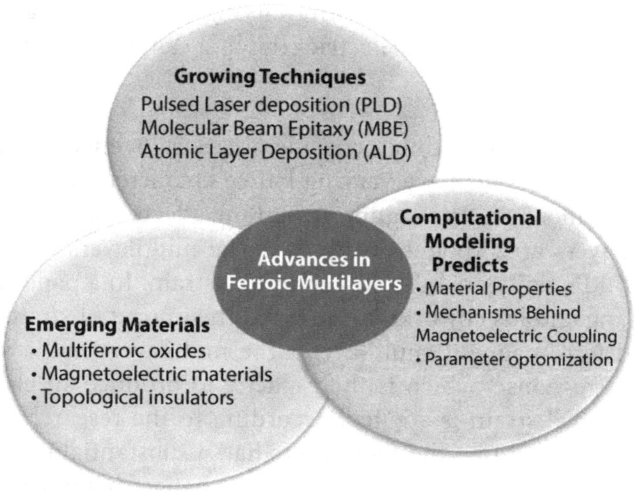

Figure 8.4 Advances in ferroic materials.

New Materials

Now that new ferroic materials have been created, there are more applications for ferroic multilayers. New materials for ferroic multilayers have been developed in recent years, including topological insulators, magnetoelectric materials, and multiferroic oxides. Another study examined the operation of topological insulator/ferromagnetic multilayers by Bonell *et al.* (2020). According to the study, the multilayers featured unique magnetic and electrical characteristics that made them suitable for application in spintronics [18].

Device Applications

Devices like spintronic devices, spintronic sensors, and memory devices might all make advantage of ferroic multilayers. In 2020, Kumari *et al.* created a ferroelectric/ferromagnetic tunnel junction device using a ferroelectric/ferromagnetic multilayer. The gadget was suitable for use in magnetic memory since it had a significant tunnel magnetoresistance effect [17].

Computational Modeling

Understanding the behavior and characteristics of ferroic multilayers requires computational modeling. In order to forecast the magnetoelectric coupling in multilayers of ferroelectric and ferromagnetic materials, Wang *et al.* (2019) developed a model. The model successfully predicted the intensity of the magnetoelectric interaction from the thickness and composition of the multilayers [19].

8.1.2.5 Recent Research

Numerous studies have been conducted recently on the creation of new multilayer structures with improved properties and potential applications. These kinds of structures include artificial superlattices and nanoscale heterostructures, both of which have attracted a lot of attention.

Layers of various materials are stacked on top of one another to create artificial superlattices. Each layer typically ranges in thickness from a few nanometers to a few micrometers. By varying the thickness and makeup of the layers, researchers can alter the superlattice's properties to suit various demands. For instance, researchers recently described how they created a new artificial superlattice using materials that are both ferroelectric and ferromagnetic in a report published in *Nano Letters*. Because of the superlattice's strong magnetoelectric interaction, spintronic devices and memory can benefit from its utilization [20].

Nanoscale heterostructures are made of different materials with different crystal structures. These heterostructures can have better properties because there exist contacts between the materials. Researchers recently claimed that they have created nanoscale heterostructures out of materials that were both ferroelectric and piezoelectric in a report published in *Nature Communications*. For uses like energy harvesting and sensing, the heterostructures are a viable choice because of their improved piezoelectric sensitivity.

A multiferroic structure is a different kind of multilayer structure that has attracted a lot of attention lately. These structures are constructed of magnetic and electrically charged materials. A thin ferromagnetic coating on top of a ferroelectric substrate was used to create a distinctive multiferroic structure, according to a recent publication in *Applied Physics Letters*. Because of the structure's improved magnetoelectric coupling, spintronic devices and memory can utilize it.

Making novel multilayer structures with improved features and potential applications, such as synthetic superlattices, nanoscale heterostructures, and multiferroic structures, has been the focus of recent study. Innovative techniques to describe these structures have allowed for the development of fresh and promising applications [21]. Making novel multilayer structures with improved features and potential applications, such as synthetic superlattices, nanoscale heterostructures, and multiferroic structures, has been the focus of recent study. Innovative techniques to describe these structures have allowed for the development of fresh and promising applications [21].

8.1.2.6 Characterization Techniques

Some of the most popular methods for describing ferroic multilayers include magnetometry, reflectometry, and ellipsometry.

Magnetometry

The magnetic properties of multilayer ferroics are studied using magnetometry. The magnetic hysteresis loops of ferroic multilayers may be found using vibrating sample magnetometry (VSM), a widely used technique. VSM was used by Jiang and his colleagues to examine the magnetic characteristics of FeNi/Cu/FeNi multilayers. They discovered that the thickness of the Cu spacer layer had a significant impact on the magnetic characteristics of the multilayers.

Reflectometry

A crucial technique for researching ferroic multilayers is reflectometry. A common technique for determining the thickness and roughness of layers in a multilayer stack is X-ray reflectometry (XRR). Wang *et al.* (2021) employed XRR to analyze the structural characteristics of CoFeB/MgO multilayers. The scientists were able to learn more about the magnetic behavior of the multilayers by measuring the thickness and roughness of each layer in a multilayer stack.

Ellipsometry

Ellipsometry is a key technique for analyzing the optical properties of ferroic multilayers. Spectroscopic ellipsometry may be used to calculate the refractive index and extinction coefficient of each layer in a multilayer stack.

Transmission Electron Microscopy (TEM)

The exceptional spatial resolution of the transmission electron microscope (TEM) makes it a suitable tool for examining the microstructure of multilayer ferroics. By using a powerful electron beam to pierce the material, it does this. This results in a visual depiction of the geometry and crystal structure of each layer in a multilayer stack. The atomic-scale geometry of interfaces and defects, such as stacking faults and dislocations, which can alter how magnetic a multilayer is, can be examined using TEM. Multilayer ferroic materials' microstructure and magnetic behavior may be thoroughly examined using TEM.

Atomic Force Microscopy (AFM)

The surface of the ferroic multilayers may be examined using AFM. It requires measuring the forces operating on the surface while a small, pointed tip is traced across it. The surface's texture and features may be described in detail using this technique. AFM may be used to get details on the mechanical properties, such as adhesion and elasticity, as well as high-resolution topographical photographs of the surface. Ferroic multilayers' surface shape and roughness may be examined using AFM; these factors may affect the multilayer's functionality and magnetic properties.

X-ray Diffraction (XRD)

In multilayer ferroic materials, crystal structure and orientation can be studied using X-ray diffraction (XRD). How well it functions is determined by the diffraction pattern of the X-rays that are dispersed by the crystal lattice in the sample. The magnetic anisotropy and domain structure of a multilayer are both influenced by the lattice parameter, crystal symmetry, and orientation of each layer, which can all be determined using XRD. Understanding the crystallographic characteristics of multilayer ferroics can also be done well by XRD.

Time-Resolved Magneto-Optical Kerr Effect (TR-MOKE)

Using sub-picosecond laser pulses, the TR-MOKE technique examines the dynamic magnetic behavior of ferroic multilayers. The instrument monitors how the polarization of the light reflected from the surface varies when a magnetic field is applied to a sample. Each layer in a multilayer stack can have its ultrafast magnetization dynamics, such as precession and damping, revealed via TR-MOKE. Understanding the magnetic characteristics and functionality of the multilayer can benefit from knowing this.

8.1.3 Heterostructures

Ferroic heterostructures are made of many layers of different ferroic materials that interact with each other at their interfaces to give them unique and controllable properties. There are three types of ferroic materials: ferromagnetic, ferroelectric, and ferroelastic. All of them have magnetic, electric, or mechanical properties that naturally order themselves.

In a ferroic heterostructure, the properties of each layer can interact and couple with each other. This can lead to emergent behavior and new properties that are not present in the individual layers alone. A magnetoelectric effect is made when an outside magnetic field causes an electric polarization in the ferroelectric layer and an outside electric field causes a magnetization in the ferromagnetic layer.

Ferroic heterostructures might be utilized in a variety of applications, including energy conversion, data storage, sensors, actuators, and actuators. It is possible to create materials with customizable qualities by managing how the various layers react and interact. This could result in more advanced technology.

8.1.3.1 Types of Ferroic Heterostructures

Heterostructures made of both ferroelectric and ferromagnetic materials are known as ferroelectric–ferromagnetic heterostructures. They have been investigated for use in magnetic tunnel junctions and other spintronic devices where the ferroelectric layer can modify the ferromagnetic layer's spin polarization [21].

Ferroelectric–Ferroelastic Heterostructures

A ferroelectric substance and a ferroelastic material combine to form these heterostructures. The ferroelastic material deforms in a way that can be undone when an electric field is introduced. This kind of construction has been investigated for potential energy harvesting applications since the ferroelastic layer can convert mechanical energy into electrical energy.

Ferromagnetic–Ferroelastic Heterostructures

These are heterostructures composed of ferroelastic and ferromagnetic materials. The ferroelastic material deforms in a reversible manner when exposed to an external magnetic field. This type of device has been investigated for potential use in magnetic sensing and actuation because the ferromagnetic layer can influence how the ferroelastic layer changes form.

Ferroelectric–Ferrimagnetic Heterostructures

This type of heterostructure is made up of two different kinds of materials: a ferroelectric material and a ferrimagnetic material. The ferrimagnetic material has a net magnetic moment that is not zero even when there is no magnetic field applied. The idea for these structures came from multiferroic devices, in which the magnetization of a ferrimagnetic layer can be changed by a ferroelectric layer next to it.

Ferroelectric–Antiferromagnetic Heterostructures

These are heterostructures composed of an antiferromagnetic and a ferroelectric substance. Despite having no net magnetic moment, the antiferromagnetic material possesses regularly ordered magnetic moments. Their application in spintronics and magnetonics, where the ferroelectric layer may regulate the dynamics of the antiferromagnetic layer, has motivated research into them.

8.1.3.2 Historical Development

The study of ferroic heterostructures entered a long and fruitful phase after the discoveries of ferroelectricity and ferromagnetism in the late 1800s and early 1900s. In the 1950s and 1960s, the development of thin-film deposition techniques reignited interest in ferroic heterostructures. This is because it is now feasible to explore the characteristics of ferroic materials when coupled with other materials by growing thin films of those materials.

Liu *et al.* examined the properties of ferromagnetic and ferroelectric bilayers [22]. They found that by applying a magnetic field to the ferromagnetic layer, it was possible to change the polarization of the ferroelectric layer. This proved that the interaction between the two materials was not trivial.

In the 1990s, the invention of molecular beam epitaxy (MBE) and other cutting-edge thin-film deposition techniques allowed for the formation of highly ordered and precise heterostructures. Because of this, a great deal of study has been done on ferroic heterostructures, with a focus on multiferroic materials, which have several ferroic properties, such as being simultaneously ferromagnetic and ferroelectric.

A thin sheet of $BiFeO_3$ was discovered to exhibit multiferroic behavior in 2003. One of the most significant findings in the area was this. This material is a strong contender for usage in next memory and computer technologies since it possesses both ferroelectric and antiferromagnetic characteristics.

Numerous investigations on the growth and appearance of ferroic heterostructures have been conducted since then. Understanding how they function has been one of the key aims of these investigations. For instance, scientists have studied how flaws and interfacial strain alter the characteristics of ferroic heterostructures. The effects of magnetic and electric fields on the characteristics of ferroic heterostructures have also been studied.

Ferroic heterostructures have taught us a lot about the fundamental physics of ferroic materials and about the potential applications for them in a variety of domains, including spintronics, energy storage, and electronics [3]. The study of ferroic heterostructures entered a long and fruitful phase after the discoveries of ferroelectricity and ferromagnetism in the late 1800s and early 1900s. In the 1950s and 1960s, the development of thin-film deposition techniques reignited interest in ferroic heterostructures. This is because it is now possible to develop thin films of ferroic materials in order to investigate the properties of such materials when combined with other materials. The study of ferroic heterostructures entered a long and fruitful phase after the discoveries of ferroelectricity and ferromagnetism in the late 1800s and early 1900s. In the 1950s and 1960s, the development of thin-film deposition techniques reignited interest in ferroic heterostructures. This is because it is now possible to develop thin films of ferroic materials in order to investigate the properties of such materials when combined with other materials.

Liu *et al.* examined the properties of ferromagnetic and ferroelectric bilayers [22]. They found that by applying a magnetic field to the ferromagnetic layer, it was possible to change the polarization of the ferroelectric layer. This proved that the interaction between the two materials was not trivial.

In the 1990s, the invention of molecular beam epitaxy (MBE) and other cutting-edge thin-film deposition techniques allowed for the formation of highly ordered and precise heterostructures. Due to this, a great deal of study has been done on ferroic heterostructures, with a focus on multiferroic materials, which have several ferroic properties, such as being simultaneously ferromagnetic and ferroelectric.

A thin sheet of $BiFeO_3$ was discovered to exhibit multiferroic behavior in 2003. One of the most significant findings in the area was this. This material is a strong contender for usage in next memory and computer technologies since it possesses both ferroelectric and antiferromagnetic characteristics.

Numerous investigations on the growth and appearance of ferroic heterostructures have been conducted since then. Understanding how they function has been one of the key aims of these investigations. For instance, scientists have studied how flaws and interfacial strain alter the characteristics of ferroic heterostructures. The effects of magnetic and electric fields on the characteristics of ferroic heterostructures have also been studied.

We have learned a lot about the basic physics of ferroic materials from ferroic heterostructures, as well as about the prospective uses for these materials in spintronics, energy storage, and electronics [3]. The study of ferroic heterostructures entered a long and fruitful period after the discoveries of ferroelectricity and ferromagnetism in the late 1800s and early 1900s. In the 1950s and 1960s, the development of thin-film deposition techniques reignited interest in ferroic heterostructures. This is because it is now possible to develop thin films of ferroic materials in order to investigate the properties of such materials when combined with other materials.

Ferroelectric and ferromagnetic bilayers' characteristics were studied by Liu *et al.* [22]. They discovered that the polarization of the ferroelectric layer could be altered by applying a magnetic field to the ferromagnetic layer. This demonstrated the nontrivial nature of the two materials' interaction.

In the 1990s, the invention of molecular beam epitaxy (MBE) and other cutting-edge thin-film deposition techniques allowed for the formation of highly ordered and precise heterostructures. Because of this, a great deal of study has been done on ferroic heterostructures, with a focus on multiferroic materials, which have several ferroic properties, such as being simultaneously ferromagnetic and ferroelectric.

A thin sheet of $BiFeO_3$ was discovered to exhibit multiferroic behavior in 2003. One of the most significant findings in the area was this. This material is a strong contender for usage in next memory and computer technologies since it possesses both ferroelectric and antiferromagnetic characteristics.

Numerous investigations on the growth and appearance of ferroic heterostructures have been conducted since then. Understanding how they function has been one of the key aims of these investigations. For instance, scientists have studied how flaws and interfacial strain alter the characteristics of ferroic heterostructures. The effects of magnetic and electric fields on the characteristics of ferroic heterostructures have also been studied.

Ferroic heterostructures have taught us a lot about the fundamental physics of ferroic materials and about the potential applications for them in a variety of domains, including spintronics, energy storage, and electronics [3].

8.1.3.3 Properties of Heterostructures

When different ferroic materials are put together in heterostructures, they can take on new properties that are not found in the single materials. For example, the combination of ferromagnetic and ferroelectric materials in multiferroic heterostructures can lead to the coupling of magnetic and electric fields, which can be used for things like magnetic field sensors.

Some of the specific properties of ferroic heterostructures are as follows:

Ferromagnetic Properties

When two or more ferromagnetic materials are put together to make a composite material, the composite material has ferromagnetic properties.

The crystal structure of ferromagnetic materials causes them to become magnetic on their own, and magnetic fields from the outside can change how magnetic they are.

In ferroic heterostructures, the combination of different ferromagnetic materials can lead to new magnetic properties like exchange coupling, magnetic anisotropy, and magnetoresistance. Exchange coupling is when the spins of two different magnetic materials interact at the interface. This makes the magnetic properties of both materials better.

Some fields that could benefit from ferromagnetic ferroic heterostructures are spintronics, magnetic storage systems, and magnetic sensors. In MRAM technology, for example, magnetic tunnel junctions (MTJs) are used a lot. MTJs are made up of two ferromagnetic layers separated by a thin layer of insulation. When ferroic heterostructures are used in MTJs, their magnetic properties can be improved and their performance can be raised. Also, ferroic heterostructures with ferromagnetic properties could be used in magnetic field sensors, which are used in many industries, including the automotive and aerospace ones.

Antiferromagnetism

Antiferromagnetic ferroic heterostructures are made up of two or more antiferromagnetic materials in a single material. When antiferromagnetic materials are put together, they can make new magnetic properties like exchange bias, spin frustration, and magnetic anisotropy.

Exchange bias is the change in the magnetic hysteresis loop that happens when a ferromagnetic layer touches an antiferromagnetic layer. This effect happens because the spins of the two materials exchange coupling across the interface. It can be used in magnetic sensors and magnetic recording devices.

Because various antiferromagnetic materials may be combined to provide novel magnetic characteristics that can be tuned for particular purposes in domains like magnetic sensors, spintronics, and quantum computing, antiferromagnetism is a potential area of study for ferroic heterostructures.

Ferroelasticity

Ferroelastic heterostructures can have significant recoverable stresses, which makes them valuable for actuators, sensors, and energy-harvesting devices. The composition, thickness, and orientation of the component pieces may all be altered to alter the properties of ferroelastic heterostructures. This enables the production of materials with certain mechanical characteristics. For instance, it has been demonstrated that the piezoelectricity and strain responses of ferroelastic heterostructures based on perovskite oxide may change. They may therefore be used for processes including energy harvesting, sensing, and actuation.

According to a study, ferroelastic heterostructures can also experience reversible form changes. New material possibilities, such memory storage and mobile equipment, are now conceivable. As a result of research on ferroelasticity in heterostructures, novel materials with distinctive mechanical characteristics may be produced.

Piezoelectricity

By mixing different materials with different piezoelectric coefficients, heterostructures can have a wide variety of piezoelectric properties that can be changed. New piezoelectric materials are created as a result, and piezoelectricity is improved. According to a study, piezoelectric heterostructures can display higher piezoelectricity than the component materials. Additionally, the materials may be made to have different piezoelectric properties by modifying their composition, thickness, and orientation.

For instance, it has been shown that piezoelectric heterostructures constructed of ferroelectric oxides, such as $Pb(Zr,Ti)O_3$ (PZT), are more efficient piezoelectric devices due to the strain coupling between the multiple layers. As a result, they may be applied to energy collecting, actuation, and sensing. The piezoelectric response of the heterostructure may also be changed by changing the crystal orientation and domain structure of the materials that make up the heterostructure, according to research. The development of innovative materials with specialized piezoelectric characteristics that perform better in applications like energy harvesting, sensing, and actuation may come from a study into piezoelectricity in heterostructures.

Magnetostriction

By mixing several magnetic materials with differing magnetostrictive coefficients, heterostructures can have a wide variety of magnetostrictive properties that can be changed. Researchers have shown that magnetostrictive heterostructures can display magnetostriction that is more powerful than the individual component materials. The magnetostriction characteristics of the materials may also be varied by changing their composition, thickness, and orientation.

For instance, it has been shown that magnetostrictive heterostructures made of ferromagnetic and magnetostrictive materials, such as FeGa and FeCo, display considerable magnetostriction due to the strain interaction between the multiple layers. As a result, applications including energy harvesting, actuation, and magnetic field sensing are made possible.

The magnetostrictive response of the heterostructure may also be changed by changing the crystal orientation and domain structure of the materials that make up the heterostructure, according to research. In conclusion, studies on magnetostriction in heterostructures may lead to the development of innovative materials with specialized magnetostrictive characteristics and enhanced utility in fields including sensing, actuation, and energy harvesting.

8.1.3.4 Characterization Techniques

There are several approaches to comprehend ferroic heterostructures, each having pros and cons. The specific traits and characteristics of a material define the appropriate technique to apply when characterizing it.

X-ray Diffraction

X-ray diffraction may be used to discover more about ferroic heterostructures (XRD). It offers significant information on the lattice parameters, orientation of the component minerals, and crystal structure. The sample is exposed to X-rays, and the angles at which they bend are then calculated. The sample's crystal structure and the positioning of the crystalline planes are revealed by the diffraction pattern.

XRD may be used to investigate the strain, makeup, and crystallographic orientation of the different layers in ferroic heterostructures. XRD may also be used to check for agreement between the lattices of different layers and to look at the production of interfacial defects such as dislocations and misfit dislocations. Ferroic heterostructures' texture, which has a big impact on their magnetic and piezoelectric characteristics, may be studied using XRD.

XRD may also be used to study the domain structures and phase transitions of ferroic heterostructures, such as the change from ferromagnetic to paramagnetic in magnetic thin films or from ferroelectric to paraelectric in ferroelectric thin films. The diffraction patterns at different temperatures may be used to determine the critical temperature for these phase transitions. X-ray diffraction may be used to discover more about ferroic heterostructures (XRD). It offers significant information on the lattice parameters, orientation of the component minerals, and crystal structure. The sample is exposed to X-rays, and the angles at which they bend are then calculated. The sample's crystal structure and the positioning of the crystalline planes are revealed by the diffraction pattern.

XRD may be used to investigate the strain, makeup, and crystallographic orientation of the different layers in ferroic heterostructures. XRD may also be used to check for agreement between the lattices of different layers and to look at the production of interfacial defects such dislocations and misfit dislocations. Ferroic heterostructures' texture, which has a big impact on their magnetic and piezoelectric characteristics, may be studied using XRD.

XRD may also be used to study the domain structures and phase transitions of ferroic heterostructures, such as the change from ferromagnetic to paramagnetic in magnetic thin films or from ferroelectric to paraelectric in ferroelectric thin films. The diffraction patterns at different temperatures may be used to determine the critical temperature for these phase transitions.

Scanning Electron Microscopy (SEM)

Scanning electron microscopy (SEM), which produces high-resolution pictures of the surface shape, microstructure, and chemical composition of ferroic heterostructures, is commonly used in research. A focused electron beam from the SEM examines a sample's surface in search of secondary electrons generated by the surface. The generated picture shows the surface shape and look of the sample.

The thickness, surface abrasion, and layer shape of a ferroic heterostructure may also be determined by SEM. SEM, for instance, may be used to investigate the surface morphology of ferroic heterostructures, including the emergence of fractures, defects, and surface roughness. The microstructure of the different layers and the areas where they converge, which has a substantial influence on the physical properties of heterostructures, may also be examined using SEM [23].

Atomic Force Microscopy

AFM analyzes the topography and geometry of the layers' surfaces as well as their interfaces in ferroic heterostructures. For example, AFM may be utilized to investigate surface roughness and the emergence of nanoscale features like islands and pits.

The ferroelectric and piezoelectric properties of ferroic heterostructures may be analyzed using a combination of atomic force microscopy (AFM), kelvin probe force microscopy (KPFM), and PFM. The AFM tip receives an AC voltage, and the sample's movement as a result is observed. On the sample's ferroelectric and piezoelectric domains, information is provided by this. The surface potential of the sample is what KPFM uses to function. This reveals details about the sample's electrical properties in that area.

Transmission Electron Microscopy (TEM)

Transmission electron microscopy (TEM), which may also find flaws including dislocations, stacking faults, and grain boundaries, can measure the thickness of a layer.

Transmission electron microscopy (TEM), energy-dispersive X-ray spectroscopy (EDS), and electron energy loss spectroscopy (EELS) can be used to analyze the electrical and chemical structure of a material's interfaces and layers (TEM). You may use EDS to identify the elements contained in a sample by looking at the X-rays that it releases when the electron beam impacts it. Energy-loss spectroscopy (EELS) is a method that tracks the energy of electrons as they interact with a material to comprehend the electronic structure and bonding of that sample.

X-ray Photoelectron Spectroscopy

XPS may be used to find out more about the chemical makeup and chemical bonding of the different layers and the gaps between them in ferroic heterostructures. The elemental makeup of the layers, their level of oxidation, and the presence of pollutants or contaminants may all be determined using XPS. By using XPS, we can also look at the chemical bonds at the interfaces, which helps us understand how the layers are joined together and how the chemistry at the interfaces affects the properties of the heterostructures.

XPS may also be used in combination with other methods like depth profiling and angle-resolved XPS to analyze the chemical composition and chemical bonding as a function of depth and angle. This offers information on the distribution of chemical species in depth as well as the orientation of chemical bonding at surfaces. XPS may be used to find out more about the chemical makeup and chemical bonding of the different layers and the gaps between them in ferroic heterostructures. The elemental makeup of the layers, their level of oxidation, and the presence of pollutants or contaminants may all be determined using XPS. By using XPS, we can also look at the chemical bonds at the interfaces, which helps us understand how the layers are joined together and how the chemistry at the interfaces affects the properties of the heterostructures.

XPS may also be used in combination with other methods like depth profiling and angle-resolved XPS to analyze the chemical composition and chemical bonding as a function of depth and angle. This gives information on the depth distribution of chemical species as well as the orientation of chemical bonding at surfaces.

Magneto-Optical Kerr Effect (MOKE)

A useful method for identifying ferroic heterostructures that can reveal their magnetic characteristics is the magneto-optical Kerr effect (MOKE). MOKE looks for changes in the polarization of light that is reflected from a material when a magnetic field is present.

MOKE may be used to investigate the magnetic properties of the different layers and the gaps between them in ferroic heterostructures. For example, MOKE can be used to study the magnetic anisotropy and magnetization reversal behavior of ferromagnetic thin films. This gives information on the domain structure, magnetic domain wall motion, and spin texture. In multilayered ferromagnetic and antiferromagnetic heterostructures, MOKE may be utilized to analyze the magnetic coupling between various layers, providing details on the interlayer exchange coupling and magnetic interlayer coupling [24]. A useful method for identifying ferroic heterostructures that can reveal their magnetic properties is the magneto-optical Kerr effect (MOKE). MOKE looks for changes in the polarization of light that is reflected from a material when a magnetic field is present.

MOKE may be used to investigate the magnetic properties of the different layers and the gaps between them in ferroic heterostructures. For instance, MOKE may be used to analyze the behavior of magnetic anisotropy and magnetization reversal in ferromagnetic thin films. This gives information on the domain structure, magnetic domain wall motion, and spin texture. In multilayered ferromagnetic and antiferromagnetic heterostructures, MOKE can be utilized to analyze the magnetic coupling between various layers, providing details on the interlayer exchange coupling and magnetic interlayer coupling [24]. A useful method for identifying ferroic heterostructures that can reveal their magnetic characteristics is the magneto-optical Kerr effect (MOKE). MOKE looks for changes in the polarization of light that is reflected from a material when a magnetic field is present.

MOKE may be used to investigate the magnetic properties of the different layers and the gaps between them in ferroic heterostructures. MOKE, for instance, may be used to examine the behavior of ferromagnetic thin films' magnetic anisotropy and magnetization reversal. This gives information on the domain structure, magnetic domain wall motion, and spin texture. In multilayered ferromagnetic and antiferromagnetic heterostructures, MOKE may be utilized to analyze the magnetic coupling between various layers, providing details on the interlayer exchange coupling and magnetic interlayer coupling [24]. The magneto-optical Kerr effect is a helpful technique for locating ferroic heterostructures that can show their magnetic characteristics (MOKE). MOKE looks for changes in the polarization of light that is reflected from a material when a magnetic field is present.

MOKE may be used to investigate the magnetic properties of the different layers and the gaps between them in ferroic heterostructures. For instance, MOKE may be used to analyze the behavior of magnetic anisotropy and magnetization reversal in ferromagnetic thin films. This gives information on the domain structure, magnetic domain wall motion, and spin texture. MOKE may be used to analyze the magnetic coupling between different layers in multilayered ferromagnetic and antiferromagnetic heterostructures, giving information on the interlayer exchange coupling and magnetic interlayer coupling [24].

Vibrating Sample Magnetometry (VSM)
The magnetic properties of the different layers and their intersections in ferroic heterostructures may be determined using VSM. VSM may be used to measure the magnetization of thin ferromagnetic films in order to learn more about their saturation magnetization, coercive field, and magnetic anisotropy. VSM may also be used to examine the magnetic properties of multilayered ferromagnetic and antiferromagnetic heterostructures. This gives information on the magnetic interlayer coupling and exchange coupling between several layers.

VSM may also be used in conjunction with other methods like MOKE and X-ray magnetic circular dichroism to better understand the magnetic properties of heterostructures.

8.2 Conclusion

Thin ferroic films, multilayers, and heterostructures are now viable for a wide range of applications due to recent advancements. Due to their unique properties of being ferroelectric, ferromagnetic, and ferroelastic, they are useful in a range of applications, including microelectronics, data storage, sensing, and energy conversion. Researchers are always pushing the limits of what is possible when it comes to ferroic materials and thin films that are similar to them. New manufacturing methods like pulsed laser deposition and molecular beam epitaxy now make it feasible to produce high-quality thin films and heterostructures that are controlled at the atomic level. Characterization techniques including X-ray diffraction, scanning probe microscopy, and magnetometry have all been very helpful in understanding the basic properties of ferroic thin films and their interfaces. This understanding makes it feasible to develop new systems and gadgets and improve their functionality. Ferroic multilayers, heterostructures, and thin films have promising futures overall. To enhance their qualities and discover new applications for them, research is continuously ongoing. It will be fascinating to observe where the next technological advances emerge from since the area is poised to significantly contribute to the creation of next-generation technology.

References

1. Wadhawan, V.K., Ferroic materials: A primer. *Resonance*, 7, 15, 2002.
2. Tagantsev, A.K., Cross, L.E., Fousek, J., Fundamentals of ferroic domain structures, in: *Domains in Ferroic Crystals and Thin Films*, A.K. Tagantsev, L.E. Cross, J. Fousek, (Eds.), pp. 11–107, Springer New York, New York, NY, 2010.
3. Chen, S., Yuan, S., Hou, Z., Tang, Y., Zhang, J., Wang, T., Li, K., Zhao, W., Liu, X., Chen, L., Martin, L.W., Chen, Z., Recent progress on topological structures in ferroic thin films and heterostructures. *Adv. Mater.*, 33, 2000857, 2021.
4. Martin, L.W., Crane, S.P., Chu, Y.H., Holcomb, M.B., Gajek, M., Huijben, M., Yang, C.H., Balke, N., Ramesh, R., Multiferroics and magnetoelectrics: Thin films and nanostructures. *J. Phys. Condens. Matter*, 20, 434220, 2008.
5. Missen, N., Bagnato, S., Wells, D.S., Crothers, L.M., Schmitt, A.J., Kolbert, J.B., Social and behavioral problems in an urban at-risk preschool population. *Int. Educ. Res.*, 3, 16, 2015.
6. Ban, Z.G., Alpay, S.P., Mantese, J.V., Fundamentals of graded ferroic materials and devices. *Phys. Rev. B - Condens. Matter Mater. Phys.*, 67, 177, 178, 2003.
7. Ganpule, C.S., Stanishevsky, A., Su, Q., Aggarwal, S., Melngailis, J., Williams, E., Ramesh, R., Scaling of ferroelectric properties in thin films. *Appl. Phys. Lett.*, 75, 409, 1999.
8. Mansilla, M.V., Gómez, J., Butera, A., Ferromagnetic resonance of disordered FePt thin films. *IEEE Trans. Magn.*, 44, 2883, 2008.
9. Yun, Q., Xing, W., Chen, J., Gao, W., Bai, Y., Zhao, S., Effect of Ho and Mn co-doping on structural, ferroelectric and ferromagnetic properties of BiFeO3 thin films. *Thin Solid Films*, 584, 103, 2015.
10. Norton, D.P., Pulsed laser deposition of complex materials: Progress toward applications, in: *Pulsed Laser Deposition of Thin Films: Applications-Led Growth of Functional Materials*, pp. 1–31, 2006.
11. Desai, J.D., Pathan, H.M., Min, S.K., Jung, K.D., Joo, O.S., FT-IR, XPS and PEC characterization of spray deposited hematite thin films. *Appl. Surf. Sci.*, 252, 1870, 2005.
12. Akai, H. and Ogura, M., Hyperfine interactions of half-metallic diluted antiferromagnetic semiconductors, in: *Hyperfine Interactions*, A. Pasquevich, M. Rentería, E.B. Saitovitch, H. Petrilli, (Eds.), pp. 21–25, Springer Berlin Heidelberg, Berlin, Heidelberg, 2007.
13. Bochenek, D., Surowiak, Z., Poltierova-Vejpravova, J., Producing the lead-free BaFe0.5Nb0.5O3 ceramics with multiferroic properties. *J. Alloys Compd.*, 487, 572, 2009.
14. Ortega, N., Kumar, A., Scott, J.F., Katiyar, R.S., Multifunctional magnetoelectric materials for device applications. *J. Phys. Condens. Matter*, 27, 504002, 2015.

15. Ma, W., Lu, J., Wan, B., Peng, D., Xu, Q., Hu, G., Peng, Y., Pan, C., Wang, Z.L., Piezoelectricity in multilayer black phosphorus for piezotronics and nanogenerators. *Adv. Mater.*, 32, 1905795, 2020.
16. Mirshekarloo, M.S., Yao, K., Sritharan, T., Ferroelastic strain induced antiferroelectric-ferroelectric phase transformation in multilayer thin film structures. *Adv. Funct. Mater.*, 22, 4159, 2012.
17. Kumari, S., Pradhan, D.K., Katiyar, R.S., Kumar, A., Ferroelectric, ferromagnetic, and multiferroic heterostructures for possible applications as tunnel junctions, in: *Magnetic, Ferroelectric, and Multiferroic Metal Oxides*, Stojanovic Ferroelectric, and Multiferroic Metal Oxides, B.D.B.T.-M. (Ed.), pp. 571–591, Elsevier, Matthew Deans, 2018.
18. Bonell, F., Goto, M., Sauthier, G., Sierra, J.F., Figueroa, A.I., Costache, M.V., Miwa, S., Suzuki, Y., Valenzuela, S.O., Control of spin-orbit torques by interface engineering in topological insulator heterostructures. *Nano Lett.*, 20, 5893, 2020.
19. Lu, X., Li, H., Wang, B., Theoretical analysis of electric, magnetic and magnetoelectric properties of nano-structured multiferroic composites. *J. Mech. Phys. Solids*, 59, 1966, 2011.
20. Stamps, R.L., Dynamic magnetic properties of ferroic films, multilayers, and patterned elements. *Adv. Funct. Mater.*, 20, 2380, 2010.
21. Becher, C., Trassin, M., Lilienblum, M., Nelson, C.T., Suresha, S.J., Yi, D., Yu, P., Ramesh, R., Fiebig, M., Meier, D., Functional ferroic heterostructures with tunable integral symmetry. *Nat. Commun.*, 5, 191, 192, 194, 2014.
22. Liu, X., Pyatakov, A.P., Ren, W., Magnetoelectric coupling in multiferroic bilayer VS2. *Phys. Rev. Lett.*, 125, 195–197, 2020.
23. Ul-Hamid, A., Specialized SEM techniques, in: *A Beginners' Guide to Scanning Electron Microscopy*, A. Ul-Hamid, (Ed.), pp. 181–232, Springer International Publishing, Cham, 2018.
24. Allwood, D.A., Xiong, G., Cooke, M.D., Cowburn, R.P., Magneto-optical Kerr effect analysis of magnetic nanostructures. *J. Phys. D. Appl. Phys.*, 36, 2175, 2003.

9

Physics of Multiferroic Materials

M. Rizwan[1]*, A. Ayub[2] and S. Ilyas[2]

[1]*School of Physical Sciences, University of the Punjab, Lahore, Pakistan*
[2]*Department of Physics, University of Gujrat, Pakistan*

Abstract
The increasing enthusiasm for multiferroic materials over the last decade is attributed to their astonishing properties such as large magnetoelectric coupling and technological applications. This chapter reviews some aspects of multiferroic materials such as their origin, history, and recent developments. Multiferroic materials can be classified based on the nature and type of coupling. Single-phase and composite multiferroics are discussed here based on the mechanism of magnetoelectric coupling. Type I and type II multiferroic materials along with their common classification are also deliberated, and well-known applications are also discussed.

Keywords: Type I multiferroics, magnetoelectric coupling, magnetic sensor, single phase multiferroics, charge order

9.1 Introduction

Over the past few years, multiferroic materials have grabbed the attention of researchers and scientists due to their unique physical properties and capacity to build multi-billion US $ market value technologies in the future.

"Multiferroics are substances in which multiple orders co-exist. These orders may include electric, magnetic, or piezo-elastic phases" [1–3].

These phases may be the consequence of various microscopic interactions; i.e., magnetic order may be the consequence of the interchanging of magnetic dipoles that may arise due to unoccupied shells of atomic orbitals. Electric order may be due to the distribution of internal electric dipoles, and

**Corresponding author*: rizwan.sps@pu.edu.pk

Inamuddin, Tariq Altalhi and Mohammad Abu Jafar Mazumder (eds.) Ferroic Materials-Based Technologies, (207–226)

elastic order may be due to changes faced by the material it bears because of external stress on atomic levels below Curie temperature [2]. The coexistence of these orders is not the only reason, as the cross-coupling effects also play a potential force behind its utilization in numerous technological applications for the purpose of storing and transmission [1].

In 1994, H. Schmidt was the first one to explore these materials that contain multiple orders at the same time, and named them "multiferroic" [4].

Based on various orders that exist simultaneously in multiferroic materials, these may be classified into various types. Some of these are as follows:

- Ferromagnets show strong piezomagnetic properties [1].
- Piezo-ferroelectrics show strong ferroelectric and ferroelastic properties [1].
- Piezomagnetic materials are the ones that show ferromagnetic and ferroelastic orders [5].
- Magnetoelectric multiferroics show strong electromagnetic properties [1].
- Elasto-magnetoelectric multiferroics have coexistence of three orders: electric, magnetic, and elastic [1].

9.2 Origin of Ferromagnetism and Antiferromagnetism

In the field of magnetism, the motion of charges especially the orbital and spin motion of electrons play a vital role in the determination of the magnetic properties of materials. Basically, it depends on the electronic configuration or the number of electrons in that specific element whether it will show a diamagnetic or ferromagnetic effect. If a material shows diamagnetic behavior in an external applied magnetic field, that means individual atoms or ions present in that material have overall canceled the effect of each other [6, 7].

On the atomic level, when an atom has completely filled a valence shell, then all electrons nullify the combined effect of each other. So, in a nutshell, all atoms and all materials show diamagnetic behavior, but this behavior may be suppressed by magnetic behavior if an atom has incomplete f or d shell orbitals and these materials show some net magnetic moment that may be coupled or not. The charge and spin of electrons play a vital role in determining the magnetic properties of materials [5].

If a material shows enhanced magnetic properties due to the addition of coupling of moments, it is termed ferromagnetism. If a material cancels the effect of coupling the moment on an atomic level, the behavior is coined antiferromagnetism, but if this effect is partially canceled out, then as a result, materials show slight magnetic behavior, and this is coined as ferri-magnetism. Free moments giving rise to magnetic properties may be termed para-magnetism. So, the arrangement of magnetic moments is also a turning point in determining the magnetic properties of materials. Various transition metal ionic salts show paramagnetic behavior because of considerable fractionation. Besides that, temperature also affects the interrelationship of magnetic moments and hence they become temperature dependent.

Basically, the origin of ferromagnetism is linked with quantum-mechanical-exchange interaction among atoms [8], and it is well depicted by two famous theories: the Curie–Weiss localized theory [9] and the Stoner-band theory [10]. In both of the models, strongly bound, restrained electrons, i.e., d shelled electrons of transition metals or f-shelled electrons of transition metals and rare-earth metals respectively play vital roles for magnetism [8]. Reported by Weiss, the arrangement of magnetic moment of ions are led by internal molecular field [9]. The internal molecular field is enhanced by the thermal energy beneath a characteristic temperature (Tc, the transition temperature) and therefore the alignment of spins (magnetic moments) exists even in the absence of enforced magnetic field for ferromagnetic materials. Weiss theory successfully explains ferromagnetic behavior, but it is unable to explain experimental results of magnetic moments. Stoner then presents the idea of band theory, in which magnetic properties of materials are determined by the modification in band energy because of the interrelationships of the spins [8]. The parallel arrangement of spins lowers the energy while oppositely oriented spins enlarge the energy of the ferromagnetic system

9.3 Origin of Ferroelectric Materials

Just like ferromagnetism, when an atom is placed in the vicinity of an electric field, it exhibits various types of electric polarization, i.e., ionic, dipole, and electronic. Materials where individual dipoles lead to the addition of dipole moments are coined ferroelectrics, and where they negate the effect of each other, they are named antiferroelectrics.

Basically, when an insulator is positioned in an electric field, dipoles are created in that insulator, forming it negatively charged at one end and positively charged on the other. The sum of the product of these charges and the distance between them defines the term dipole moment and the dipole moment present in a unit volume gives us the polarization [5].

Some materials do show dipole moment even in the absence of electric field so when those materials are placed in an external field, they orient their internal field in a direction where it enhances the overall electric field and, hence, the dipole moment, making them ferroelectric materials.

The overall electric displacement D in ferroelectrics in an external applied electric field E is given by

$$D = \varepsilon_0 E + P \tag{9.1}$$

where ε_0 represents the vacuum permittivity and P is the polarizability.

Also, even if they are removed from the vicinity of electric field, internal electric fields do not disappear completely and they exhibit remanent polarization just like ferromagnetic materials. Similar to ferromagnetic materials, ferroelectric materials also show small domains in which dipoles are oriented randomly in various directions. Ferroelectric materials also show temperature-influenced behavior such that, under certain temperatures, their behavior varies from antiferroelectric to ferroelectric behavior.

Basically, ferroelectric behavior was observed long ago in Rochelle salts. Then, its discovery in KH_2PO_4 and $BaTiO_3$ expanded interests in it. While investigating materials for a capacitor with high dielectric constant $BaTiO_3$, the first non-hydrogen-bonded ferroelectric material was explored during World War II. Helen D. Megaw explained this behavior in detail and Miyake and Ueda verified its existence [5].

Based on spontaneous polarization shown by different materials, they can be classified as:

- Ferroelectrics: Materials that exhibit spontaneous polarization in an electric field under an ordering temperature. Inversion symmetry is absent in them and they do not have a proper rotation axis. Most of all, the polarization in these materials is easily reoriented by application of an external electric field. Diversion of polarization is connected to reorientation of the polarization direction of domains through rotation. The redirection of polarization domains in ferroelectrics is similar to that present in ferromagnets [2].
- Pyroelectrics: Materials that show spontaneous polarization beneath a certain ordering temperature are called pyroelectrics. Ferroelectrics belongs to the class of pyroelectrics. While the difference between these two class of materials

is that in pyroelectrics, the exterior electric field, which is essential for polarization reorientation, can surpass the electric breakdown field, arising from catastrophic impact ionization. This implies that polarization cannot be reoriented by an externally applied electric field. The polarization can only be reoriented under a specific temperature in an external field by ordering transition. Twinned samples may be the consequence of cooling in zero field, like inversion twins.

- Proper ferroelectrics: In these materials, the ferroelectric ordering is controlled by phase-transition basic mode, which can be mapped out via the alterations in the consonance via the transitions. Most of the time, the value of polarization is greater than 1 μC cm^{-2}.
- Improper ferroelectrics: In such materials, ferroelectric order is controlled by secondary order parameters. An example of ferroelectric ordering is in $TbMnO_3$, where electric orientation is found as a result of magnetic orientation. The polarization here is mostly less than 1 μC cm^{-2}.
- Antiferroelectrics: In these materials, the local electric dipoles are oriented in a direction opposite to applied field. For example: $PbZrO_3$ shows antiferroelectricity. Antiferroelectric materials basically belong to a group of antipolar materials (these materials possess anti-parallel electric ordering but are not polarizable [11]).

Figure 9.1 describes the spin configuration for ferromagnetic and ferroelectric materials. Some materials can switch from one state to another, for example, when materials switch from paraelectric state to ferroelectric

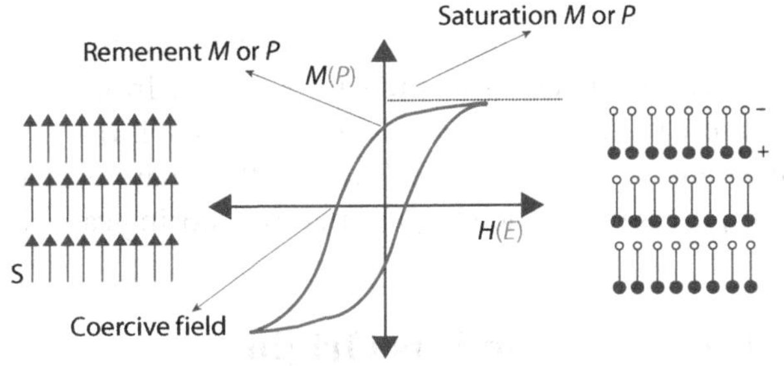

Figure 9.1 Hysteresis loop and ordering of spins for ferromagnetic and ferroelectrics materials where M = magnetization, P = electric polarization, H = magnetic field, and E = electric field [5].

state; this is known as ferroic transition. Besides this, there are many materials that exhibit two states at a time.

9.4 Historical Background and Present

Schmidt was the first one who used the term "multiferroicity" in 1994 [4]. Although this term was firstly used in 1994, the concept of "multiferroicity" dated back to 1894, when Curie described the concept of magnetoelectric coupling grounded on crystal symmetry [12]. The term "electro-magneto coupling" was used after that by "Debye" in 1926 [13].

After the exploration of ferroelectric properties in $BaTiO_3$ in 1940, multiferroic materials grasped much attention [14]. Dzyaloshinskii for the first time reported magnetoelectric effect in antiferromagnetic dielectrics such as $Cr2O_3$ in 1959 [15]. In 1970, O'Dell summed up all of the research data in the field of magnetoelectric coupling in a book [16]. In 1973, Wood and Austin published a review based on applications of multiferroics [17]. In 1982, Smolenskii and Chupis published a review enlisting 50 multiferroic materials [14].

After that, this research topic lost its importance for a short span of time. The main reasons behind that were that only a small number of multiferroic materials were discovered. Other than that, multiferroic materials discovered till that time were not having strong magneto-electro coupling, and above all, cryogenic temperatures were required for their operation. However, after 2000, this topic again grabbed the attention of researchers. This was particularly due to the industrial demand of these materials in various applications as well as the development in methodology of preparing materials especially using nano-technology and discovery of new materials with special reference to nano-materials [1].

Nowadays, "multiferroic materials" are considered to be one of the top emerging fields of research and technological applications. This is mainly because of their appealing properties and their vast use in technologies that make them potential materials to increase revenue. In fact, it is predicted that merging these materials with semiconductor or spintronic materials can bring advancement in the field of technology that may replace the existing RAM and computers with much higher memory storing capacity and speed.

9.5 Multiferroicity and Its Origin

As mentioned earlier, "coupling" of different orders is one of the basic characteristics of multiferroic materials, as shown in Figure 9.2.

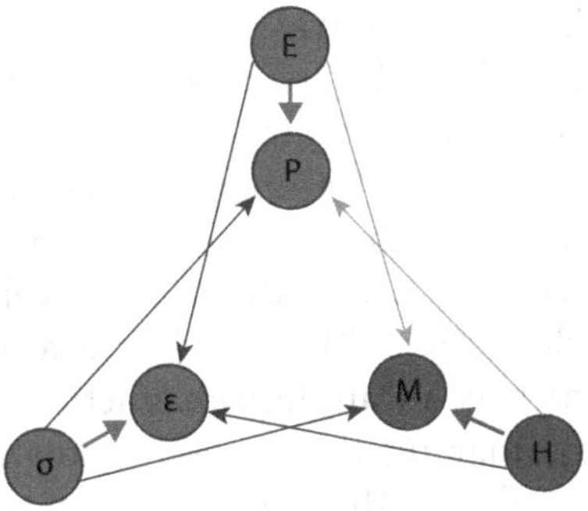

Figure 9.2 Possible cross-couplings in multiferroics where ε = strain, E = electric field, H = magnetic field, P = electric polarization, M = magnetization, and σ = stress [1].

Consider "piezo-ferro coupling" where coupling of ferro phases and piezo-elastic materials help us to control the ferromagnetic and ferroelectric characteristics using externally enforced stress/force. Also, there exist materials where the external magnetic and electric field applies stress on the material and changes the shape of materials.

Another type of coupling that may be present in multiferroic materials is "magneto-electro coupling", where magnetic as well as electric phases exist simultaneously. This effect is based on the interconversion of energy between electric and magnetic field, well explained by Maxwell's equations. This also creates a link between insulators and spin of the electrons as ferroelectric materials are mostly insulators and unpaired spin is required for ferromagnetic properties to show.

Maxwell's equations basically provide a mathematical form of differential equations where electric and magnetic field varies with time. Thus, these equations explain only time-varying fields. As regards static electric and magnetic fields in solids. Pierre Curie presented the concept in 1894 based on the symmetry of crystals [1].

"Magneto-electro coupling" may be of two types:

- Direct magneto-electro coupling: A directly applied external magnetic or electric field changes either the magnetization or the electric polarization of the material. These types of materials are basically known as "single-phase multiferroics" [18].
- Indirect magneto-electro coupling: This type of coupling is mostly observed in "elasto-magnetoelastic multiferroics", where an externally applied field does not have a direct impact on the magnetization or polarization but it puts a stress on material that results in change [19, 20].

As mentioned earlier, ferroelectric properties are mainly observed in insulators as they do not allow flow of charge while ferromagnetic behavior is observed in materials having unpaired spin electrons, and these do not have any link with electrical properties. Magnetic materials having good electrical properties are commonly observed but magnetic insulators have also been discovered. Basically, this fact is explained by Jahn-Teller distortion. For example, $YTiO_3$ [21], which is a ferromagnetic material, and $LaMnO_3$ [22], which is an antiferromagnetic material, both exhibit ignorable conductivity, making them good insulators. Magnetoelectric and magneto-dielectric effect in these materials can be depicted because of powerful electron–phonon couplings and polar-defect formation [23–26], while magnetic conductors like transition-metal-ion-based doped $BaTiO_3$ can be explained via lattice defects [27].

Thus, multiferroicity can be basically explained by these two properties. According to Schmidt, proper symmetrical conditions are required by materials to show more than one ferroic properties at a time or multiferroicity [28].

For the detailed explanation of this symmetry, Heesch developed some point groups [29], and later on, Shubinikov and Belov [30] also explained this concept based on time (1`) and space (1) inversion or their

Table 9.1 Magnetic–electric polarization compatibility Heesch–Shubnikov point groups [5].

P_s (Electric polarization)	**1**	m		m'		$1'$	$m1'$
	2	mm2	$2'$	$m'm'2$	$m'm2'$	$21'$	$mm21'$
	3	3m		$3m'$		$31'$	$3m1'$
	4	4mm	$4'$	$4m'm'$	$4'mm'$	$41'$	$4mm1'$
	6	6mm	$6'$	$6m'm'$	$6'mm'$	$61'$	$6mm1'$
M_s (Magnetic polarization)	**1**		$\underline{1}$	$2'$	m'	$m'm2'$	$2'/m'$
	2	m	2/m	$2'2'2$	$m'm'2$		$M'm'm'$
	3		$\underline{3}$	$32'$	$3m'$		$\underline{3}m'$
	4	$\underline{4}$	4/m	$42'2'$	$4m'm'$	$\underline{4}2'm'$	$4/mm'm'$
	6	$\underline{6}$	6/m	$62'2'$	$6m'm'$	$\underline{6}m'2$	$6/mm'm'$

Bold symbols are suitable for electric as well as magnetic polarization.

combination vector (1`) accompanied with crystallographic point groups known as Heesch–Shubnikov point groups, which are given in Table 9.1. In total, 122 Heesch–Shubnikov point groups are found, of which 31 are good ferroelectrics and 31 are good ferromagnetics [31], and only 13 show spontaneous electric as well as magnetic polarization, making them common in both [32–35]. Hence, materials that show multiferroicity based on Heesch–Shubnikov point groups are very rare [36]. Later on, Hill pointed out that only on the basis of symmetry can these materials not be explained [27]. As there are so many Heesch–Shubnikov point groups that do not have ferroelectric or ferromagnetic properties, these can be explained well along with the occupancy of d-shelled orbitals in materials. In a nutshell, we may also say that solids that are multiferroic crystals and that possess electric and magnetic order should possess both time and spatial inversion symmetry [1].

9.6 Multiferroic Materials

Multiferroic or ferro-magnetoelectric materials belong to a class of solid compounds that possess a unique property that enables the coexistence of at least two magnetic, electric, or piezo elastic phases. The coexistence of these phases makes materials such as ferromagnetic materials having significant piezo magneticity and piezo-ferroelectric materials important multiferroics. These multiferroic materials are extremely valuable for applications because of the ease of controlling their ferroelectric and ferromagnetic properties by manipulating the coupling of ferro and piezo phases via external stress. The shapes of multiferroics can be manipulated by the reverse process of inducing external magnetic or electric fields. The coupling between the ferro phase and piezo phase gives rise to piezoelectric and piezo magnetic properties, and further additional cross-coupling can give rise to magnetoelectric multiferroics and elasto-magnetoelectric multiferroics, making these materials quite valuable for research and technological applications [37].

The coupling between ferro order and piezo state is well understood, but the additional cross-coupling leading to magnetoelectric coupling in multiferroics is not quite simple. The relationship between electricity and magnetism has been established by Maxwell's equations. These equations establish a relationship between time-varying fields to their gradients. However, Maxwell's equations are based on a time-varying field; the real question is whether static electric and magnetic fields have any relationship in solids. Pierre Curie established the relationship of linear magnetoelectric coupling in 1984. The properties of multiferroic materials are determined by symmetry since the ferroic properties are affected by space

Table 9.2 Types of multiferroic materials and their general features.

Type	Nature/ mechanism	Nature of coupling	Origin of coupling	Examples
Type I	Independent ferroelectricity and magnetism	Weak coupling	Lone pair	$BiFeO_3BiMnO_3PbVO_3$
			Charge ordered	RFe_2O_4 $RFeMO_4RGaCu_2O_4$ where R = Yb, Lu and M = Co, Cu
Type II	Magnetism results in ferroelectricity	Strong coupling	Spiral or collinear magnetic ordering	Ortho-$RMnO_3$ (R = Tb, Tm, Yb) Hex-$RMnO_3$ (where R = Y, Er, Tm) RMn_2O_5 (where R = Yb, Er, Tm, Y)

and time inversion. Electric polarization is invariant under the time inversion symmetry operator, while magnetization remains invariant under a special inversion symmetry operator.

A combination of the two phases of ferroic properties is very hard, but a large sum of research has contributed to understanding the multiferroic properties of such materials. Table 9.2 gives a list of well-known multiferroic materials and their properties [38]. In recent years, many materials have been studied due to their diverse multiferroic properties. These materials are reported to have these properties in bulk, thin film, and heterostructure as well. In the following section, these multiferroic materials are classified, and their properties are explained briefly [39].

9.7 Classification of Multiferroic Materials

Based on the mechanism of ferroelectricity possessed by materials and their nature, multiferroic materials are organized into two subgroups: heterostructure/single phase and composite multiferroic materials [40]. Furthermore, based on the nature and associated mechanism of multiferroicity, single-phase materials are divided into type I and type II multiferroic materials as shown in Table 9.2. Type I multiferroic materials consist of materials in which ferric order are independent, while in type II materials, ferroelectric polarization is a result of another driving order. Type I multiferroics have high polarization and high ordering temperatures, but the coupling between phases is weak [41].

9.7.1 Single-Phase Multiferroics

In these multiferroic materials, there is the coexistence of electric and magnetic order at any point, and these materials are homogeneous and isotropic. These materials could also possess elastic or piezo-elastic properties, but that is not compulsory in all materials. A very firm symmetry condition is required for the appearance of coexistence of magnetic and electric order in the same phase demands. Electric-ordered crystals that possess electric polarization should remain unchanged under the action of time inversion symmetry operator; a crystal that allows magnetic polarization should be invariant when acted by a space inversion symmetry operation. Figure 9.3 shows the time inversion and special inversion symmetry operator mechanism.

The total number of structures that are possible are 122 [42]; out of these structures, 31 possess electric order and 31 point groups have magnetic order, while 13 phases possess both electric and magnetic phase at the same time. The 13 point groups that allow multiferroic phase are tabulated in [43]. Table 9.3 shows some of the well-known single-phase multiferroics (SPM).

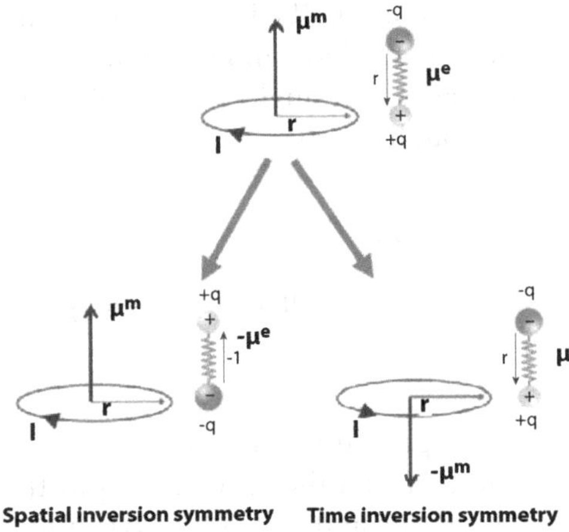

Figure 9.3 Schematic figure of time and spatial symmetry operator on a magnetic and an electric moment [42].

Table 9.3 Single-phase and composite multiferroics [42].

Single-phase multiferroics	Ref.	Composite multiferroics	
Cr_2O_3	[44]	$La_{0.7}Sr_{0.3}MnO_3$/PZT laminated layers	[45]
$Cr_3B_7O_{13}Cl$	[46]	$Tb_{1-x}Dy_xFe_2$-PZT in polymer matrix	[47]
$GaFeO_3$	[48]	$BaTiO_{3-}$ $CoFe_2O_4$ composite fiber	[49]
$LiCoPO_4$	[50]	$Tb_{1-x}Dy_xFe_2$/PZT laminated layers	[51]
$NdCrTiO_5$	[52, 53]	$CoFe_2O_4$ in $BaTiO_3$ matrix	[54]
$TbPo_4$	[52, 53]	$Tb_{1-x}Dy_xFe_2$/PZT/ $Tb_{1-x}Dy_xFe_2$ laminates	[55]

In single-phase multiferroics, most materials are dielectric oxides since one of the contributing components is electric, which is usually provided by dielectric materials. Single-phase multiferroics are very hard to synthesize and do not occur abundantly in nature. The reason behind the scarcity of these materials is linked to symmetry restrictions due to only 13-point groups exhibiting multiferroicity.

SPM materials are further divided into type I and type II multiferric materials. These materials are classified based on the mechanism of electric polarization. Type I SPMs have independent electric and magnetic phase coexistence, whereas type II single-phase multiferroic materials have strong magnetoelectric coupling [56].

9.7.1.1 *Type I Multiferroics*

These multiferroics are further classified, and some type I multiferroics are discussed here.

9.7.1.1.1 Perovskite-Type Multiferroics

Perovskites are materials that have the general composition of ABX_3, where A and B are said to be cations and X can be oxygen or any halogen. The cations are decided by coordination polyhedral and charge on the anion. Perovskites posing multiferroic properties have been explored for a long time. The electrical conductivity and magnetism in rare earth manganates are due to the existence of giant magnetoresistance.

Some perovskites also belong to the family of multiferroic materials such as barium titanate. In barium titanate, ferroelectric moment is due to the lone pair ion that is situated on the A-site such as presented in Table 9.3, and partly from the transition metal that is situated on the B-site having d-electron configuration. Magnetism in these materials arises due to the transition metal occupied on the B-site or the rare earth element sitting on the A-site. Controlling the compositions in these materials can lead to multiferroicity.

Bismuth ferrite is one of the few type I materials existing at room temperature. In this material, the multiferroicity arises from the line pair of bismuth that is situated on the A-site, and magnetism arises from Fe, with unpaired ions existing on the B-site as mentioned earlier. The antiferromagnetic and ferroelectric transition temperatures are 643 K and 1143 K, respectively. A lot of research has been performed on this material due to reports of large ferroelectric polarization in thin films of bismuth ferrite. The bulk counterpart of bismuth ferrite did not have large polarization previously. In this material, since ferroelectricity and ferromagneticity coexist, the magnetoelectric coupling is very diminished [57].

$EuTiO_3$ is another perovskite type I multiferroic material that belongs to a class of quantum paraelectric. In such materials, at 0 K, the initial ferroelectric transition is stopped and quantum fluctuations arise since thermal vibrations stop [58].

9.7.1.1.2 Organic–Inorganic Hybrids

Organic–inorganic hybrid materials were proposed to be an alternative multiferroic material during research to find new materials of this class. In these hybrid materials, magnetism is provided by a transition metal halide, and organic materials provide structural flexibility. This transition metal halide mainly determines the electron properties of these hybrids further. Organic–inorganic hybrids belong to a subclass of metal organic framework materials (MOFs). These MOFs belong to a very researched class of materials due to their astonishing applications. Organic–inorganic hybrids such as $((CH_3)_2\ NH_2)Mn\ (HCOO)_3$ do exhibit multiferroism, and show antiferromagnetic order at 8.4 K [59].

9.7.1.1.3 Charge Ordered

As mentioned in Table 9.3, the origin of weak coupling could be either the lone pair present in perovskites or the charge order. Here, we will discuss how the charge order will produce multiferroic properties. Materials like nearly half-doped manganites such as $R_{1-x}Ca_xMnO_3$ possess charge order, which can be considered as site based; here, Mn_4^+ and Pb_3^+ are arranged in a checkerboard pattern. In this pattern, the Mn sites make the bond inequivalent and thus charge density concentrates between nuclei. These charge order configurations can retain the inversion symmetry. A charge order conformation that happens on a site-centered and bond-centered charge order has electric polarization [60].

$LuFe_2O_4$ is a material in which ferroelectricity is determined by charge order. It is synthesized by stacking layers of LuO_2 and Fe_2O_2. $LuFe_2O_4$ exhibits charge order under 320 K and intermediate layers possess net electric polarization [61].

9.7.1.2 *Type II Multiferroics*

In type II multiferroics, there is strong coupling, which arises due to ferroelectricity. Magnetic order in a material can break the inversion symmetry possessed by a crystal. Type II materials are based on a microscopic magnetoelectric coupling mechanism. Materials in which ferroelectricity arises as a result of spin order are classified into three subgroups. The first subgroup possesses collinear magnetic order, and symmetric exchange striction provides electric polarization. The materials having non-collinear magnetic order and magnetization arising from anti-symmetric exchange striction belongs to the second group. Spin dependent p-d hybridization produces electric polarization in the third subgroup [62].

9.7.2 Composite Multiferroics

Even though single-phase multiferroic materials are a group of multiferroic materials that are extensively researched, they do not exhibit strong room temperature magnetoelectric coupling. The absence of the aforementioned property makes single-phase multiferroics materials unattractive for future applications. The research in multiferroic materials has been diverted towards finding alternative of these materials such as multiferroic composites. In comparison to single-phase materials, these composites have coexistence of electric, magnetic, and piezo order; furthermore, these materials are chemically isotropic and are also homogeneous [63]. The classification of composite multiferroic materials is based on the fact that three existing phases are independent of each other. Multiferroic composites offered design flexibility and magnetoelectric coupling for technological applications. Multiferroic composites have stronger coupling than single-phase materials and even exist above room temperature. These materials do not

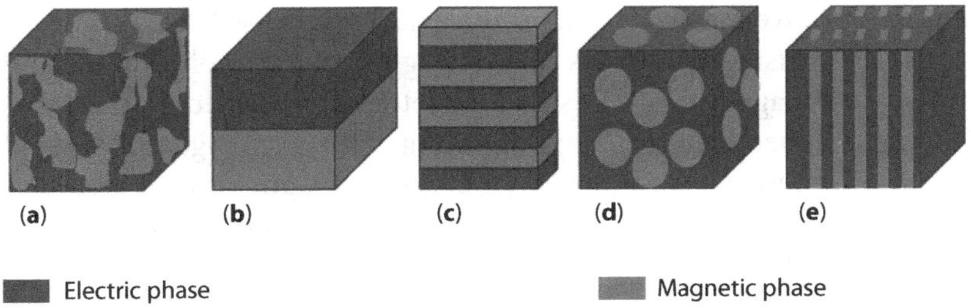

Figure 9.4 Possible phase mixing of electric and magnetic order in multiferroic composites [42].

need to be dielectric due to separated ferroic phases. The electric phase in multiferroic composites can be dielectric, but there is no restriction on magnetic phase to be dielectric. This makes these materials widely versatile, since there are large materials to choose from. These multiferroics do not also have to deal with simultaneous breakup of time and spatial inversion operation, since it automatically fulfills this condition [64]. Figure 9.4 shows possible mixing of electric and magnetic phases in these composites.

Figure 9.4 also reveals that bilayer and multi-layer composites are promising multiferroic composites. These multiferric composite configurations can be developed in blk and laminated thin films as well, which makes them suitable for nanofabrication as well. Nanofabrication merged with lithography makes it possible for novel nanoscale multiferroic composites to be integrated with electronic circuits, thus revolutionizing the field of multiferroics. Nanoscale fabrication of multiferroic composites could introduce new quantum effects at such scale. In multiferroic composites, there is an indirect magnetoelectric coupling that is not affected by electric and magnetic field application, but is modified via strain application [65].

9.8 Applications of Multiferroics

Ever since their discovery, multiferroics have become extremely important for technological applications. Barium titanate, a type I multiferroic, is used in high-energy-density capacitors, transducers, and many more [66]. Ferroelectric thin film integration with semiconductors further increased their scope in electrocaloric, FeRAM devices. Similarly, magnetic materials also had a wide range of applications in various fields [67]. The modern-day application of multiferroic materials is based on information storage via electric and magnetic polarization. Such application has an edge over conventional ferroelectric or ferromagnetic memory devices, which only use magneto-resistive materials that produce heat and noise [68]. This heat can be avoided by using capacitors under magnetic field application that will also increase magnetic bit density. Magnetoelectric random access memories use magnetoelectric coupling along with exchange coupling that exist on multiferroics and ferromagnets [69]. Magnetic field sensors are also another application of such magnetoelectric materials. It is very easy to measure a minute single of voltage at the point of no current, in comparison to resistivity or magnetization. Magnetoelectric materials can also produce terahertz radiations, such as $BiFeO_3$ [70].

9.9 Conclusion

It will not be an exaggeration to state that multiferroic materials are a great contender to revolutionize our society and existing scientific growth. In this chapter, we introduced the concept of ferromagnetism, antiferromagnetism, and ferroelectricity and the origin of these phenomena coexisting in multiferroic materials. The classification of multiferroic materials based on the internal coexistence of magnetic and electric phase is also discussed. Common applications of multiferroics are also touched upon.

References

1. Vopson, M.M., Fundamentals of multiferroic materials and their possible applications. *Crit. Rev. Solid State Mater. Sci.*, 40, 223, 2015.
2. Buurma, A.J.C., Blake, G.R., Palstra, T.T.M., Adem, U., Multiferroic materials: Physics and properties, in: *Reference Module in Materials Science and Materials Engineering*, Elsevier, Amsterdam, 2016.
3. Fiebig, M., Revival of the magnetoelectric effect. *J. Phys. D. Appl. Phys.*, 38, R123, 2005.
4. Schmid, H., Multi-ferroic magnetoelectrics. *Ferroelectrics*, 162, 317, 1994.
5. Achary, S.N., Jayakumar, O.D., Tyagi, A.K., Multiferroic materials, in: *Functional Materials: Preparation, Processing and Applications*, S. Banerjee, and A.K.B.T.-F.M. Tyagi, (Eds.), pp. 155–191, Elsevier, London, 2011.
6. Wiegand, H. and Pielou, E.C., An introduction to mathematical ecology. Wiley Interscience. John Wiley & Sons, New York 1969. VIII + 286 S., 32 Abb., Preis 140 s. *Biom. Z.*, 13, 219, 1971.
7. David Mermin, N. and Ashcroft, N.W., Hans Bethe's contributions to solid-state physics, in: *Hans Bethe and his Physics*, pp. 189–200, World Scientific, Singapore, 2006.

8. Hill, N.A., Density functional studies of multiferroic magnetoelectrics. *Annu. Rev. Mater. Sci.*, 32, 1, 2002.
9. Weiss, P., L'hypothèse du champ moléculaire et la propriété ferromagnétique. *J. Phys. Théor. Appl.*, 6, 661, 1907.
10. Geshev, J., Pereira, L.G., Schmidt, J.E., Mikhov, M., Dependence of the magnetization and remanence of single-domain particles on the second cubic anisotropy constant. *J. Appl. Phys.*, 90, 6243, 2001.
11. Lines, M.E. and Glass, A.M., Macroscopics and phenomenology, in: *Principles and Applications of Ferroelectrics and Related Materials*, M.E. Lines, and A.M. Glass, (Eds.), pp. 59–86, Oxford University Press, Walton Street, Oxford, 2010.
12. Curie, P., Sur la symétrie dans les phénomènes physiques, symétrie d'un champ électrique et d'un champ magnétique. *J. Phys. Théor. Appl.*, 3, 393, 1894.
13. Debye, P., Bemerkung zu einigen neuen Versuchen über einen magneto-elektrischen Richteffekt. *Z. Phys.*, 36, 300, 1926.
14. Smolenskii, G.A. and Chupis, I.E., Ferroelectromagnets. *Sov. Phys. - Usp.*, 25, 415, 1982.
15. Shpol'skiĭ, É. V., Problems of semiconductors. *Sov. Phys. Usp.*, 2, 628, 1959.
16. Willmore, A.P., Small rocket instrumentation techniques I. Maeda (North-Holland Publishing Company, Amsterdam, 1969, 240 pp. *Geophys. J. Int.*, 19, 210, 1970.
17. Erratum, *Wood Preserv.*, vol. 17, p. 236a, 1991.
18. Prellier, W., Singh, M.P., Murugavel, P., The single-phase multiferroic oxides: From bulk to thin film. *J. Phys. Condens. Matter*, 17, R803, 2005.
19. Srinivasan, G., Magnetoelectric composites. *Annu. Rev. Mater. Res.*, 40, 153, 2010.
20. Nan, C.W., Bichurin, M.I., Dong, S., Viehland, D., Srinivasan, G., Multiferroic magnetoelectric composites: Historical perspective, status, and future directions. *J. Appl. Phys.*, 103, 31101, 2008.
21. Fukatani, R., Yokota, H., Ogura, S., Uesu, Y., Bartasyte, A., Fukunaga, M., Noda, Y., Ferroelectricity and ferromagnetism of BaTiO 3 /BaFeO3 superlattice thin films. *Jpn. J. Appl. Phys.*, 51, 09LB01, 2012.
22. Von Helmolt, R., Wecker, J., Holzapfel, B., Schultz, L., Samwer, K., Giant negative magnetoresistance in perovskitelike La2/3Ba1/3MnOx ferromagnetic films. *Phys. Rev. Lett.*, 71, 2331, 1993.
23. Rogado, N.S., Li, J., Sleight, A.W., Subramanian, M.A., Magnetocapacitance and magnetoresistance near room temperature in a ferromagnetic semiconductor: La2NiMnO6. *Adv. Mater.*, 17, 2225, 2005.
24. Furukawa, N., Transport Properties of the Kondo Lattice Model in the Limit S= ∞ and D= ∞. *J. Phys. Soc. Japan*, 63, 3214, 1994.
25. Millis, A.J., Littlewood, P.B., Shraiman, B.I., Double exchange alone does not explain the resistivity of La1-xSrxMnO3. *Phys. Rev. Lett.*, 74, 5144, 1995.
26. Röder, H., Zang, J., Bishop, A.R., Lattice effects in the colossal-magnetoresistance manganites. *Phys. Rev. Lett.*, 76, 4987, 1996.

27. Hill, N.A., Bättig, P., Daul, C., First principles search for multiferroism in BiCrO3. *J. Phys. Chem. B*, 106, 3383, 2002.
28. Schmid, H., Some symmetry aspects of ferroics and single phase multiferroics. *J. Phys. Condens. Matter*, 20, 434201, 2008.
29. Heesch, H., VI. Zur Strukturtheorie der ebenen Symmetriegruppen. *Z. Kristallogr. - Cryst. Mater.*, 71, 95, 1929.
30. Buerger, M.J., Antisymmetry: Colored symmetry. A series of publications from the Institute of Crystallography, Academy of Sciences, U.S.S.R., 1951-1958. A. V. Shubnikov, N. V. Belov, and others. Translated from the Russian by Jack Itzkoff and Jack Gollob. William T. Holser, Ed. Pergamon, London; Macmillan, New York, 1964. xxvi + 263 pp. Illus. $9.75. *Sci. (80-.)*, 145, 804, 1964.
31. Cook, W.R., Ferroelectric, piezoelectric, and electro-optic materials, in: *Digest of Literature on Dielectrics*, vol. 38 1974, pp. 502–660, 2016.
32. Helesic, J., Leichtfried, M., Statzner, B., Schenkova, J., International review of hydrobiology: Preface. *Int. Rev. Hydrobiol.*, 92, 361, 2007.
33. Schmid, H., On the possibility of ferromagnetic, antiferromagnetic, ferroelectric, and ferroelastic domain reorientations in magnetic and electric fields. *Ferroelectrics*, 221, 9, 1999.
34. Ascher, C.S., Council manager government. *Public Adm. Rev.*, 34, 93, 1974.
35. Birss, R.R., Macroscopic symmetry in space-time. *Rep. Prog. Phys.*, 26, 307, 1963.
36. Hill, N.A., Why are there so few magnetic ferroelectrics? *J. Phys. Chem. B*, 104, 6694, 2000.
37. Rado, G.T. and Folen, V.J., Observation of the magnetically induced magnetoelectric effect and evidence for antiferromagnetic domains. *Phys. Rev. Lett.*, 7, 310, 1961.
38. Fiebig, M., Lottermoser, T., Meier, D., Trassin, M., The evolution of multiferroics. *Nat. Rev. Mater.*, 1, 16046, 2016.
39. Cross, L.E., Ferroic materials and composites: Past, present and future, in: *Advanced Ceramics III*, S. Sōmiya, (Ed.), pp. 71–102, Springer Netherlands, Dordrecht, 1990.
40. Thomas, R., Scott, J.F., Bose, D.N., Katiyar, R.S., Multiferroic thin-film integration onto semiconductor devices. *J. Phys. Condens. Matter*, 22, 423201, 2010.
41. Zhou, J.P., Zhang, Y.X., Liu, Q., Liu, P., Magnetoelectric effects on ferromagnetic and ferroelectric phase transitions in multiferroic materials. *Acta Mater.*, 76, 355, 2014.
42. Vopson, M.M., Fundamentals of multiferroic materials and their possible applications. *Crit. Rev. Solid State Mater. Sci.*, 40, 223, 2015.
43. Birss, R.R. and Fletcher, D., Space-time symmetry restrictions on transport coefficients. *J. Magn. Magn. Mater.*, 15–18, 915, 1980.
44. McGuire, T.R., Scott, E.J., Grannis, F.H., Antiferromagnetism in a Cr2O3 crystal. *Phys. Rev.*, 102, 1000, 1956.

45. Srinivasan, G., Rasmussen, E.T., Gallegos, J., Srinivasan, R., Bokhan, Y.I., Laletin, V.M., Erratum: Magnetoelectric bilayer and multilayer structures of magnetostrictive and piezoelectric oxides [Phys. Rev. B 64, 214408 (2001)]. *Phys. Rev. B - Condens. Matter Mater. Phys.*, 66, 2002.
46. Rivera, J.P., On definitions, units, measurements, tensor forms of the linear magnetoelectric effect and on a new dynamic method applied to Cr-Cl boracite. *Ferroelectrics*, 161, 165, 1994.
47. Van Den Boomgaard, J., Terrell, D.R., Born, R.A.J., Giller, H.F.J.I., An *in situ* grown eutectic magnetoelectric composite material - Part I Composition and unidirectional solidification. *J. Mater. Sci.*, 9, 1705, 1974.
48. Rado, G.T., Observation and possible mechanisms of magnetoelectric effects in a ferromagnet. *Phys. Rev. Lett.*, 13, 335, 1964.
49. Cai, N., Nan, C.W., Zhai, J., Lin, Y., Large high-frequency magnetoelectric response in laminated composites of piezoelectric ceramics, rare-earth iron alloys and polymer. *Appl. Phys. Lett.*, 84, 3516, 2004.
50. Rivera, J.P., The linear magnetoelectric effect in LiCoPO4 revisited. *Ferroelectrics*, 161, 147, 1994.
51. Ryu, J., Carazo, A.V., Uchino, K., Kim, H.E., Magnetoelectric properties in piezoelectric and magnetostrictive laminate composites. *Jpn. J. Appl. Phys., Part 1 Regul. Pap. Short Notes Rev. Pap.*, 40, 4948, 2001.
52. Hwang, J., Choi, E.S., Zhou, H.D., Lu, J., Schlottmann, P., Magnetoelectric effect in NdCrTiO 5. *Phys. Rev. B - Condens. Matter Mater. Phys.*, 85, 214408, 2012.
53. Rado, G.T., Ferrari, J.M., Maisch, W.G., Magnetoelectric susceptibility and magnetic symmetry of magnetoelectrically annealed TbPO4. *Phys. Rev. B*, 29, 4041, 1984.
54. Ma, J., Shi, Z., Nan, C.W., Magnetoelectric properties of composites of single Pb(Zr,Ti)O3 rods and terfenol-D/epoxy with a single-period of 1-3-type structure. *Adv. Mater.*, 19, 2571, 2007.
55. Lee, M.K., Nath, T.K., Eom, C.B., Smoak, M.C., Tsui, F., Strain modification of epitaxial perovskite oxide thin films using structural transitions of ferroelectric BaTiO3 substrate. *Appl. Phys. Lett.*, 77, 3547, 2000.
56. Khomskii, D., Classifying multiferroics: Mechanisms and effects. *Phys. (College. Park. Md)*, 2, 20, 2009.
57. Rojac, T., Kosec, M., Budic, B., Setter, N., Damjanovic, D., Strong ferroelectric domain-wall pinning in BiFeO3 ceramics. *J. Appl. Phys.*, 108, 1–41, 2010.
58. Katsufuji, T. and Takagi, H., Coupling between magnetism and dielectric properties in quantum paraelectric (formula presented). *Phys. Rev. B - Condens. Matter Mater. Phys.*, 64, 054415, 2001.
59. Stroppa, A., Jain, P., Barone, P., Marsman, M., Perez-Mato, J.M., Cheetham, A.K., Kroto, H.W., Picozzi, S., Electric control of magnetization and interplay between orbital ordering and ferroelectricity in a multiferroic metal-organic framework. *Angew. Chem. - Int. Ed.*, 50, 5847, 2011.

60. Efremov, D.V., Van Den Brink, J., Khomskii, D.I., Bond- versus site-centred ordering and possible ferroelectricity in manganites. *Nat. Mater.*, 3, 853, 2004.
61. Ikeda, N., Ohsumi, H., Ohwada, K., Ishii, K., Inami, T., Kakurai, K., Murakami, Y., Yoshii, K., Mori, S., Horibe, Y., Kitô, H., Ferroelectricity from iron valence ordering in the charge-frustrated system LuFe2O4. *Nature*, 436, 1136, 2005.
62. Tokura, Y., Seki, S., Nagaosa, N., Multiferroics of spin origin. *Rep. Prog. Phys.*, 77, 76501, 2014.
63. Ramesh, R. and Spaldin, N.A., Multiferroics: Progress and prospects in thin films. *Nat. Mater.*, 6, 21, 2007.
64. Vopsaroiu, M., Thwaites, M.J., Rand, S., Grundy, P.J., O'Grady, K., Novel sputtering technology for grain-size control. *IEEE Trans. Magn.*, 40, 2443, 2004.
65. Prof. Dr. phil. Otto Vaupel 70 Jahre. *Mater. Test.*, 14, 37, 1972.
66. Scott, J.F., Applications of modern ferroelectrics. *Sci. (80-.)*, 315, 954, 2007.
67. Deutscher, I., *Lenin's childhood*, p. vii, 67. $4.95, Oxford University Press, New York, 1970. *Am. Hist. Rev.*, 1972.
68. Shaw, B.I., Recollections of pioneers in xenotransplantation research: Medicine and Health David K. C. Cooper, ed. New York, NY: Nova Science Publishers, Incorporated, 2018. 445 pp. *Am. J. Transplant.*, 19, 606, 2019.
69. Béa, H., Gajek, M., Bibes, M., Barthélémy, A., Spintronics with multiferroics. *J. Phys. Condens. Matter*, 20, 434221, 2008.
70. Takahashi, K., Kida, N., Tonouchi, M., Terahertz radiation by an ultrafast spontaneous polarization modulation of multiferroic BiFeO3 thin films. *Phys. Rev. Lett.*, 96, 117402, 2006.

10

Overview of Comparison Between Primary Ferroic Crystals with Secondary Ferroic Crystals

V. Renuga

Department of Chemistry, National College, Tiruchirappalli, Tamil Nadu, India

Abstract

The present chapter describes an overview of some basic concepts related to the process and formation mechanism of ferroic materials. The chapter deals with definitions of ferroics and their types, namely, primary and secondary ferroics. Further, it deals with the crucial classifications that exist in each of the ferroics and their formation, domains, domain walls, functions, phase structure, and their feasibility in a detailed manner. Also, this chapter deals with a comparison between primary and secondary ferroics. In addition to all the above facts, it also deals with multiferroics and their formations, and their relation to primary and secondary ferroics is discussed in an elaborate manner.

***Keywords*:** Primary ferroics, secondary ferroics, multiferroics, phase structure, domains and domain walls

10.1 Introduction

Ferroics, regardless of age-old concepts, are a class of materials that are aligned with the real world through their basic and advanced research studies [1]. The most distinctive nature of these materials inspired research communities and drove them to study their topological properties including magnetic [2, 3], electrical [4], skyrmions, negative capacitance [5, 6], and even resistive switching [7, 8]. The solitary nature of these ferroic materials is their ability to manifest most of their in-built electric, elastic, and magnetic properties in a long-range order. These properties provide

Inamuddin, Tariq Altalhi and Mohammad Abu Jafar Mazumder (eds.) Ferroic Materials-Based Technologies, (227–256) © 2024 Scrivener Publishing LLC

most of the fascinating nature of physics and exhibit their state of the art to modern society in emerging fields such as energy harvesting, sensor technology, and even in medical diagnosis. The launch of ferroics research was raised to an astounding level, both in theory and in practice, and even up the atomic level. In addition to this, the vast improvement in synthesis methodology and available tools for characterization techniques leads to the stabilization of this novel field and helps to engineer the ferroic properties in a desirable manner [9, 10]. The conceptual nature of these ferroic materials, opened up new ways to science and technology, and led to most of the key areas such as low-energy spintronics [11] in the field of neuromorphic computing as data storage at innumerable levels and also to the upcoming genesis nanotechnology. The word "ferroic" advocates the twinned crystals [12, 13], which usually mimics the orientation or direction in which more than one twin constituent exhibits their nature through the influence of proper external driving forces.

The difference in orientation states (twins) in the crystal governs the driving fields, which accompany the nature and orientation and are basic requirements needed to reorient between states. The domain states of ferroic crystals are shuffled between two primary and secondary states by the chosen external fields. The existence of domains such as ferroelectricity and ferromagnetism or ferroelasticity is influenced by spontaneous polarization and magnetization or strain, respectively, resulting in primary ferroics. The domain states in secondary ferroic crystals are influenced by either induced magnetization or induced polarization.

"Ferroics" can be defined as a substance in a condensed state that lowers its symmetry spontaneously under external influences such as temperature, pressure, strain, etc., resulting in collective phase transition from its original structure in an ordered manner [14, 15]. Ferroic materials' functional properties are directly related to the generation or administration of domains and domain walls, and are controlled by the materials' crystal arrangements, variations in size, stress, and dielectric conditions, which are major features that have to be considered in the field. Based on the above facts, ferroics can be defined as their electric, elastic, or magnetic order, which occurs impulsively through a non-disruptive phase transition [16].

These microscopic interactions count the driving order, which results in the existence of various primary types of ferroic order present as ferromagnetism, ferroelectricity, ferroelasticity, and ferrotoroidicity [17]. Due to this, an excess of antiferroic phases was formed, literally, and most of the magnetically ordered systems are antiferromagnetic.

Multiferroics [18] are another category of ferroic materials, in which it has a minimum of more than one type of ferroic order that exists in the identical phase and has a maximum association with magnetoelectric and magnetocapacitance effects [19]. These multiferroics are not dependent on long-range order, but there are certain parameters that describe the ferroic phase with at least two symmetrically equivalent directions. Between these directions, shifting takes place with the help of an external conjugate field. Here, regions with identical assimilation of the same order parameter are called domains and the interface that distinguishes them and that exists between them is called a domain wall [20, 21].

In general, these ferroics and multiferroics are the most distinguished objects and the most elemental part of physics, which contains complex nonlinear phenomena that are present in them at the micro and nano level. These nano-sized ferroics are the most promising among the most advanced applications due to their capability to control their size effects, influencing the most advanced applications in nano-electronics, nano-electromechanics, optoelectronics, nonlinear optics, and information technologies.

Van Aken [22] was the first person to discover the most exciting fourth type of primary ferroic crystal named ferrotoroidic crystal. The domains of this fourth type of ferroic crystal are most prominent due to a toroidal moment [23–25]. This ferrotoroidic domain may undergo switching between ferromagnetic and antiferromagnetic domains based on the symmetry as ferromagnetic/ferrotoroidic and antiferromagnetic/ferrotoroidic by either the combined influence of electric and magnetic fields or by a magnetic field alone [26].

A crystal that exhibits a spontaneous electric dipole moment is known as a ferroelectric crystal. In the absence of an external electric field, the crystals exhibit improper arrangements of the center of both positive and negative charges. The most requisite condition that governed, though not necessary, is the ferroelectric crystals, which should lack a center of symmetry. It concludes that all ferroelectrics will be piezoelectric, but the reverse, all piezoelectric, will not be ferroelectric (e.g., quartz).

In the early 1970s [27], Aizu described ferroic materials as "materials with a memory", since they have the ability to set back in various analogous forms, which may easily be shifted back and forth among the forms through the appropriate induction of fields, such as electric, magnetic, and stress. V. Wadhawan [28] describes the phase transition of ferroic systems as follows: (i) if it appears as a consequence of certain modifications of the nondisruptive phase, it is considered as the "prototypic phase"; (ii) the second one assumes this prototype system loses one or more point-symmetry operators.

The above definitions conclude that ferroelectric transitions generally involve the lack of time-inversion symmetry, and ferroelastic transitions correspond to the lack of symmetry elements such as rotation axis or

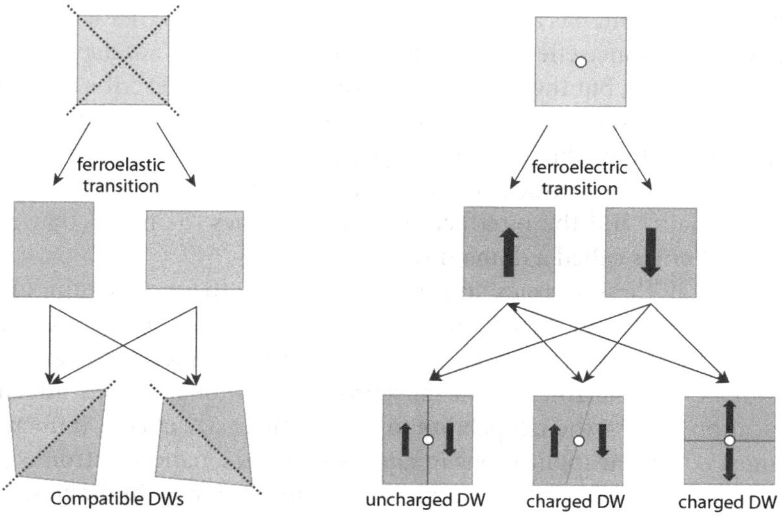

Figure 10.1 The basic illustration of ferroelastic and ferroelectric transitions is depicted here. Left: In the ferroelastic case, the two different domain pairs formed with the combination of two domains and are split by the respective compatible walls. They coincide with the mirror planes and are lost during the transition. To bring the two domains to contact each other, a small rotation is necessary. Right: In ferroelectric, examples of two different charged and uncharged domain walls are depicted. The wall's orientation is not fixed either by symmetry or strain. Adopted from Ref. [29].

mirror planes. Some of the fundamental demonstrations of the formation of a ferroelastic and a ferroelectric transition are depicted in Figure 10.1.

10.2 Formation of Ferroic Domains and Domain Boundaries

In ferroic crystals, domain formations take place, due to ferroic transitions, which accompany the regions that have different orientations, but have equivalent symmetry in space. The point groups present in the phases of prototypic and ferroic are represented as G and F, respectively. The number of available domains corresponding to this point group is simply given by the index of F in G. Through the symmetry operations, these domains are interconnected to each other and lost at the phase transition.

The domain formation is based on symmetry operation and might be obtained through the applications of the Curie principle. In accordance with this principle, the cause and effect of the symmetry almost assume the same symmetry and follow scalar change, and it follows that most of the symmetry frameworks such as temperature or hydrostatic pressure are symmetric in nature and maintain the crystal to retain its symmetry and never allow to lower its symmetry. The domain formation in the crystal is related to this principle, and indeed it maintains its symmetry on an average when considered to account for all possible domains.

There are a number of different types of domains existing in the crystals, which are distinguished by domain walls or otherwise known as domain partitions. These two domain walls and domain partitions act as transformation spaces. The symmetry properties related to this transition region do not influence the shape or orientation in space, but they have important classifications.

In ferroelastic materials, the disclosure of some of the spontaneous strains that decide the transition and formation of domain walls is formed by minimizing the elastic energy of the crystal.

The boundaries that are formed in planes with both domains are strained equally, in relation to the prototypic phase, and such kinds of walls are deemed compatible as depicted in Figure 10.1.

The feasible orientations of domain walls are derived from all possible symmetry descents [30–32]. The two types of domain walls are named W, whose orientations stand on crystallographic symmetry, and W', whose orientations mainly depend on the coefficient standards of the spontaneous strain tensor.

The ferroelectric materials, which follow one more notion, are based on the difference among charged domains, whether they are uncharged or even neutral to charged domain walls. For example, the unconstrained polarizations in domains 1 and 2 are labeled as P1 and P2, respectively, and the corresponding inclination of the domain wall is given by a vector n, and the corresponding charges present over the exterior of the domain walls are given as (P1 – P2).n. If the charge goes to zero, the domain walls are considered neutral, but in all other cases, it is charged as shown in Figure 10.1. However, in general, domain walls that are charged are considered as energetically less favorable.

Based on internal structure, domain walls are classified as Ising walls, Bloch walls, and Neel walls. In the case of Ising walls, all the other parameters go to zero at the domain walls, but they maintain a constant direction (Figure 10.2a). Bloch walls maintain constant magnitude and rotate in a plane parallel to the wall, and exist with two direct chiralities (Figure 10.2b).

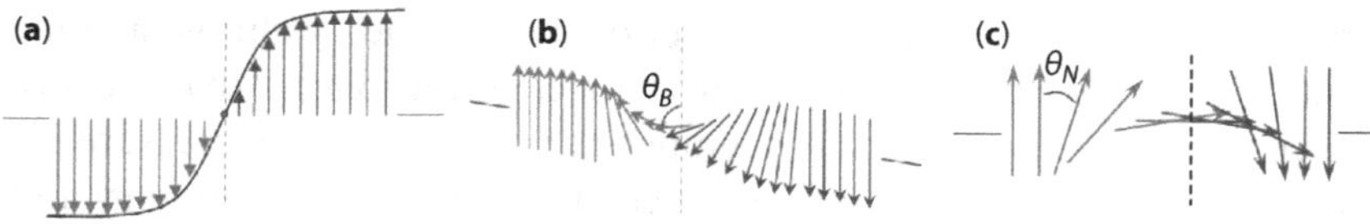

Figure 10.2 Representation of domain walls with respect to internal structure: (a) Ising, (b) Bloch, and (c) Néel. Adapted from Ref. [33].

The parameter that describes Neel Walls is known as the order parameter, which oscillates perpendicular to the plane of the wall (Figure 10.2c). Both Bloch and Neel walls initially exist in magnetic materials, but in recent times, researchers found that they belong to ferroelectric domain walls as well. A detailed description of all these types of walls is depicted in Figure 10.2.

There are a variety of microscopic techniques available to anticipate the formations of domains and domain walls in both ferroics and multiferroics. Some intriguing parameters such as magnetic order in an exceptional arrangement and enhanced conductivity also help to analyze the nature of domains and domain walls.

Primary ferroic crystals that arise due to the spontaneous magnetization, polarization, and strain are named ferromagnetic, ferroelectric, and ferroelastic crystals, respectively, due to the influence of magnetic, electric, and mechanical stress, and are discussed elaborately in the following section.

10.3 Description of Ferroelectricity—Phenomenological Way

Based on Landau's theory [39], ferroelectrics are further classified phenomenologically as proper, improper, and pseudo-proper ferroelectrics.

These three eminent classifications are connected with Landau theories' phase transition point with an effective thermodynamic potential in terms of a primary parameter [34–38], which governs symmetry breaking and is associated with new physical properties.

10.3.1 Proper Ferroelectrics

They are formed when the primary order parameter is directly connected with electric polarization P. Considering the consistent polarization in an infinite system results in the formation of an unbounded system formed in the near phase transition point, and the corresponding Landau free energy formation is shown in Equation 10.1

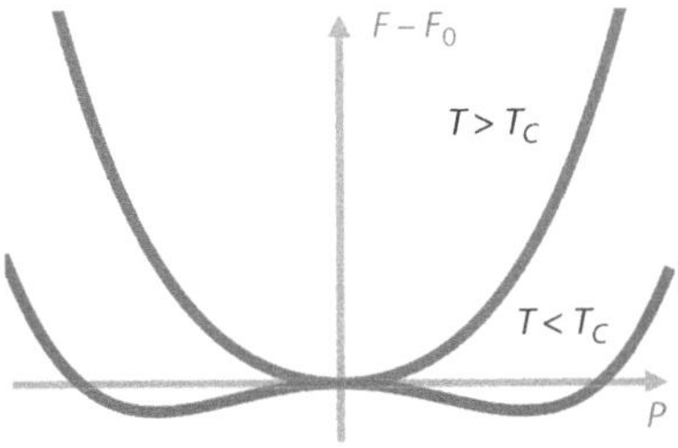

Figure 10.3 Expression of Landau free energy with respect to the order specification of P at T > Tc and T < Tc.

$$F(T,P) = F_0(T) + \frac{1}{2}a(T)P^2 + \frac{1}{4}b(T)P^4. \tag{10.1}$$

Here, F_0 is high—the symmetry state's free energy, and T is temperature—a control parameter.

The scalar quantity free energy is uniform during the space-inversion symmetry operation and its expansion never contains the eccentric power of P.

The second-order transition takes place at T = Tc in a continuous way and assumes a positive sign at co-efficient "a" and it stays positive at the coefficient "b". The existence of proper ferroelectrics based on Landau-free energy is shown in Figure 10.3.

Hence, it is sufficient to consider the point near Tc as the first-order terms, which appeared by expanding coefficients as a = a'(T-Tc) with (a'>0) and b(T) = b(Tc) = constant.

Figure 10.3 depicts the responsibility of free energy, with respect to the order parameter P. To relate P with second order transition at T = Tc, it is customary to assume the coefficient of sign changes from positive to negative in a continuous manner , so to dissipate at Tc, by keeping the coefficient of b as positive.

The existence of non-equilibrium in free energy state with respect to P can be correlated if it meets the following conditions, such as,

- If T > Tc (a > 0), the energy exhibits only one minimum that corresponds to P = 0. Whereas,
- If T < Tc (a < 0), the free energy exhibits two symmetric minima that correspond to P = 0.

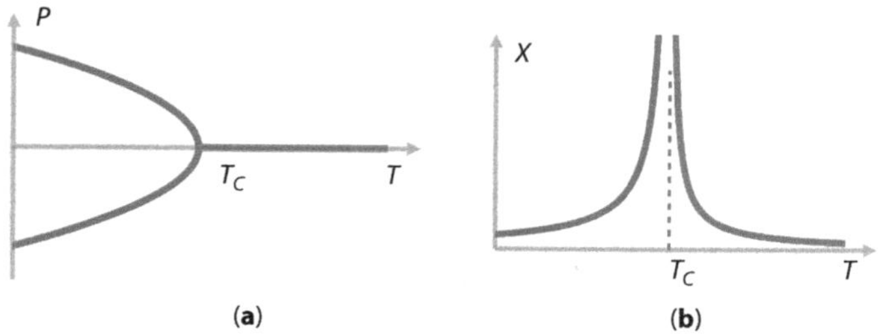

Figure 10.4 Influence of temperature on the parameters such as (a) P related to order and (b) electric susceptibility related to proper ferroelectrics.

The two parameters—(a) order P and (b) electric susceptibility—that belong to proper ferroelectrics and are obtained mainly depending on temperature are depicted in Figure 10.4.

10.3.2 Improper Ferroelectrics

According to Landau free energy [39], improper ferroelectrics means having a primary order parameter with different variables, such as Q, and the corresponding electric polarization is the by-product of it [40]; it happens in magnetically induced ferroelectric materials.

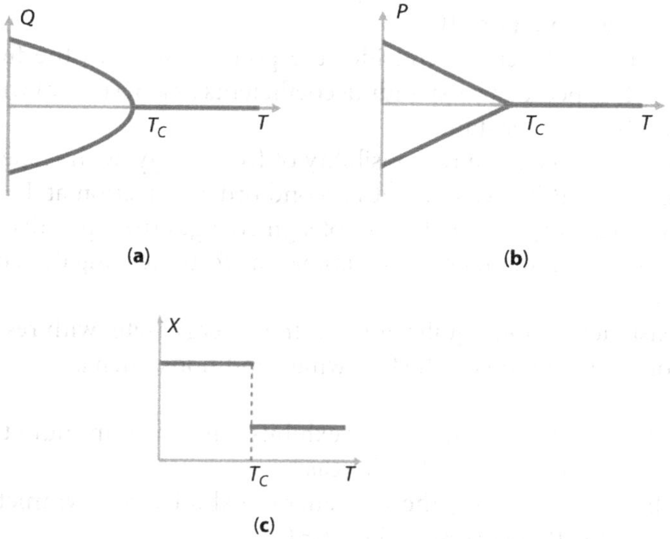

Figure 10.5 The order parameters directly related to temperature are (a) Q, (b) P, and s(c) temperature dependence of electric susceptibility—for improper ferroelectrics.

The equation that compares the basic and standard forms of ferroelectrics on the basis of Landau free energy is given in Equation 10.2, and the corresponding existence is depicted in Figure 10.5.

$$F(T,P) = F_0(T) + \frac{1}{2}aP^2 + \frac{1}{2}AQ^2 + \frac{1}{4}BQ^4 - \lambda PQ^2. \quad (10.2)$$

10.3.3 Pseudo-Proper Ferroelectrics

The third type of ferroelectrics is pseudo-proper ferroelectrics; it has intermediate properties of proper and improper ferroelectrics.

Based on symmetry consideration, the present case consists of the parameter with primary order P, which is observed as a "qualified" one. Q, the second primary order parameter, has the same symmetry property but differs in physical meaning compared to P. This usually happens in spin-spiral ferroelectrics systems. According to Landau's theory, in a spin-spiral ferroelectrics systems system, the magnetic free energy will be written as shown in Equation 10.3

$$F_m = \sum_i \frac{a_i}{2} M_i^2 + \frac{b}{4} M^4 + \frac{c}{2} \mathrm{M}(q^2 + \nabla^2)^2 \mathrm{M}. \quad (10.3)$$

From the above equations, the Landau coefficients of second-order, fourth-order, and gradient forms are described as a, b, and c, respectively. Here, M represents the distribution of magnetization.

In pseudo-proper ferroelectrics, the Landau free energy can be considered in the form as shown in Equation 10.4. The corresponding temperature dependence of the order parameters (a) Q and (b) P, and (c) the electric susceptibility for pseudo-proper ferroelectrics are depicted in Figure 10.6.

$$F(T,P) = F_0(T) + \frac{1}{2}aP^2 + \frac{1}{2}AQ^2 + \frac{1}{4}BQ^4 - \lambda PQ. \quad (10.4)$$

The occurrence and engineering of domains and domain walls in ferroic materials are directly related to their functional properties. Thus, properties like size effects, crystallographic structure, strain, and electrostatic conditions are directly correlated with this domain's nature.

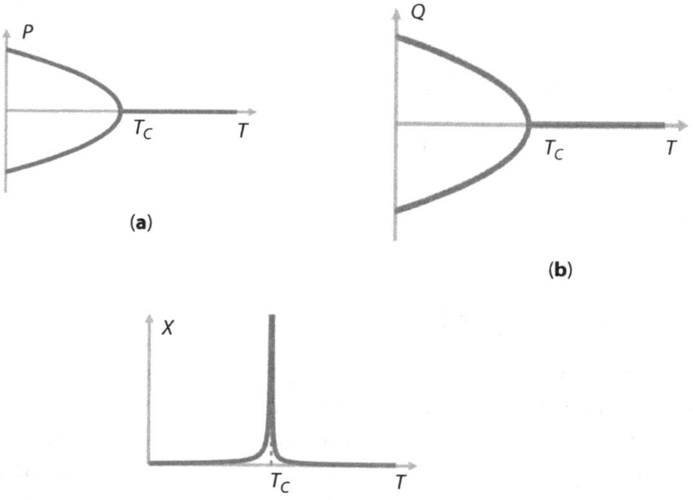

Figure 10.6 The order parameters directly related to temperature are (a) Q, (b) P, and (c) the electric susceptibility—with respect to temperature mentioned for improper ferroelectrics.

From the definition we came across, ferroics' properties such as elastic, electric, or magnetic order, which were formed spontaneously through their electric dipoles, were aligned in a way to provide polarization instinctively in a non-disruptive phase transition manner [41]. The order depends on microscopic interactions, leading to a more complex structure, which results in a structure beyond the primary type of ferroic order, and has a plethora of antiferroic phases. In general, most of the existing antiferromagnetic systems are magnetically arranged in an ordered manner. These ferroelectric systems also exist in nanoparticles, such as nanotubes, nanowires, and nanodots. There are a number of research available [42–52] to prepare ferroelectrics based on nanomaterials. Scientists fabricated novel materials [53, 54] using the above material.

Particularly, ferroelectric-based nanotubes and nanoshells are very interesting, due to their specific topology and are used for engineering functionalities, which are related to most of the technological applications [42, 43, 55]. However, the limiting factor of the above concepts is that the nanosystems in ferroelectrics have limited stability. The emergence of a nontrivial explanation of equations corresponds to linearized direct control over equations for polarization, which are related to electrostatic equations, at the boundary conditions [47]. Due to the finite-size effect, its corresponding transition temperature Tc drops drastically, and it has been studied both theoretically and experimentally [43, 47, 56–63]. Levanyuk *et al.* [36] studied the ferroelectric phase transition in both 2D and 3D nanostructures.

10.4 Important Term in Primary Ferroics

The term ferroic is defined as materials that relatively possess macroscopic properties with long-range order and also have the ability to develop a dipole having a conjugated field. The four types of primary ferroics are ferroelectric, ferromagnetic, ferroelastic, and ferrotoroidicity; their description and the corresponding figures are explained in the following section and in Figure 10.7, respectively.

10.4.1 Ferroelectric Materials

Ferroelectric materials are materials that have a stable spontaneous polarization P, and it will be transferred through an external applied electric field E.

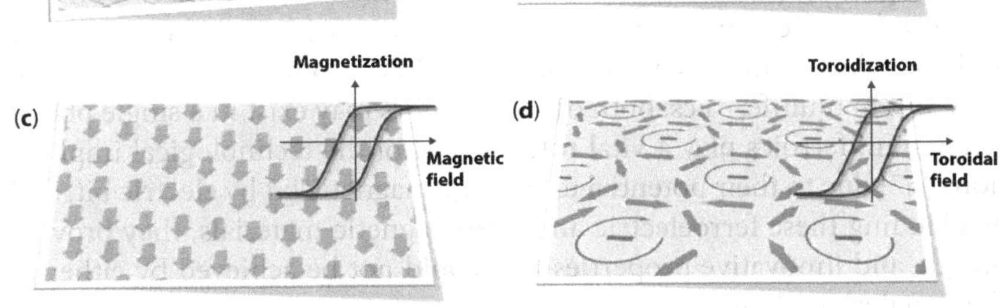

Figure 10.7 Exhibition of ferroic orders in four types. (a) Ferroelasticity, (b) ferroelectricity, (c) ferromagnetism, and (d) ferrotoroidicity. (a) Ferroelasticity describes impulsive strain σ, as a result of stress, (b) Spontaneous polarization, P, due to applied electric field, (c) spontaneous magnetization, M, due to magnetic field, and (d) ferrotoroidics exhibit a vortex-like alignment of spins with a toroidization due to toroidic field [65].

10.4.2 Ferromagnetic Materials

Ferromagnetic materials are materials that have stable spontaneous magnetization M, and they will be transferred through an externally applied magnetic field H.

10.4.3 Ferroelastic Materials

Ferroelastic materials are materials that undergo stable spontaneous deformation and transformation with applied stress.

10.4.4 Ferrotoroidic Materials

A ferrotoroidic state is a state having a domain that may undergo domain transformation in space–time in the antisymmetric external field, which is also known as toroidal field. Due to this toroidal field, they exhibit spins of vortex-like alignment with respect to the periodization that exists in the system [64].

10.5 Multiferroics

Multiferroics are nothing but materials that simultaneously possess a combination of both ferroelectric and ferromagnetic properties. Hans Schmid [66] was the first person, in 1994, to introduce the concept of multiferroics. These are the materials that possess all the ferroic properties like ferroelectric, ferromagnetic, and ferroelastic, and in which two or more ferroic properties are co-existing in an ordered manner and are responsible to control magnetism with the help of an electric field. Both ferromagnetic and ferroelectric in primary ferroics may differ from a technical point of view, but in multiferroics, both of these factors may exist in a single phase. These multiferroics play a vital role in most of the technological implications [67] due to their potential to manage magnetic bit by electric bit, and by blending these ferroelectric and ferromagnetic materials, they provide unique and innovative properties that could not be achieved by either of these materials alone.

An electric voltage itself is sufficient to read and write magnetic bit; hence, the energy loss is avoided; otherwise, high energy is needed. These multiferroics are very small, work fast, and provide energy-efficient data storage, which will be the most desired technology in the future. The characteristic requirement of a material to be multiferroics is that the ferroelectric material should be non-centrosymmetric and has the ability to pass spontaneous electric polarization, due to the presence of 31 point groups in it. Out of this, 13 point groups possess both of these properties simultaneously. Based on their formation mechanism and the order of combinations of ferroelectric and ferromagnetic properties, multiferroics are further divided into two kinds, Type 1 and Type 2, and are discussed in the following section.

10.5.1 Type 1 and Type 2 Multiferroics

In Type 1, the ordering of both magnetic and ferroelectric properties takes place independently, due to their corresponding source being different [2], as depicted in Figure 10.8. Here, the blending of magnetism and ferroelectricity is generally weak. The driving mechanisms in Type 1 multiferroics formations correspond to three different mechanisms: (i) charge ordering, (ii) lone pair, and (iii) geometric ferroelectricity. The Type 2 multiferroics are the combinations of ferroelectricity and magnetic transition and are depicted in Figure 10.2. Here, the coupling of the magnetism and ferroelectricity is generally larger than Type 1 and it follows a spin-driven mechanism. L. Liang *et al.* [69], in their paper "Characterization of Multiferroic Domain Structures in Multiferroic Oxides," state that, in multiferroics, because of the existence of 2 or 3 primary ferroic orders, it concurrently exists in the same plane. Further, it leads to intricate coupling

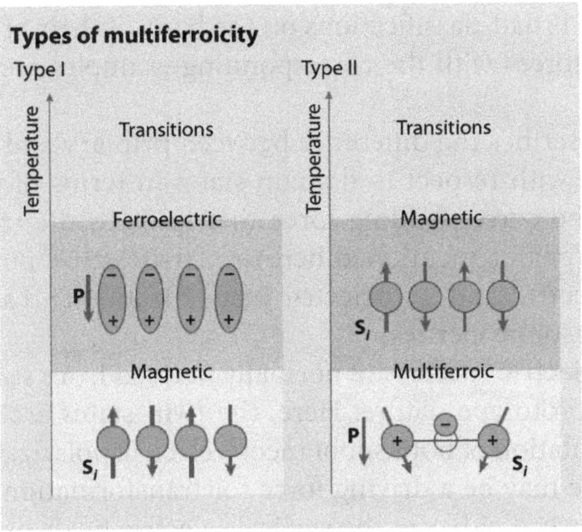

Figure 10.8 Formation of Type 1 and Type 2 multiferroics. Here, P is polarization, and Si represents spin at site [68].

taking place between them and new interesting multiferroic properties emerge; they are perceived by the amalgamation of a unique nanometer range with charge-oriented ordered domains.

10.6 Secondary Ferroics

In secondary ferroics, the suitably chosen external force defines the domain states [70, 71], whereas in the case of primary ferroics, the domain states are defined by spontaneous transformation. The domains in secondary ferroics are defined as induced strain, induced magnetization, or induced polarization. In other words, secondary ferroics [72] are described as the existence of all types of echoic twinned crystals, where the orientation of one or many twin components are affected suitably by the applications of chosen external driving forces.

Examples of secondary ferroics are ferrobielectric, ferromagnetic, ferrobielastic, ferroelastoelectric, ferromagnetoelastic, and ferromagnetoelectric. In the secondary ferroics, the differences in piezoelectric, piezomagnetic, and magnetoelectric coefficients occur between domains leading to ferroelastoelectricity, ferromagnetoelasticity, and ferromagnetoelectricity, respectively. The broad classification of ferroics falls on the nature of the tensor's properties [73], which usually differ amid states of orientation (twins) in the crystal. The difference in orientation states will regulate the type and driving force for orientations, which are essential to realizing reorientation among states. The primary and secondary ferroics order and their broad classifications on the basis of their orientation states and switching forces with the corresponding examples are given in Table 10.1.

The table describes the difference between primary and secondary ferroic states [74] with respect to domain states in terms of the orientation of states and the corresponding force that leads to the transformations. The primary domains mentioned here are ferroelectric polarization [48], ferroelectric electric, and ferroelectric magnetization and are described as lower-order tensor properties.

These ferroelectric species are normally derived from structures having a non-centric prototype nature. Here, the twin states are normally definite in the orientation of both spontaneous electric polarization and elastic strain and there may be a driving force for transformation that may have taken place between either of the states as electric fields or elastic stresses [75]. It is interesting that ferroic systems with higher-order tensor properties that are derived recently are secondary ferroics, which are obtained by

Table 10.1 Primary and secondary ferroics classifications.

S. no.	Ferroic class	Orientation states differ in	Switching force	Example
Primary Ferroics				
1.	Ferroelectric	Spontaneous polarization	Electric field	$BaTiO_3$
2.	Ferroelastic	Spontaneous strain	Mechanical stress	$Pb_3(PO_4)_2$
3.	Ferromagnetic	Spontaneous magnetization	Magnetic field	Fe_3O_4
Secondary Ferroics				
1.	Ferrobielectric	Dielectric susceptibility	Electric field	$SrTiO_3$
2.	Ferrobimagnetic	Magnetic susceptibility	Magnetic field	NiO
3.	Ferrobielastic	Elastic compliance	Mechanical stress	SiO_2
4.	Ferroelastoelectric	Piezoelectric coefficients	Electric field and mechanical stress	NH_4Cl
5.	Ferromagnetoelastic	Piezomagnetic coefficients	Magnetic field and mechanical stress	CoF_2
6.	Ferromagnetoelectric	Magnetoelectric coefficients	Magnetic field and electric field	Cr_2O_3

induced forces and are named ferrobielectric, ferrobimagnetic, ferrobielastic, ferroelastoelectric, ferromagnetoelastic, and ferromagnetoelectric, and are summarized in the second part of the table [76]. Poling is a phenomenon [77] that is more precise in secondary ferroics. In the case of polycrystalline poled secondaries, remanent magnetization and polarization are generally absent. In the present case, some class of secondaries undergo small changes, whereas in some cases, certain properties do not exist till they are poled.

For example, in the case of polycrystalline quartz, an example of ferroelastic, the presence of poling generates a few percent of elastic anisotropy. However, the circumstance is entirely different in other cases like ferroelectrics; polycrystalline ferroelastoelectrics are not piezoelectric unless poled.

Out of six types of secondaries, three secondaries such as ferroelastoelectrics, ferromagnetoelectrics, and ferromagnetoelastics depend on null properties, whereas the remaining three—ferroelastics, ferroelectrics, and ferromagnetics—are not.

10.6.1 Ferrobielectrics and Ferrobimagnetics—Secondary Ferroic Systems

The difference between ferroelectric and ferroelectricity is that in the former, electric polarization occurs spontaneously, known as primary electrics, whereas in the latter, electric polarization occurs through induced polarization known as secondary ferroics. In secondary ferroic systems, changes in orientation states occur due to the existence of contrast in the dielectric permittivity tensor. It usually occurs in crystals with a non-polar nature, with echoic twinning, and possessing substantial dielectric anisotropy [78].

Ferroelectric domains, an example of secondary ferroics, occur due to the existence of dielectric constant differences in the corresponding domains. Similarly, ferromagnetic domains occur due to differences in magnetic susceptibility.

10.6.1.1 Ferrobielectrics

Ferrobielectricity belongs to secondary ferroics and follows a phenomenon called field-induced electric polarization, which exhibits different orientation states, due to the extant contrast arrangements in the dielectric permittivity tensor. They generally occur in echoic twining and substantial dielectric anisotropy with the non-polar crystal structure.

In the case of primary ferroics, antiferroic materials exist, which is opposite to ferroelectric materials;, examples of such classes are $NaNbO_3$ [79] and $SrTiO_3$, which are considered as the most effective materials due to their large dielectric permittivities. However, the same materials, $SrTiO_3$, which also act as ferroelastic, exhibit a ferroelectric nature below 110°C. At this temperature, phase transition takes place with a change in symmetry structure from cubic (class m3m) to tetragonal (4/mmm).

The evidence for ferroelectric changes is very minimal. Sometimes, antiferroelectric materials may reveal dielectric permittivity in the form of substantial anisotropy, which is considered as the most important condition for the formation of ferroelectricity.

$NaNbO_3$, an example of orthorhombic material, manifests the respective atmospheric dielectric constant and is aligned to the b axis, around 600, which is usually greater than six times the corresponding existence with respect to perpendicular directions [79].

Influenced by the strong electric fields, sodium niobate domain wall motion will undergo antiferroelectric formation. The endurance of individual domains that have dielectric anisotropy possess the maximum dielectric constant contributing to the direction of the field, which is known as ferroelectric switching.

An example of this ferroelectricity is $NaNbO_3$ (10) and $SrTiO_3$. However, there is no promising evidence for the existence of ferroelectricity in both single-crystal and polycrystalline material formation.

In general, both ferroelectric and ferroelastic crystals have equal potential to contribute to the secondary ferroelectric systems. The existence of pure ferrobielectric is not possible, but it resides either as a primary or as a secondary ferroics.

The most important and distinct domain switching of polycrystalline ferrobielectrics is in accord either with the increase of the dielectric constant or with the corresponding electric field. The same circumstance happened in manganese oxide, which adopts ferromagnetic effects.

Besides ferroelectricity, all kinds of secondary ferroics possess more than one example. Some of the typical examples considered for secondary ferroics are ammonium chloride, chromium oxide, nickel oxide, iron carbonate, and quartz.

10.6.1.2 Ferrobimagnetism

Ferromagnetism is analogous to anisotropic magnetic susceptibility. Here, magnetic fields, which correspond to the various kinds of magnetization related to the domain states, are significantly formed through induced mechanism and are considered as a ferromagnet. The existence of relatively low χij mostly appears in antiferromagnetic crystals, and also the presence of isotropic effect in both diamagnetic and paramagnetic solids leads to ferromagnetism. For example, the metal oxide NiO has both ferroelastic and ferromagnetic properties. Above Neel temperature, at 523°K, NiO exists paramagnetically with a cubic rock salt structure, but below that temperature, Ni2+ spins align in a rhombohedral disordered nature with antiferromagnetic orientation.

In the present case, unit cell contraction takes place rather than any one of the <111> body diagonals corresponding to the angle between 90° and 90°4'. Obviously, the system's contraction occurs in any of the body diagonals and every domain present in optically uniaxial directions, and they are parallel to the existing contraction directions. Here, domain walls align themselves in the direction, which possesses the highest susceptibility coefficient and also coincides with respect to the applied field, thereby enhancing the magnetic field by an induced magnetization. There are a number of evidence available for ferromagnetic conversion and are found in several polycrystalline antiferromagnets. According to Bizette, Squire, and Tsai [78] who measured magnetic susceptibility on manganese oxide, they found evidence for antiferromagnetism, and such material exhibits cusp-shaped temperature dependence, which is analyzed through molecular field theory. From the measurements, they found possible domain motion due to changes in ferromagnetic systems.

The experimental findings of polycrystalline MnO curves and their typical single-crystal susceptibilities were measured and are depicted in Figure 10.9. To estimate the susceptibility of the particular systems, a small magnetic field (6000 oersteds) required in the ratio of χ (O)//χ (T N) of 0.69. These magnetic fields are almost similar to the prescribed values as 2///3. The susceptibility values mostly depend on the applied field, rather than the Neel temperature, which corresponds to the field strength of 24,000 oersteds; χ (O)χ(TN) rises to 0.8.

The magnetic susceptibility in an arbitrary direction for a uniaxial antiferromagnet [80] is represented in Equation 10.5.

$$\chi = \chi_{11} \sin^2 \theta + \chi_{33} \cos^2 \theta \tag{10.5}$$

Here, the axis of spin is represented as χ3. The appearance of the angle in the middle of χ3 and to the applied magnetic field is mentioned as 0. The mean was estimated to calculate the susceptibility of a polycrystalline sample. Similarly, the magnetic susceptibility for an unpoled specimen in the randomly oriented grains that exist independent of direction is given by (2χ11 + χ33)/3. Compared to spontaneous primary ferroelectric materials like ferroelectric or ferromagnetic, moments occur generally

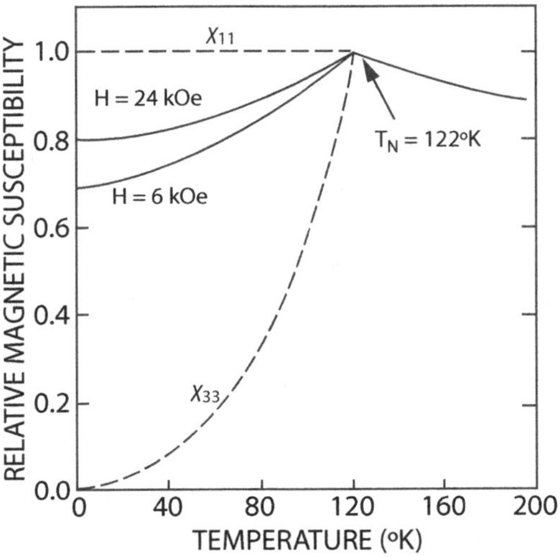

Figure 10.9 Polycrystalline MnO's magnetic susceptibility was plotted against temperature. The rise in susceptibility corresponds to the fields obtained due to the occurrence of domain switching that happened in ferrobimagnetic systems. In the above diagram, the dashed line represents the existence of single crystal susceptibility coefficients.

greater than the induced secondary ferroic systems like ferroelectric and ferromagnetic. Out of the existing 94 fully ferroelastic species, only 42 are ferroelectric.

Here, the commencement of ferroelectric character begins with the help of an electric field compared to ferroelectric, an example of secondary ferroelectric system. The search for ferroelectric materials endures 52 ferroelastic species. To eradicate ferromagnetic behavior in ferromagnetism, the existing 31 gyromagnetic classes (81) have to be eliminated. In ferroelastic materials, the hexagonal-to-monoclinic crystal system transformation involves the conversion of prototype symmetry to its corresponding ferroic derivative. For example, $SrTiO_3$, a ferroelastic system, at low temperatures has three orientation states with the space group of m3mF4/mmm. Here, upon cooling the existing prototype group with the m3m space group, it undergoes changes to tetragonal 4/mmm through symmetry changes from hexagonal to monoclinic. Compared to primary ferroics, the induced secondary ferroics arises due to the twins in them conflicting within their elastic constants or some other physical properties. In both ferroelectric and ferrobimagnetic, switching between domains occurs due to the difference in electric and magnetic fields, respectively.

10.6.1.3 Ferroelastoelectricity

Ferroelastoelectric results in simultaneous applications of electric fields and mechanical stresses over domain walls. One of the most interesting types of secondary ferroics is ferroelastoelectricity. Here, piezoelectrics are made by multiply-twinned material through poling in an electric field in the presence of mechanical stress, and they are applied simultaneously. Mohler and Pitka [82] are the first to discover the presence of ferroelasto-electricity in ammonium chloride (NH_4C1), as shown in Figure 10.10, and quartz (SiO_2) is considered as the best example of ferroelastoelectricity. Ammonium chloride has a ferroic group of m3mF43m, and when cooled at −30°C, (111) plates of NH_4C1 undergo order–disorder transformation [82] under the impact of the combined effect of mechanical stress and electric field, but the same has not been observed in quartz. The poling experiments of ferroelastoelectric for polycrystalline materials have not been illustrated yet by any of the scientists. The point group 43m of ammonium chloride has one unconstrained piezoelectric coefficient, d123, which relates to electric polarization through the cubic (100) direction to a shearing stress of about (100).

10.6.1.4 Ferrobielasticity

In ferroelastics, secondary ferroics have the domains contradicted in stiffness coefficients, which leads to strain difference due to the influence of enforced mechanical stress. The produced stress acts upon the strain

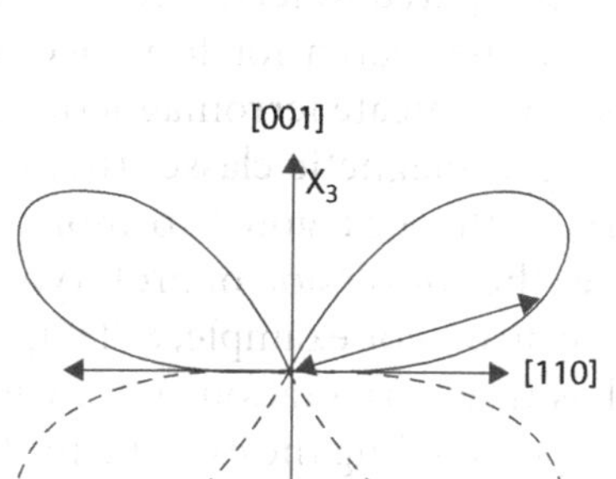

Figure 10.10 The existence of low -temperature longitudinal piezoelectric surface.

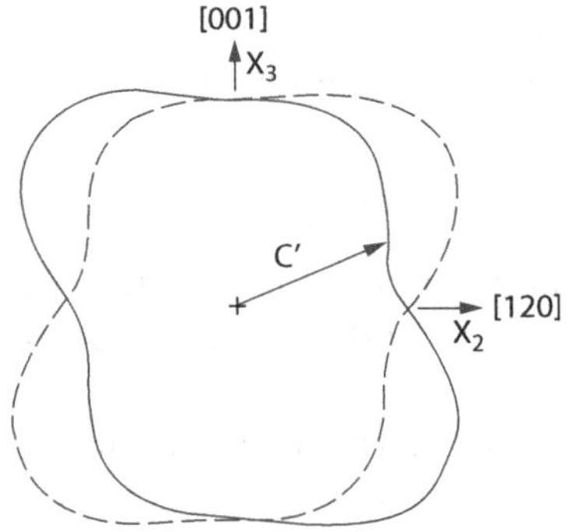

Figure 10.11 The existence of direction of two twin states of quartz surface with respect to longitudinal stiffness.

and creates an energy difference and gives rise to domain instability and wall motion. α-Quartz, SiO_2, is the best example of ferroelastic material. Compared to natural quartz, the most common [83] twinning present in α-quartz is Dauphin. Here, a class of deceptive twinning occurs, where more than one inclination state exists in 180° spinning through the optic axis, and a symmetry operation undoes the sign along the polar a-axis. The quartz's surface with longitudinal elastic stiffness is depicted in Figure 10.11. Here, the tensile stresses are applied in the directions, where they promote large variations among the surfaces and are more active in promoting twin wall motions.

The mechanical stress present in quartz will be controlled by the Dauphine twins, present in them, due to the elastic constants present in twins being different. In the Dauphine twins, the stiffness coefficient present in them is c1123, which has a contradictory sign for the corresponding two twin states, whereas all the other five independent coefficients present in common and Dauphine twins are c1111, c1122, c1133, c2323, and c3333. Due to the difference in c1123 between common and Dauphine twins, certain stress (18) combinations are utilized to derive twin walls. The two twin states of longitudinal stiffness of quartz are depicted in Figure 10.11. The single crystal of quartz possesses controlled Dauphin~ twins, due to the influence of laser-induced thermal stresses [77] and static stresses [84, 85]. The conformations of Dauphin~ twins and twin-wall motion are restricted in polycrystalline quartz formation. Elastic constant measurements reveal that polycrystalline quartz delivers some reasonable evidence for the meager presence of twins in polycrystalline quartz materials.

10.6.1.5 Ferromagnetoelectricity

Ferromagnetoelastic and ferromagnetoelectric domain wall formation requires both magnetic fields and mechanical stress and electric and magnetic fields, respectively.

Chromium oxide, an antiferromagnetic material with a point group $\underline{\bar{3}m}$ below the Neel temperature of 34°C, is considered the best example of a ferromagnetoelectric material. Cr203 has domains in 180° type with a similar structure for spin inversion. According to Foner and Hanabusa [86], magnetoelectric effects were absent in Cr203 powders due to their improperly poled nature. Due to the random arrangement of powder material, and if they present in large size, they behave independently, but below Neel's point, the crystalline structure proceeds freely to ordering their arrangements and results in an equal number of time-inversion-related domains. Hence, the magnetoelectric effect was not present in powder specimens. Shtrikman and Treves [87] proposed the method of preparation of ferromagnetic electricity material, which operates below the Neel point through the time-inversion process. Here, ferromagnetoelectric switching helps to separate the energy degeneracy that exists between the two spin-reversed magnetic structures, which promotes the extension of one of the domains at the expense of the other. Under the electric field, the induced magnetization influences one structure, and the converse one assumes a different time-reversed structure. Out of these structures, the difference in free energy occurs in the magnetic field and has an energy twofold that of reversed domains compared to the more probable one.

The principle behind the magnetoelectric effect formation is directly related to the combined applications of both electric and magnetic fields at temperatures below T N to a powder sample feasible to their formations. In reality, in lab practice, the energy barriers, which separate the time-reversed states and utilize the available small energies to proceed with the process, are usually slow. Shtrikman and Treves [87], who discovered the methodology to improve the equilibrium rate, conducted the experiments in the applied fields, which showed simultaneous slow cooling at the Neel point. The induced electric and magnetic fields are responsible for any magnetoelectric properties of the powder sample, in which the corresponding powder tensor possesses a symmetry structure and it either has identical or greater symmetry than the fields they produce. The final conclusion arrived here is that the symmetry formed due to the induced magnetoelectric field is not obviously connected to the symmetry of the single crystal tensor.

10.6.1.6 Ferromagnetoelasticity

CoF_2 [88] and MnF_2 [89] are the best examples of ferromagnetoelastic materials. In a typical experiment, the samples are placed in a mechanical vise and the obtained changes were recorded. Out of 90 magnetic points, 66 points exhibit a null effect. However, these materials have very low transition temperatures. There were no poling experiments conducted for the corresponding polycrystalline specimens, but they expect similar results for magnetoelectric Cr203 materials. In both cases, the existence of domain states related to spin inversion difference leads to easier domain wall movement.

10.7 Applications of Ferroic Materials

The most common features existing in both primary and secondary ferroics, along with multiferroics, due to their formation of domains and domain walls, bring an unbelievable variety of applications of these ferroic materials to the world. Ravez [90] coined all such above categories into common terms as ferroic materials or ferroics for all such materials. In addition to the domain structures of ferroics, a second important attribute of the present system is its large contribution to certain functions of these materials, which can be beneficial to ferroic transformation. Another major feature of this application is that most of the macroscopic properties depend on temperature variations that exist either along or in the vicinity of ferroic phase transformation during domain formations. Further, there exists a strong interaction between these properties with other properties, leading to a variety of applications including deriving device formation and its applications as ferroics [28] both in physics and chemistry, as exhibited in Figure 10.12. The major applications of these ferroics as electronic devices are as transducers, sensors, amplifiers, and phase shifters, as depicted in Figure 10.12 in tree form.

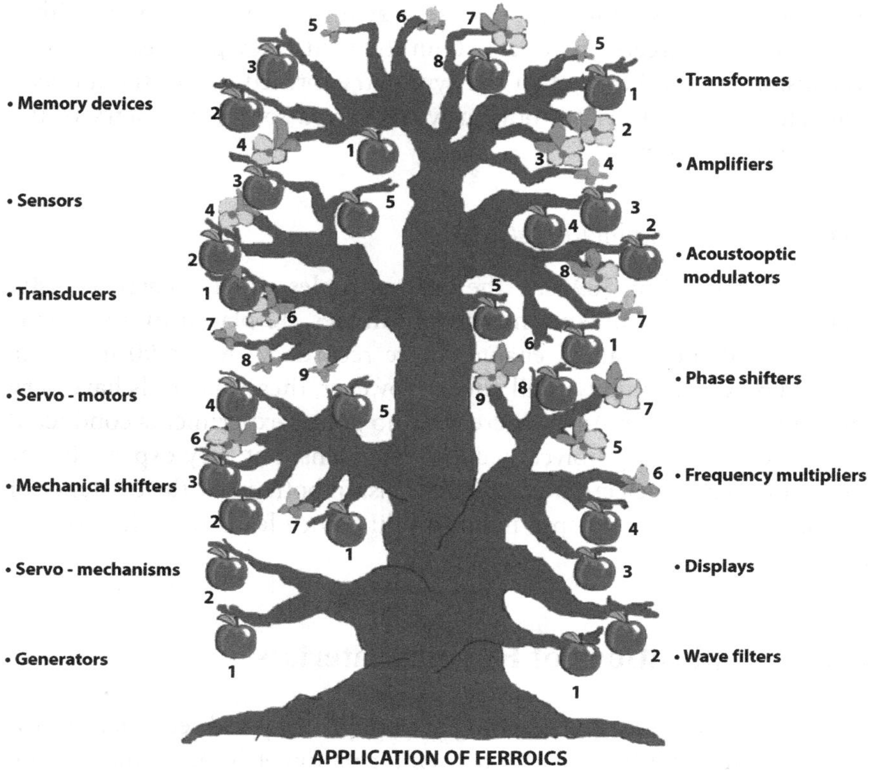

Figure 10.12 Applications of ferroics in various fields adopted from Ref. [91].

10.8 Conclusions

Ferroic materials are found in a variety of physical and chemical systems, and they are considered an important material, due to their applications in society both in the scientific and industrial world. Considering the current trends in this field, ferroics paved the way toward a generation of ferroic-based electronics. Hence, in this chapter, ferroic materials' definitions, formation mechanisms, classifications as primary and secondary ferroics, functions, characterizations, and their applications are discussed elaborately. Further, the inner classifications of each of these ferroics are also discussed in detail. The ferroics and functional properties are directly related to the generation of domains and domain walls, which are further controlled by the material's crystal structure, size effects, strain, and electrostatic conditions. Multiferroics are another class of materials that have properties in between those of primary and secondary ferroics, and their differences were discussed in this chapter in detail. In addition, the comparison between primary and secondary ferroics is also featured here. Finally, applications of these ferroic materials in various fields are discussed. The utility of ferroic materials in the nano fields is also pointed out, and their pros and cons are also elaborated. At the outset, ferroic materials' research leads to sustainable growth, though it is an age-old concept, and their revived investigation brings new avenues, which are tied with technological advancements. In the future, researchers will have new outlook in this field, and they will provide new output and will have to revise their thoughts and scientific strategies to generate more unique functional ferroic materials for a more sustainable future.

References

1. Balzer, C., *Kommunale Ordnungsdienste*, KSV Verwaltungspraxis, Wiesbaden, 2019.
2. Dicloralurea 116-52-9, in: *Sax's Dangerous Properties of Industrial Materials*, John Wiley & Sons, Inc., Hoboken, New Jersey, USA, 2012.
3. Meier, D., Valanoor, N., Zhang, Q., Lee, D., Domains and domain walls in ferroic materials. *J. Appl. Phys.*, 129, 219901-1&219901-2, 2021.
4. Doi, E., Ohtsuru, C., Matoba, T., Carboxypeptidase in commercial bromelain powder: Partial purification and some properties of pineapple carboxypeptidase. *J. Biochem.*, 75, 1063, 1974.
5. Mühlbauer, S., Binz, B., Jonietz, F., Pfleiderer, C., Rosch, A., Neubauer, A., Georgii, R., Böni, P., Skyrmion lattice in a chiral magnet. *Sci. (80-.)*, 323, 915, 2009.
6. Other books received. *Int. Aff.*, 85, 915, 2009.
7. Fert, A., Reyren, N., Cros, V., Magnetic skyrmions: Advances in physics and potential applications. *Nat. Rev. Mater.*, 2, 1–15, 2017.
8. Claridge, J.B. *et al.*, ChemInform abstract: A polar corundum oxide displaying weak ferromagnetism at room temperature. *ChemInform*, 43, 2012. *J. Am. Chem. Soc.*, 134, 8, 3737–3747, 2012.
9. Khan, A.I., Chatterjee, K., Wang, B., Drapcho, S., You, L., Serrao, C., Bakaul, S.R., Ramesh, R., Salahuddin, S., Negative capacitance in a ferroelectric capacitor. *Nat. Mater.*, 14, 182, 2015.
10. Íñiguez, J., Zubko, P., Luk'yanchuk, I., Cano, A., Ferroelectric negative capacitance. *Nat. Rev. Mater.*, 4, 243, 2019.
11. Rodriguez Contreras, J., Kohlstedt, H., Poppe, U., Waser, R., Buchal, C., Pertsev, N.A., Resistive switching in metal-ferroelectric-metal junctions. *Appl. Phys. Lett.*, 83, 4595, 2003.

12. Chanthbouala, A., Garcia, V., Cherifi, R.O., Bouzehouane, K., Fusil, S., Moya, X., Xavier, S., Yamada, H., Deranlot, C., Mathur, N.D., Bibes, M., Barthélémy, A., Grollier, J., A ferroelectric memristor. *Nat. Mater.*, 11, 860, 2012.
13. Han, J.-C., Dutta, S., Ekkad, S., *Gas turbine heat transfer and cooling technology*, 2nd ed., CRC Press, 2013. CRC Press Taylor & Francis Group 6000 Broken Sound Parkway NW, Suite 300 Boca Raton, FL 33487-2742.
14. Martin, L.W., Chu, Y.H., Ramesh, R., Advances in the growth and characterization of magnetic, ferroelectric, and multiferroic oxide thin films. *Mater. Sci. Eng. R Rep.*, 68, 89, 2010.
15. Martin, L.W. and Rappe, A.M., Thin-film ferroelectric materials and their applications. *Nat. Rev. Mater.*, 2, 1–15, 2016.
16. Back, C., Cros, V., Ebert, H., Everschor-Sitte, K., Fert, A., Garst, M., Ma, T., Mankovsky, S., Monchesky, T.L., Mostovoy, M., Nagaosa, N., Parkin, S.S.P., Pfleiderer, C., Reyren, N., Rosch, A., Taguchi, Y., Tokura, Y., Von Bergmann, K., Zang, J., The 2020 skyrmionics roadmap. *J. Phys. D. Appl. Phys.*, 53, 363001, 2020.
17. Demkowicz-Dobrzański, R., Górecki, W., Guţă, M., Multi-parameter estimation beyond quantum Fisher information. *J. Phys. A Math. Theor.*, 53, 363001, 2020.
18. Hayashi, S., Takahashi, S., Yamamoto, M., Magneto-plastic effect in nickel single crystals. *J. Phys. Soc. Japan*, 30, 381, 1971.
19. National Academy of Engineering, *Engineer and the City: A Symposium Sponsored by the National Academy of Engineering at Its Fifth Autumn Meeting*, October 22-23, 1969, The National Academies Press, Washington, DC, 1969.
20. Contents of Vol. 37, 1969. *J. Electrochem. Soc. Jpn.*, 37, 228, 1969.
21. Spaldin, N.A. and Fiebig, M., The renaissance of magnetoelectric multiferroics. *Sci. (80-.)*, 309, 391, 2005.
22. Ramesh, R. and Spaldin, N.A., Multiferroics: Progress and prospects in thin films. *Nat. Mater.*, 6, 21, 2007.
23. Wadhawan, V., *Introduction to ferroic materials*, CRC Press, London, 2000.
24. Van Aken, B.B., Rivera, J.P., Schmid, H., Fiebig, M., Observation of ferrotoroidic domains. *Nature*, 449, 702, 2007.
25. Fiebig, M., Lottermoser, T., Meier, D., Trassin, M., The evolution of multiferroics. *Nat. Rev. Mater.*, 1, 1–14, 2016.
26. Kimura, T., Goto, T., Shintani, H., Ishizaka, K., Arima, T., Tokura, Y., Magnetic control of ferroelectric polarization. *Nature*, 426, 55, 2003.
27. Tagantsev, A.K., Cross, L.E., Fousek, J., A preview of concepts and phenomena, in: *Domains in Ferroic Crystals and Thin Films*, A.K. Tagantsev, L.E. Cross, J. Fousek, (Eds.), pp. 1–10, Springer New York, New York, NY, 2010.
28. Meier, D., Seidel, J., Gregg, M., Ramesh, R., *Domain walls: From fundamental properties to nanotechnology concepts*, Oxford University Press, Walton Street, Oxford, opposite Somerville College, in the inner suburb of Jericho, 2020.
29. Irkhin, V.Y. and Skryabin, Y.N., Topological phase transitions in strongly correlated systems: Application to Co3Sn2S2. Письма В Журнал Экспериментальной И Теоретической Физики, 114, 625, 2021.

30. Schmid, H., On ferrotoroidics and electrotoroidic, magnetotoroidic and piezotoroidic effects. *Ferroelectrics*, 252, 41, 2001.
31. Windsch, W., Freeman, A.J., Schmid, H., Magnetoelectric interaction phenomena in crystals. Gordon and Breach Science Publishers London 1975. £ 10.90. *Krist. Tech.*, 11, K45, 1976.
32. Schmid, H., Some symmetry aspects of ferroics and single phase multiferroics. *J. Phys. Condens. Matter*, 20, 434201, 2008.
33. Aizu, K., Possible species of ferromagnetic, ferroelectric, and ferroelastic crystals. *Phys. Rev. B*, 2, 754, 1970.
34. Nataf, G.F., Barrett, N., Kreisel, J., Guennou, M., Raman signatures of ferroic domain walls captured by principal component analysis. *J. Phys. Condens. Matter*, 30, 35902, 2018.
35. Sapriel, J., Domain-wall orientations in ferroelastics. *Phys. Rev. B*, 12, 5128, 1975.
36. Janovec, V. and Přívratská, J., Domain structures, in: *International Tables for Crystallography*, pp. 449–505, 2006.
37. Lee, D., Behera, R.K., Wu, P., Xu, H., Li, Y.L., Sinnott, S.B., Phillpot, S.R., Chen, L.Q., Gopalan, V., Publisher's note: Mixed Bloch-Néel-Ising character of 180° ferroelectric domain walls [Phys. Rev. B 80, 0601029 (1–4) (2009)]. *Phys. Rev. B*, 80, 2009.
38. Rabe, K.M., Dawber, M., Lichtensteiger, C., Ahn, C.H., Triscone, J.M., Modern physics of ferroelectrics: Essential background, in: *Topics in Applied Physics*, pp. 1–30, Springer Berlin Heidelberg, Berlin, Heidelberg, 2007.
39. Rothman, M.A., General physics. Mechanics and molecular physics. L. D. Landau, A. I. Akhiezer, and E. M. Lifshitz. Translated from the Russian edition (Moscow, 1965) by J. B. Sykes, A. D. Petford, and C. L. Petford. Pergamon, New York, 1967. x + 372 pp., illus. $8. *Sci. (80-.)*, 160, 667, 1968.
40. Ginzburg, V.L. and Levanyuk, A.P., Light scattering near second-order phase-transition and Curie points. *J. Phys. Chem. Solids*, 6, 51, 1958.
41. Ginzburg, V.L., Phase transitions in ferroelectrics: Some historical and other remarks, in: *Ferroelectricity: The Fundamentals Collection*, pp. 131–136, 2007.
42. Bratkovsky, A.M. and Levanyuk, A.P., Continuous theory of ferroelectric states in ultrathin films with real electrodes. *J. Comput. Theor. Nanosci.*, 6, 465, 2009.
43. Strukov, B.A. and Levanyuk, A.P., General characteristics of structural phase transitions in crystals, in: *Ferroelectric Phenomena in Crystals*, B.A. Strukov, and A.P. Levanyuk, (Eds.), pp. 1–29, Springer Berlin Heidelberg, Berlin, Heidelberg, 1998.
44. Valasek, J., Piezo-electric and allied phenomena in Rochelle salt. *Phys. Rev.*, 17, 475, 1921.
45. Bernal, A., Tselev, A., Kalinin, S., Bassiri-Gharb, N., Ferroelectric nanotubes: Free-standing ferroelectric nanotubes processed via soft-template infiltration (Adv. Mater. 9/2012). *Adv. Mater.*, 24, 1159, 2012.

46. Gruverman, A. and Kholkin, A., Nanoscale ferroelectrics: Processing, characterization and future trends. *Rep. Prog. Phys.*, 69, 2443, 2006.
47. Scott, J.F., Applications of modern ferroelectrics. *Sci. (80-.)*, 315, 954, 2007.
48. Morozovska, A.N., Glinchuk, M.D., Eliseev, E.A., Phase transitions induced by confinement of ferroic nanoparticles. *Phys. Rev. B - Condens. Matter Mater. Phys.*, 76, 1–33, 2007.
49. Morozovska, A.N., Glinchuk, M.D., Eliseev, E.A., Ferroelectricity enhancement in ferroelectric nanotubes. *Phase Transitions*, 80, 71, 2007.
50. Levanyuk, A.P. and Blinc, R., Ferroelectric phase transitions in small particles and local regions. *Phys. Rev. Lett.*, 111, 1–5, 2013.
51. Anoufa, M., Kiat, J.M., Kornev, I., Bogicevic, C., Vortices of polarization in BaTiO3 core-shell nanoceramics: Calculations based on *ab initio* derived Hamiltonian versus Landau theory. *Phys. Rev. B - Condens. Matter Mater. Phys.*, 88, 144106–(1–12), 2013.
52. Herchig, R., Schultz, K., McCash, K., Ponomareva, I., Terahertz sensing using ferroelectric nanowires. *Nanotechnology*, 24, 45501, 2013.
53. Lee, B., Nakhmanson, S.M., Heinonen, O., Strain induced vortex-to-uniform polarization transitions in soft-ferroelectric nanoparticles. *Appl. Phys. Lett.*, 104, 262906, 2014.
54. Baudry, L., Sené, A., Luk'Yanchuk, I.A., Lahoche, L., Scott, J.F., Polarization vortex domains induced by switching electric field in ferroelectric films with circular electrodes. *Phys. Rev. B - Condens. Matter Mater. Phys.*, 90, 024102-(1-7), 2014.
55. Gui, Z., Wang, L.W., Bellaiche, L., Electronic properties of electrical vortices in ferroelectric nanocomposites from large-scale ab initio computations. *Nano Lett.*, 15, 3224, 2015.
56. Philippot, G., Elissalde, C., Maglione, M., Aymonier, C., Supercritical fluid technology: A reliable process for high quality BaTiO3 based nanomaterials. *Adv. Powder Technol.*, 25, 1415, 2014.
57. Liu, M. and Wang, J., Giant electrocaloric effect in ferroelectric nanotubes near room temperature. *Sci. Rep.*, 5, 1–7, 2015.
58. Bernal, A., Tselev, A., Kalinin, S., Bassiri-Gharb, N., Free-standing ferroelectric nanotubes processed via soft-template infiltration. *Adv. Mater.*, 24, 1160, 2012.
59. Fu, H. and Bellaiche, L., Ferroelectricity in barium titanate quantum dots and wires. *Phys. Rev. Lett.*, 91, 257601–(1–4), 2003.
60. Naumov, I.I., Bellaiche, L., Fu, H., Unusual phase transitions in ferroelectric nanodisks and nanorods. *Nature*, 432, 737, 2004.
61. Geneste, G., Bousquet, E., Junquera, J., Ghosez, P., Finite-size effects in BaTiO3 nanowires. *Appl. Phys. Lett.*, 88, 1–4, 2006.
62. Miyata, N., Ota, H., Ichikawa, M., Correlation between scanning-probe-induced spots and fixed positive charges in thin HfO2 films. *Appl. Phys. Lett.*, 86, 1, 2005.

63. Durgun, E., Ghosez, P., Shaltaf, R., Gonze, X., Raty, J.Y., Polarization vortices in germanium telluride nanoplatelets: A theoretical study. *Phys. Rev. Lett.*, 103, 247601-(1-4), 2009.
64. Tagantsev, A.K., Gerra, G., Setter, N., Short-range and long-range contributions to the size effect in metal-ferroelectric-metal heterostructures. *Phys. Rev. B - Condens. Matter Mater. Phys.*, 77, 174111-(1-9), 2008.
65. Hong, J., Catalan, G., Fang, D.N., Artacho, E., Scott, J.F., Topology of the polarization field in ferroelectric nanowires from first principles. *Phys. Rev. B - Condens. Matter Mater. Phys.*, 81, 172101-(1-4), 2010.
66. Lottermoser, T. and Meier, D., A short history of multiferroics. *Phys. Sci. Rev.*, 6, 1-11, 2021.
67. Hoffman, J., Pan, X., Reiner, J.W., Walker, F.J., Han, J.P., Ahn, C.H., Ma, T.P., Ferroelectric field effect transistors for memory applications. *Adv. Mater.*, 22, 2957, 2010.
68. Schmid, H., Multi-ferroic magnetoelectrics. *Ferroelectrics*, 162, 317, 1994.
69. Nataf, G.F., Guennou, M., Gregg, J.M., Meier, D., Hlinka, J., Salje, E.K.H., Kreisel, J., Domain-wall engineering and topological defects in ferroelectric and ferroelastic materials. *Nat. Rev. Phys.*, 2, 634, 2020.
70. Liang, L., Wu, H., Li, L., Zhu, X., Characterization of multiferroic domain structures in multiferroic oxides. *J. Nanomater.*, 2015, 1, 2015.
71. Aizu, K., Electrical, mechanical and electromechanical orders of state shifts in nonmagnetic ferroic crystals. *J. Phys. Soc. Jpn.*, 32, 1287, 1972.
72. Newnham, R.E. and Cross, L.E., Secondary ferroics and domain-divided piezoelectrics. *Ferroelectrics*, 10, 269, 1976.
73. Aizu, K., Possible species of "ferroelastic" crystals and of simultaneously ferroelectric and ferroelastic crystals. *J. Phys. Soc. Jpn.*, 27, 387, 1969.
74. Aizu, K., Kumada, A., Yumoto, H., Ashida, S., Simultaneous ferroelectricity and ferroelasticity of Gd2(MoO4)3. *J. Phys. Soc. Jpn.*, 27, 511, 1969.
75. Cross, L.E., Ferroic materials and composites: Past, present and future, in: *Advanced Ceramics III*, S. Sōmiya, (Ed.), pp. 71–102, Springer Netherlands, Dordrecht, 1990.
76. Lumpkin, G.R. and Ewing, R.C., Geochemical alteration of pyrochlore group minerals: Betafite subgroup. *Am. Min.*, 81, 1237, 1996.
77. Newnham, R.E. and Skinner, D.P., Polycrystalline secondary ferroics. *Mater. Res. Bull.*, 11, 1273, 1976.
78. Bidaux, R., Conte, R., Nasser, J.A., More on spin orientation in MnO at 0 K and zero applied field - theoretical predictions. *J. Physiol. Paris*, 41, 1317, 1980.
79. Kao, K.C., Ferroelectrics, piezoelectrics, and pyroelectrics, in: *Dielectric Phenomena in Solids*, pp. 213–282, 2004.
80. Shull, C.G., Strauser, W.A., Wollan, E.O., Neutron diffraction by paramagnetic and antiferromagnetic substances. *Phys. Rev.*, 83, 333, 1951.
81. Małuszyńska, H., Tylczyński, Z., Cousson, A., Ferroelastoelectric ordering in (NH4)2CuBr 4·2H2O single crystal. *CrystEngComm*, 15, 7498, 2013.

82. Buerger, M.J., Mineralogy: Dana's system of mineralogy . By Charles Palache, Harry Berman and Clifford Frondel. Seventh edition, volume I, elements, sulfides, sulfosalts, oxides. New York: John Wiley and Sons, Inc., 1944. $10.00. *Sci. (80-.)*, 101, 650, 1945.
83. Aizu, K., Second-order ferroic state shifts. *J. Phys. Soc. Japan*, 34, 121, 1973.
84. Wooster, W.A., Wooster, N., Rycroft, J.L., Thomas, L.A., The control and elimination of electrical (Dauphiné) twinning in quartz. *J. Inst. Electr. Eng. - Part IIIA Radiocommun.*, 94, 927, 1947.
85. Foner, S. and Hanabusa, M., Magnetoelectric effects in Cr2O3 and (Cr 2O3)0.8·(Al2O3)0.2. *J. Appl. Phys.*, 34, 1246, 1963.
86. Schuele, W.J., Shtrikman, S., Treves, D., Observation of superparamagnetism by the Mössbauer effect. *J. Appl. Phys.*, 36, 1010, 1965.
87. Anzus Pacific security pact. *Int. Organ.*, 17, 986, 1963.
88. Astrov, D.N., Ermakov, N.B., Borovik-Romanov, A.S., Kolevatov, E.G., Nizhankovskiĭ, V.I., External quadrupole magnetic field of antiferromagnetic Cr2O3. *JETP Lett.*, 63, 745, 1996.
89. YEN, W.M., Experimental studies of energy transfer in rare earth ions in crystals, in: *Spectroscopy of Solids Containing Rare Earth Ions*, A.A. Kaplyanskii, R.M.B.T.-M.P. Macfarlane, C.M.S. (Eds.), pp. 185–249, Elsevier, Radarweg 29, 1043 NX Amsterdam, The Netherlands, 1987.
90. Ravez, J., Ferroelectricity in solid state chemistry. *C. R. Acad. Sci. - Ser. IIc Chem.*, 3, 267, 2000.
91. Bochenek, D., Surowiak, Z., Poltierova-Vejpravova, J., Producing the lead-free BaFe0.5Nb0.5O3 ceramics with multiferroic properties. *J. Alloys Compd.*, 487, 572, 2009.

11

Robust Domain Boundary Engineering of Ferroic and Multiferroic Materials

Uzma Hira*, Ayeza Arshad and Abdul Sattar

School of Physical Sciences (SPS), University of the Punjab, New Campus, Lahore, Pakistan

Abstract

Multiferroic materials are a group of materials that simultaneously show many ferroic orders, like ferromagnetism and ferroelectricity, in the same material. The interaction between various ferroic properties is controlled by the domain structure and domain boundary properties in multiferroic materials. This chapter discusses the importance of domain boundary engineering in multiferroic materials and how it affects the properties of these materials. We first provide a definition of multiferroic materials and explain how their properties are related to the interaction between various ferroic orders. We then describe how domain boundary engineering can be used to manipulate the domain structure and affect the interaction between various ferroic orders. We will also highlight the advantages and challenges of domain boundary engineering, including the effects of defects and strain on domain boundary properties in multiferroic materials. Overall, the chapter emphasizes the importance of domain boundary engineering for developing new functional materials with unique properties. By controlling the domain structure and domain boundary properties in multiferroic materials, it is possible to create materials having personalized properties for use in many types of applications such as data storage, sensors, and energy conversion.

Keywords: Ferroic materials, domain structures, ferroelectricity, multiferroicity, ferromagnetism and domain boundary engineering

**Corresponding author*: uzma.sps@pu.edu.pk

Inamuddin, Tariq Altalhi and Mohammad Abu Jafar Mazumder (eds.) Ferroic Materials-Based Technologies, (257–278) © 2024 Scrivener Publishing LLC

11.1 Introduction

The presence of different order parameters, such as elasticity, electricity, or magnetism, during a reversible phase transition is the main characteristic of ferroic materials [1]. The swapping of the spontaneous polarization, the strain, or the magnetization by the suitable external factors, such as electric field, mechanical stress, or magnetic field, respectively, is related to these ferroic properties, such as ferroelectricity, ferroelasticity, and ferromagnetism [2]. These materials show a regular arrangement of their spontaneous moments in nearby unit cells, referred to as "ferro". These ferroic properties are very useful for usage in microelectronic devices [3].

Ferroic phase transitions are related to domain structures. These transitions result in domain structures, when boundary circumstances are suitable, which frequently predominate in the material's macroscopic susceptibility. In these transitions, when domain structures alter due to external stimuli, the ferroic materials exhibit very significant levels of switching in their magnetic moment, polarizations, and spontaneous strain [4]. These order parameters have at least two polarities that are symmetrically equal [5]. By using an external field, these polarities can be changed. Regions having the same polarity inside these materials are known as domains while domain walls are the interfaces present in between these regions. Inside a domain, spontaneous polarizations have the same direction, but two adjacent domains have different directions [6]. Domain boundaries are expected to be softer than the bulk because they have less chemical bonding with many structural components [7].

Multiferroic materials show multiple primary ferroic orders (ferroelectric, ferromagnet, and ferroelastic) [5]. These multiferroic materials draw more consideration among the scientists because they allow for either "electric switching of magnetic domains" or "magnetic switching of ferroelectric domains", which will open new opportunities in the manufacture of data storage systems [7]. The advantages of the localized wall properties are the transfer of information with very high resolution and high information density. Due to extremely high storage density of information, the devices related to wall properties can perform better than bulk devices significantly [7].

In multiferroics, for the coupling association between ferroic orders, the magnetic as well as dielectric properties are correlated. This correlation between magnetic and electric effect is known as electromagnetic effect [5].

In multiferroicity, at least two of the three ferroic properties are combined. Its investigation started in the 1960s with significant work on boracites and has continued with research on ceramics that have perovskite-like structures. Moreover, it was observed that "martensites" and "ferroelastics" describe the similar properties of material but have a different history for their nomenclature (like "ferroelastic" alloys are known as "martensites" while materials having "martensitic" properties, like ceramics and minerals, are called as "ferroelastics"). Ferroelastics show stepwise phase transitions frequently, but not always. For example, magnetic martensites are multiferroics and operate in the same way as many ferroelastic cum ferromagnetic ceramics [7].

Ferroic properties are associated with the surfaces and interfaces rather than the bulk of the material. During the last decade, this idea has become the basis for the greatest advancement in designs of ferroic devices. When only domain boundaries have required functionalities, electrical wiring on a much finer scale can be achieved and high memory capacitors can be manufactured, more than these properties achievable with bulk technologies. This is also true in the case of electrical conductivity as in the case of ferroelectricity and ferromagnetism. It is already found that domain boundaries carry high currents, but in engineering of domain boundaries, design of the domain boundary is primarily dependent on changing suitable domains, and the recognition of twin boundaries with superconducting properties created a variety of possibilities. Moreover, it was observed that ferroelectric switching is restricted to domain walls; i.e., it was observed that electric dipole moments were retained in domain walls, which were found incompatible with additional domain switching and depolarization fields in the bulk. Therefore, the active device length scale is reduced to the domain wall's size. The main aim is to concentrate high memories along wall lengths for many applications [3].

In recent years, simultaneous usage of magnetic and electric properties is of great interest, i.e., in magnetoelectric materials. These materials have attained so much interest not only because of their appealing physics but also because of their magnetoelectric capacities and significant part in the production of multiferroic devices. Multiferroic substances could be used in gas sensors, resistance change memories, data storage systems, and spintronic devices due to their magnetoelectric properties. Bismuth ferrite ($BiFeO_3$), the most familiar single-phase multiferroic substance, shows multiferroic properties at ambient temperatures, and this material has been under broad investigations [8].

11.2 Ferroic Materials

A unique group of solid-state compounds, having mostly two of the order parameters like electric, piezo-elastic, or magnetic phases, are called multiferroic materials. As a result, piezo-ferroelectrics and ferromagnets having great piezo-magnetic properties are a special group of electro-elastic/magneto-elastic multiferroics. These materials are critical for innovative applications in both circumstances, because the connection between the piezo-elastic properties and the ferro-phases makes it possible to immediately control ferroelectric and ferromagnetic properties by means of applied stress. Moreover, inverse effects also exist, in which mechanical strain caused by external electric and magnetic fields can alter the material's shape. This intricate relationship between electric polarization and strain in electric field, magnetization and strain in the magnetic field, and strain and stress in the electric and magnetic polarization is depicted graphically in Figure 11.1 [9].

11.3 Multiferroic Domain Boundaries

Domain boundaries have less chemical bonding with many structural components (particularly twin boundaries). One would expect that domain boundaries are softer than the bulk when it comes to elastic, magnetic, or electric susceptibilities; however, because the proportional volume of interfaces to the bulk is so small, such softening is hard to measure macroscopically. Relaxor materials, which have relaxor regions, exhibit effects of strong finite size and soft susceptibilities, and are an exception in particular.

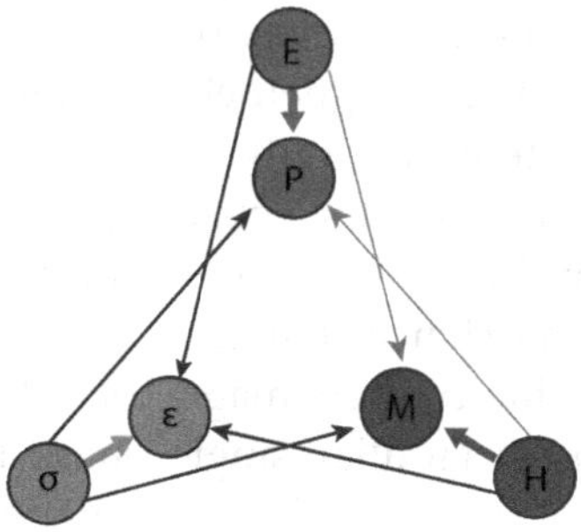

Figure 11.1 Possible cross-couplings in multiferroics. **E** = electric field; **σ** = applied mechanical stress; **P** = electric polarization; **ϵ** = strain; **M** = magnetization; **H** = magnetic field.

Internal ferroic properties of twin boundaries can outperform bulk macroscopic properties because volume of these boundaries can extend some parts per million of volume of bulk. The high information density is a benefit of the localized wall properties: They are retained within very thin twin wall sheets and may be handled spatially with very high resolution. This indicates that information inscribed in twin walls has more storage density, making it possible for wall-related devices to perform better than bulk devices.

An additional aspect to consider is the presence of twin wall structural gradients that span several interatomic distances. As the transition point approaches, there is an increase in the thickness of the walls. In accordance with Landau-Ginzburg theory's predictions for a phase transition of second-order mechanism, the wall thickness in $LaAlO_3$ increases when diffusely diffracted signals related to the walls are observed. This has been demonstrated through careful investigations [10].

Identification of ferroic features on surfaces plays a crucial role in computer simulations for material designing and theoretically studies of great physical properties of solids. This research now encompasses not only inorganic materials but also biological samples in the field of life sciences. To illustrate, let us consider an example from solid-state physics involving the material $CaTiO_3$, which is categorically nonpolar. Theoretical predictions have indicated the presence of significant spontaneous polarization in the 100-g twin walls of $CaTiO_3$. The local polarization in and around these walls exhibits a diverse texture, as described by theoretical models. Notable local distortions have nonzero contributions and antiferroelectric distortions perpendicular to the plane of the walls. However, these contributions do not affect the overall macroscopic dipole moment. Specifically, individual displacements of Ti by 2 pm away from the center of the octahedron result in a net polarization equal to about 0.6 pm displacement along the twin angle bisector. This phenomenon appears to be inherently linked to the presence of twin boundaries within the matrix, which have previously been associated with the positioning of oxygen vacancies in $CaTiO_3$.

Although there have been previous indirect reports of the twin wall's polarity, as well as in grain boundaries and anti-phase boundaries (APBs), the findings in $CaTiO_3$ provide valuable insights as they represent the first conclusive evidence of polar properties of twin walls and the structural mechanism for correlation between dipole moments and strain. $CaTiO_3$ is a purely ferroelastic material and exhibits an orthorhombic structure in its low-temperature state. Ferroelectric properties are not observed in this material. However, the TiO_6 octahedron is known to have the potential to form polar structures when the position of Ti is off-centered with regard to the geometric center of oxygen atoms surrounding it. This type of structure is present among various substances like $BaTiO_3$, $PbTiO_3$, etc. In both orthorhombic and tetragonal phases of $CaTiO_3$, there is octahedral rotation. However, this known competition suppresses the off-centering effect. Nevertheless, this effect is restored within the twin walls when material density decreases or the rotation angle becomes zero. Considering that both of these conditions have met within the twin walls, it is unsurprising to say that these walls exhibit dipole moments. The specific arrangement of the polarization field and the magnitude of the original polarization within the walls were previously unknown.

For investigation of polar arrangement in the $CaTiO_3$ ferroelastic walls, several experiments have been conducted depending on atomic-scale interpretation of these walls, considering interatomic interactions defined empirically. Three-dimensional periodic boundary conditions were utilized, while open boundary conditions were applied perpendicular to the walls to account for the presence of surfaces. Two twin walls were included for satisfaction of the periodic boundary conditions. When a prototypical cubic structure unit cell was employed, a supercell was constructed, consisting of 10 unit cells along the plane where twin angle (y) bisects, 6 unit cells along the plane perpendicular to twin angle (z), and 26 unit cells directing perpendicular to walls (x). This configuration resulted in a total of 7,800 atoms.

In addition to the development of wall polar characteristics, an increase was also expected in the presence of O_2 vacancies. When a vacancy of oxygen is transferred from the bulk to the wall, it acquires an energy of approximately 1.1 eV. While this effect is anticipated, as geologically twin walls are said to be decorated with defects, such calculations reveal that the geometrical importance of these defects may appear insignificant, resulting in only a modest increase of approximately 1.0% in the lattice spacing of $CaTiO_3$. These minor changes are distinct for interfaces and twin boundaries, explaining why dopant materials tend to concentrate in these regions. Consequently, the attributes of the walls, such as their polarity or conductivity, can be systematically altered by introducing localized dopants. Doping with magnetic ions, for example, can improve the magnetic properties of the walls without necessarily affecting bulk material. The expansion of unit cells at the interfaces can result in reduction in the elastic deformation. It indicates that the wall's compressibility is higher than the bulk of the material. However, such an effect has been observed, and it appears to be less significant than initially predicted based solely on the density reduction inside the wall. Actually, the decrease in distance between the next closest neighbors partially offsets the expected decrease in the relevant elastic modulus caused by the increment in distance between nearest neighbors. As a result, the structural relaxation greatly compensates for the elastic softening caused by expansion of the interface [9].

11.4 Highly Conducting Interfaces

11.4.1 Interfacial Magnetism

For an interface, the space-inversion symmetry inherently breaks up even if our basic material is not polar. It will be highly promising to access the magnetoelectric interaction if any interfacial magnetism is taking part in breaking the time-reversal symmetry since both of these would be violated at the same time. Carrier mediation may be used to realize this interfacial magnetism in a variety of straightforward ways. Here, four of these ways are discussed.

11.4.1.1 Spin-Dependent Oxidization, Screening, and Bonding

The most straightforward framework of the model in this class is the ferromagnetic metal's surface, breaking both space-inversion and time-reversal symmetries. The magnetoelectric effect causes noticeable changes in the magnetocrystalline anisotropy and magnetization of the surface [10]. This type of magnetoelectric effect could be significantly increased at the interfaces in-between a high-κ dielectric material and a ferromagnetic film, for instance, $SrRuO_3/SrTiO_3$ [11].

The carrier-mediated magnetoelectric effect can also be upgraded in ferromagnetic hetero-structures. A measure of surface charges is identical to ferroelectric polarization close to an interface. From the value of polarization, the charge density can be calculated; for instance, 10 μC/cm^2 corresponds to ~0.1 electron/unit cell in the event of the lattice constant of pseudo-cubic phase values ~4 Å [12]. This consequence can be considered as the ferroelectric effect that was anticipated to show up in the $SrRuO_3/BaTiO_3/SrRuO_3$ hetero-structure [13]. Subsequently, the magnetization of $SrRuO_3$ at interfaces becomes significantly different because of the ferroelectric behavior in $BaTiO_3$. The attained anomaly of magnetic moment brought about through polarization reversal per Ru is 0.31 μ_B per interface. In the interim, Cai *et al.* suggested a technique to understand the magnetoelectric impact of ambient temperature in a tri-component ferroelectric/ferromagnet/typical-metal superlattice [14]. Spin-dependent screening gathered at the ferroelectric/ferromagnet interface results in an extra magnetization that is not cancelled due to depletion at the ferroelectric/

typical-metal interface because the inversion symmetry in-between normal metal/ferroelectric and ferromagnetic/ferroelectric interfaces is broken. A huge magnetization over the entire superlattices can be imparted. In view of the following explanation, Lee *et al.* anticipated a vigorous magnetoelectric interaction in Pt/$BaTiO_3$/Fe and Pt/$PbTiO_3$/Fe superlattices [15].

11.4.1.2 Electrical Tuning of Magneto-Crystalline Anisotropy

A solid piezo strain effect can adjust the magneto-crystalline aeolotropy. Actually, there is another way for electrically tuning the magneto-crystalline aeolotropies, that is to say, by means of altering interfacial carrier density. This mechanism is a selective way to deal with the electrical regulation of magnetism because this anisotropy decides the preferred ordering of magnetization. A control of magneto-crystalline anisotropic behavior by the application of electric field has been examined recently and acknowledged in various hetero-structures and thin films with metallic–ferromagnetic interfaces [16–20] as well as at dilute magnetic semiconductor interfaces [21]. On the other hand, the interfacial magneto-crystalline anisotropic behavior can be more effectively tuned by a linked ferroelectric film through its ferroelectric polarization, as shown by a bunch of *ab initio* method and trials [20, 22]. For instance, Lukashev *et al.* concentrated on multilayers of $BaTiO_3$/Fe_4 and tracked down a significant shift of around 30% in the perpendicular interfacial magnetic anisotropy energy in response to the reverse process of ferroelectric polarization of $BaTiO_3$ [23]. Furthermore, this ferroelectric polarization is able to tune the manganite's magnetic anisotropy, as described by the magneto-resistance anisotropy of ferroelectric/manganite hetero-structures [24].

The basic physics behind this can be perceived by investigation of spin–orbital-resolved density of different phases and charge density. The reallocation of electrons in 3*d* orbital at interfaces brought about by reverse polarization is liable for change in magneto-crystalline anisotropy. As expressed earlier, the magneto-crystalline anisotropic behavior is a consequence of spin–orbit coupling among empty conducting and filled valence bands. The change in orbital occupancy upsets the spin–orbit coupling, and consequently, the magneto-crystalline anisotropy is also disturbed [24]. Moreover, ferroelectrics of organic compounds with ferroelectrics of traditional transition metal oxide are embraced to change the magneto-crystalline anisotropic behavior in applicable organic magnetic/ferroelectric hetero-structures. In these hetero-structures, the significant varieties of magneto-crystalline anisotropic energy were noticed that were determined by inversion of ferroelectric polarization. A few of them even came up to half. These perceptions provide more insight into potential outcomes to use the magnetoelectric coupling system at natural ferroelectric/ferromagnetic interfaces for adaptable and inexpensive gadgets.

11.5 Ferroelectric Management of Magnetic Phase

In addition to the already mentioned persistent adjustment of magnetic properties, more sensational magnetoelectric effect was found in hetero-structures with a ferroelectric part and a magnetic corresponding electronic part like manganite $La_{1-x}A_xMnO_3$ (A= Sr, Ca, or Ba). The electronic material that is related shows phase graphs relying on the density, which is doped carrier, consisting of plethoric phases that are competing with various resistivities, magnetic orders, and structures [25]. Specifically, if the value of doped carrier density is near the phase boundary, understanding of switching connections will become possible, which exist among the concerned two phases (like antiferromagnetic versus ferromagnetic) by applying electric field from an external source. In this manner, an enormous magnetoelectric effect can be expected. Albeit charge screening affects the driving thrust, the fundamental mechanism does not depend solely on the simple adding of polarized spin carriers, but depend more on the doping alteration of magnetic ground state. Without a doubt, sensational magnetoelectric coupling was anticipated in $La_{0.5}A_{0.5}MnO_3/BaTiO_3$ hetero-structures by the first-principles calculation [26]. The direction of polarization in $BaTiO_3$ was utilized to modify the density of hole carrier in $La_{0.5}$-$A_{0.}$5-MnO_3 electrostatically and subsequently the magnetism of $La_{0.5}A_{0.5}MnO_3$ between the antiferromagnetic and ferromagnetic state at interfaces. Assuming that polarization points are at a large distance from the interface of material, an obvious uphill shifting of local density of the state occurs, inferring reduction of e^- population at interfaces, which correspond to the hole aggregation state. However, an opposite effect happens if polarization points are toward the interfaces, which leads to a downward shifting of local density, and there occurs a decrease in hole state. Subsequently, interfacial layers favor either ferromagnetic leads to hole depletion or antiferromagnetic state leads to hole accumulation, but all depends on the orientation of polarization.

11.6 Ferroelectric–Magnetic Tunneling Junctions

The magnetoelectric effect is conspicuous in $La_{0.8}Sr_{0.2}MnO_3/PbZr_{0.2}Ti_{0.8}O_3$ hetero-structures driven from the interfacial magnetic transitions, which are charge-driven. In reality, an obvious variety of in-plane conductivity is also possible in magnetic hetero-structures due to magnetic layers. Two polarization states empower a great variety of resistivity in a $La_{0.8}Sr_{0.2}MnO_3$ layer by a mechanism that is called interfacial mediation, and it was previously discussed. At the end of the day, the remarkable variety in behavior of transportation is brought about by the ferroelectric layer, as a nonferroelectric covering layer ($LaAlO_3$ or $SrTiO_3$) has no difference at all in behavior of transportation of the neighboring $La_{0.8}Sr_{0.2}MnO_3$ layer from that of an exposed $La_{0.8}Sr_{0.2}MnO_3$ layer. Moreover, with observation of conductance modulation, which is in plane, a tunneling electro-resistance effect is also found in ferroelectric junctions. To put it plainly, there are three elements that decide the function of interface transmission and, thus, tunneling electro-resistance, which are as follows: [27]

i. Electrostatic potential that occurs across the junction
ii. Bonding strength of interface
iii. Strain that is connected to the piezo-electric effect

Inadequate screening of polarization charge at the interfaces produces an electrostatic effect that, in turn, produces an electrostatic potential that superimposes on contact potential that is present in tunnel junctions, developing an unequal potential profile across two ends of electrodes. The ferroelectric shifting at boundaries between ferroelectric layers and two electrodes is what causes the interfacial effects [28]. Orbital hybridization at the interfaces and transmission across it are caused by atomic displacements [29]. All ferroelectrics have a piezo-electric action, which causes the strain effect. The transportation properties of the barrier, such as attenuation constant and barrier thickness, are twisted by a piezo-electric distortion that acts along the junction axis.

11.7 Highly Conducting Interfaces in Case of WO_3

Twin walls have a tendency to draw flaws, which makes it possible to selectively dope twin boundaries, or inject flaws into the boundaries, but it cannot be done in case of bulk. In the 18th century, with the addition of sodium and O vacancy in twin walls, that possibility was initially employed to alter a material's conductivity containing WO_3 [27, 30, 31]. The twin walls' chemical makeup was slightly changed (for example, from WO_3 to $WO_{2.95}$), and it resulted in a shift from a metal to an insulator at low temperatures, which, in turn, induced superconductivity. Since dopants adhere to the twin wall's pathways, it is possible to change the nano patterning of superconducting materials by first patterning the twin boundaries and then doping.

This study is particularly well suited to tungsten oxide, which is WO_3, and its derivative, particularly WO_{3-x}; this is due to the fact that they exhibit a shift from metal to insulator while still showing thermodynamic stability. WO_3 can easily release O and incorporate hydrogen and alkali ions. WO_3 is well recognized for its use in solar cells, electrochromic devices, and catalysis. It also exhibits exceptional superconducting capabilities. Chemically adding Na to the walls or removal of oxygen from walls marginally reduces the superconducting twin wall in WO_3, which has a superconducting transition temperature close to 3 K; chemically altered walls are then superconducting. This configuration of superconducting twin borders with the production of domain junction and needle domain may hold the key to developing Josephson junction arrays and very sensitive magnetic scanners since the surrounding matrix maintains its own insulation [30].

11.8 Case of LNO ($LaNiO_3$) and LCMO ($La_{2/3}Ca_{1/3}MnO_3$)

In complicated oxide superlattices, interface engineering is a growing subject that allows for the management of the exceptional features of these substances as well as the admission of new phase and emerging physical trends. Now, we are going to demonstrate how a bulk paramagnetic material can develop a complicated charge and spin structure as a result of interfacial interactions. We researched on a superlattice that can be said to be SLs composed of two types of materials, one is paramagnetic while the other is ferromagnetic. $LaNiO_3$ is a paramagnetic material, which can be said to be LNO, and $La_{2/3}Ca_{1/3}MnO_3$ is a ferromagnetic material, which is highly spin polarized and can be called LCMO. This superlattice was grown on a substrate called $SrTiO_3$ (001). The magnetism in LNO can be demonstrated by using an exchange bias operation that occurs at the interfaces and can be observed in magnetic X-ray reflectivity at resonance.

Unsymmetrically induced magnetization patterns were found at the interfaces in LCMO and LNO; we correlate them to a recurring complicated charge and spin super-structure. The high-resolution STEM scans uncover that there are no obvious structural differences between lower and upper interfaces. A wide range of magnetic symmetries arising in the LNO layer exhibit immense possibilities of interfacial remodeling, which will serve as a tool for customized electrical characteristics [32].

Now, we are going to describe the perception of a spontaneously occurring magnetic depth profile using XRMR. Various electronic structures of interfaces in LNO and LCMO superlattice (see Figure 11.2) have been seen as a result of growth order. Magnetic and nonmagnetic resonance scattering characteristics are explained qualitatively by using spin and charge profile using a uniform simple charge model. The side of sharper interface valence for Mn is enhanced by a broad range of about half valence e^-, but for the nickel value of valence that reduces with respect to the same number of e^-, it happens to compensate for the charge at both interfaces. It brings a ferromagnetic order in interfaces of LNO and LCMO accompanying an increased moment in Mn. For LCMO and LNO, however, the valence remains constant at their bulk levels at the larger, more powerful intermixed interface. Due to the absence of coupling double exchange in this situation, the transfer of charge negates the Mn magnetic polarization [33].

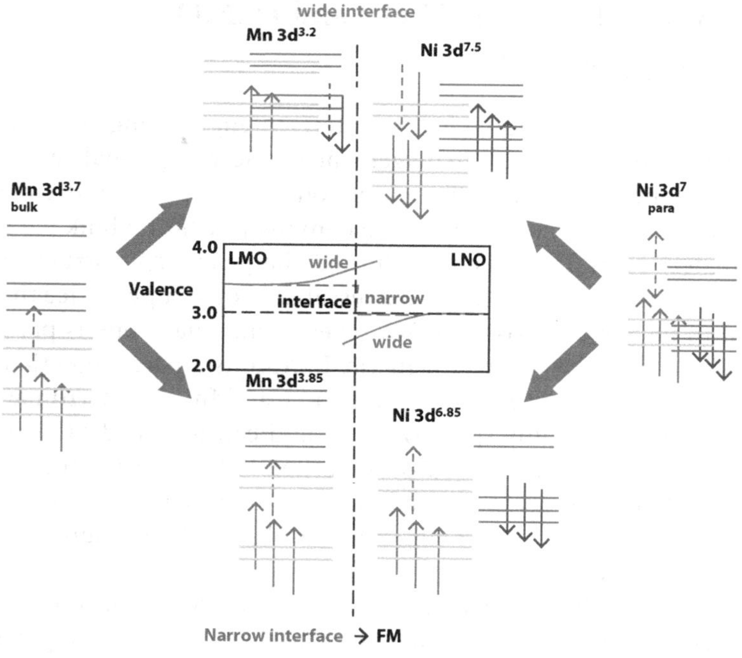

Figure 11.2 The streamlined electronic structures of bulk Mn in LCMO and Ni in LNO are seen on the left and right, respectively. The electrical structure of both in contact at the wide and narrow interfaces is shown in the center section.

11.9 Applications

The principles of magnetic interface, structures, transportation of electrons, and spin over the ferromagnet or non-ferromagnet barrier [34–37], as well as prospective applications of hybrid oxides of FM/non-FM structures present in spintronic devices, are of particular interest to investigate.

There has only been paramagnetism discovered for samples of powder LNO. In its bulk form, it is the only member of family perovskite nickelates that lack magnetic order [38]. It has been claimed that antiferromagnetic ground states exist in LNO single crystals. However, these outcomes are debatable and disputed due to stoichiometry of oxygen [38]. The mechanics of spin transportation, for instance, in spintronic devices, is improved by magnetic remodeling at interfaces among FM and non-FM oxides. For such interfaces, non-ferromagnetically arranged layers may show some net FM moments. Magnetic moments that are present in barrier substances can also affect magnetic switching. As a result, a complicated nano magnetic state is generated at the interface, while a spin transport response is already present, which is exotic. It has been suggested that the rebuilt chemical bonds cause the formation of an induced-magnetic state, which will be present at a barrier that is diamagnetic. Until now, only ¼ ¼ ¼ antiferromagnetic order has been observed in the case of LNO and RXO superlattices, which contain only 2 u.c. of LNO [34].

11.10 Dynamics of Domain Movements and Ferroic Switching

11.10.1 Dynamics of Domain Movements

If the interface is considered an active component of a substance, then the question is that either they are impacted by outside forces or a change in position will determine its applications. A single mobile interface will alter the size of its adjacent domains and change the conducting properties of the domain wall. Conducting properties of the domain wall play an important role in neuromorphic computation [39]. Experimentally, these dynamic behaviors are studied by the statistical analysis of avalanches like "jerk", signals such as measurement from acoustic emission (AE) spectra while stress drops. Traditionally, power law-distributed jerks in space and time [40, 41] are an indication of wide-scale constant avalanches.

This chapter focuses on ferroelastic domain switching that is induced by the fluctuation of ferroelastic twin boundaries [42]; because of their potential importance and rich physics, many researchers focus on their deformation mechanism [43–45].

11.10.1.1 Analysis of Avalanche Formation

The cause of this phenomenon is commonly seen in shape-memory alloy because motion of the austenite and martensitic interface results in acoustic emission, which has similarity with an avalanche behavior called Barkhausen. In classic instances of jerky spreading of one interface, several additional interfaces are introduced, and eventually an avalanche that consists of propagating phase front can be observed [46]. Such an avalanche can be considered close to a self-organized pressure point and is theoretically predicted to follow power law distributions. Although the clarity of this notion makes it attractive, it is impossible to imagine that the martensite's numerous pinning centers' unpredictability will not go all the way down to values where pinning is only possible at very low temperature [47]. Actually, it appears that the majority of investigations show that athermal dynamics govern depinning, pinning, and acoustic emission, or are not thermally generated. Recent calorimetric examination of transitions in shape memory alloys consisting of Al (15:65), Zn (16:71), and Cu (67:64), which limited the thermal driving force, was an important experiment [48]. Each avalanche was seen as a distinct peak.

In the latent heat of scanning, the transition rate was a few mK/h. The resulting DTA curve shows jerk and a continuous background comprising two parts. The jerks have no effect on the transition entropy, which remains constant during cooling or heating. For the individual jerks, no memory impact was noted, with the exception of the strongest avalanches. The statistical analysis of jerks shows that a power law of energy of jerk is being followed, and it is the same as that of AE spectra [49].

The short range of the applied forces determines whether the jerks in elastic measurements can be directly observed. In a meticulous dynamical mechanical analyzer (DMA) experiment, the study of transitions in Cu (74:08), Al (23:13), and Be (2:79) was done. The three-point-bending excitation was selected to have a frequency of 0.1 Hz, lower applied forces (maximum amplitude of 50 mN), and a cooling or heating rate of 0.140 K/h. Mechanical loss exhibits spiky avalanche behavior and $\sim \tan\theta$.

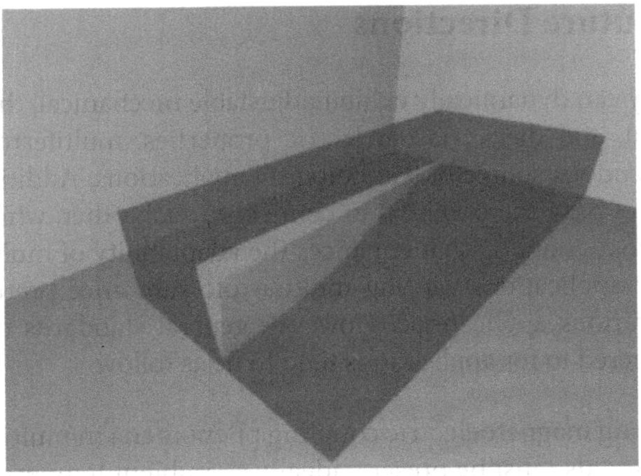

Figure 11.3 Inside the ferro-elastic matrix; the shape of the needle domain.

In Figure 11.3, the pathway of needle domains in $LaAlO_3$ is observed. Advancement or reversal of needle domains is pointed by the defects that are mainly positioned at pointing edges of hull-shaped domains. The regional fixing of lines in three dimensions, where d=; 3D=1, is thus described as pinning. Larkin measurement of edges in a flexible system is believed to be huge, and a quantity of robust pinning centers is needed to pin the pointing needle; hence, it is uncommon for pinning to take place in this configuration.

The newly made contributions by Proville [50] are apposite in this context. He demonstrated how one may still assume avalanches with a definite depinning force in a position where the system is bigger than Larkin length. This result puts into question the ongoing status. Inter-atomic springs among the atoms that are present in the wall work for a straightforward presentation of the flexibility of the interface.

The distinct mechanism in the ferroelastic system is not this one. It was seen that two principal forces are applied on the interface—with the wall energy itself being "anisotropy energy." Each wall segment rotation demands a lot of energy ~cosφ; here, φ is an angle that exists between low stress equilibrium order and the actual order of wall section. Another important energy is wall "bending energy". It opposes bending of walls and is distinctly high in case of compact walls. All mentioned energies will work to restrict the gradual roughness of the wall and ensure that bi-fold walls continue to be planar smooth throughout to influence doping states. It will casually contribute toward uneven walls in magnetic structures [51].

11.11 Future Directions

Because they can dynamically exhibit adjustable mechanical, thermal, magnetic, optical, and dielectric or electric properties, multiferroic materials offer a tremendous source of technological applications. Additionally, these characteristics have the potential to connect to each other, which results in appealing cross-coupling that enhances the adaptability of multiferroics for cutting-edge application. Varying multiferroic substance prerequisites are applied to various applications. However, general standards that must be properly adhered to for applications tend to be as follows:

a) Strong magnetoelectric coupling phenomena in multiferroic materials must be present, ideally at ambient temperature
b) Multiferroic components and material should be easy to produce and ideally made of inexpensive substances
c) Multiferroic systems need to be capable of chip-level integration and mass manufacture
d) Multiferroic devices need to be able to be manufactured at micro- and nanoscale levels without losing their ability to function properly.

In reality, there is experimental proof that multiferroic magnetoelectric coupling persists even at the nanoscale, indicating that it is certainly feasible to develop nanotechnologies based on these materials.

The majority of the aforementioned requirements are currently satisfied by composites of multiferroic materials. In fact, elastic PVDF polymer-based multiferroic composites and their applications have recently emerged, indicating that the study field of these materials continues to expand steadily. Despite significant progress in these materials, a single-phase multiferroic material that is optically and piezo active exhibits electric and magnetic order parameters, and has high cross-coupling coefficients at ambient temperatures, which represents a long-awaited scientific achievement. These substances would have unmatched versatility for high-tech applications, with the ability of mechanical operations, diverse memory phases, sensing, logical operations, and photoactive capabilities all in a single multiferroic multi-functional solid-state entity. The possibility of discovering such materials with focused study efforts is extremely real.

The idea that multiferroic substances have the capacity to radically alter our civilization and enable the continuation of the recent technological and scientific advancements is not exaggerated. Multiferroic technologies may soon achieve a market worth of many billions of US dollars, even if just some of the planned uses are implemented in actual commercial goods. In addition, the combination of multiferroics with spintronic and semiconducting materials to create multipurpose hybrid systems may help to spur the development of genuinely revolutionary technology.

11.12 Conclusion

Although it has long been known that domain boundaries carry large currents, the finding of twin boundaries that are superconducting has opened up a vast array of applications for "domain boundary engineering," in which domain boundary functions such as the device itself and a map of the device's material are heavily influenced by changing the proper domain boundaries. Because they can dynamically exhibit adjustable mechanical, thermal, magnetic, optical, and dielectric or electric properties, multiferroic materials offer a tremendous source of applications in the field of technology.

Walls of WO_3's chemical makeup underwent a small modification (for example, fromWO_3 to $WO_{2.95}$) that led to metal insulator transitions at a low temperature, which induced superconduction in twin walls. Because the dopants obey the twin walls' pathways, it is possible to nano-pattern the superconducting structure by first patterning the twin boundaries and then doping them.

There has only been paramagnetism discovered for samples of powder LNO. In its bulk form, it is the only member of family perovskite nickelates that lack magnetic order. Although the power law distribution of avalanches during ferroic switching and martensitic transformations is confirmed, we also notice some modest dynamics within the same jerk spectra. These minor occurrences are related to local lattice relaxations that take place during creep tests or after the switching process itself. Their physical genesis mostly results from kink movements within domain walls for the relaxing following switching. The effects of structural relaxations at domain wall crossings during mild occurrences during creep are considerably more negligible. We observed that shear-induced ferroelastic domain switching coexists with wild and mild dynamics in detail. Shocking shifts in domain patterns are what cause the wildness (generated by junctions and intersections of domain boundaries).

References

1. Evans, D.M., Garcia, V., Meier, D., Bibes, M., Domains and domain walls in multiferroics. *Phys. Sci. Rev.*, 5, 2, 2020.
2. Salje, E.K.H., Multiferroic domain boundaries as active memory devices: Trajectories towards domain boundary engineering. *ChemPhysChem*, 11, 940, 2010.
3. Salje, E.K.H. and Ding, X., Ferroelastic domain boundary-based multiferroicity. *Crystals*, 6, 163, 2016.
4. Salje, E. and Zhang, H., Domain boundary engineering. *Ph. Transit.*, 82, 452, 2009.
5. Wu, H., Li, L., Liang, L.Z., Liang, S., Zhu, Y.Y., Zhu, X.H., Recent progress on the structural characterizations of domain structures in ferroic and multiferroic perovskite oxides: A review. *J. Eur. Ceram. Soc.*, 35, 411, 2015.
6. Bartkowska, J.A., The magnetoelectric coupling effect in multiferroic composites based on PZT-ferrite. *J. Magn. Magn. Mater.*, 374, 703, 2015.
7. Salje, E.K.H. and Lashley, J.C., Domain boundary engineering in ferroic and multiferroic materials: A simple introduction, in: *Springer Series in Materials Science*, T. Kakeshita, T. Fukuda, A. Saxena, A. Planes, (Eds.), pp. 1–18, Springer Berlin Heidelberg, Berlin, Heidelberg, 2011.
8. Pooladi, M., Sharifi, I., Behzadipour, M., A review of the structure, magnetic and electrical properties of bismuth ferrite (Bi2Fe4O9). *Ceram. Int.*, 46, 18453, 2020.
9. Vogel, J., Kuch, W., Camarero, J., Fukumoto, K., Pennec, Y., Bonfim, M., Pizzini, S., Petroff, F., Fontaine, A., Kirschner, J., Time and layer resolved magnetic domain imagig of FeNi/Cu/Co trilayers using x-ray photoelectron emission microscopy (invited). *J. Appl. Phys.*, 95, 6533, 2004.
10. Pramanick, A., Prewitt, A.D., Forrester, J.S., Jones, J.L., Domains, domain walls and defects in perovskite ferroelectric oxides: A review of present understanding and recent contributions. *Crit. Rev. Solid State Mater. Sci.*, 37, 243, 2012.
11. Rondinelli, J.M., Stengel, M., Spaldin, N.A., Carrier-mediated magnetoelectricity in complex oxide heterostructures. *Nat. Nanotechnol.*, 3, 46, 2008.
12. Dong, S., Zhang, X., Yu, R., Liu, J.M., Dagotto, E., Microscopic model for the ferroelectric field effect in oxide heterostructures. *Phys. Rev. B - Condens. Matter Mater. Phys.*, 84, 7, 2011.
13. Niranjan, M.K., Burton, J.D., Velev, J.P., Jaswal, S.S., Tsymbal, E.Y., Magnetoelectric effect at the SrRuO3/BaTiO3 (001) interface: An *ab initio* study. *Appl. Phys. Lett.*, 95, 4, 2009.
14. Cai, T., Ju, S., Lee, J., Sai, N., Demkov, A.A., Niu, Q., Li, Z., Shi, J., Wang, E., Magnetoelectric coupling and electric control of magnetization in ferromagnet/ferroelectric/normal-metal superlattices. *Phys. Rev. B - Condens. Matter Mater. Phys.*, 80, 4, 2009.

15. Lee, J., Sai, N., Cai, T., Niu, Q., Demkov, A.A., Interfacial magnetoelectric coupling in tricomponent superlattices. *Phys. Rev. B - Condens. Matter Mater. Phys.*, 81, 6, 2010.
16. Borisov, V.S., Ostanin, S., Maznichenko, I.V., Ernst, A., Mertig, I., Magnetoelectric properties of the Co/PbZrxTi1-XO3(001) interface studied from first principles. *Phys. Rev. B - Condens. Matter Mater. Phys.*, 89, 7, 2014.
17. Fechner, M., Maznichenko, I.V., Ostanin, S., Ernst, A., Henk, J., Bruno, P., Mertig, I., Magnetic phase transition in two-phase multiferroics predicted from first principles. *Phys. Rev. B - Condens. Matter Mater. Phys.*, 78, 5, 2008.
18. Niranjan, M.K., Burton, J.D., Velev, J.P., Jaswal, S.S., Tsymbal, E.Y., Magnetoelectric effect at the SrRuO3/BaTiO3 (001) interface: An ab initio study. *Appl. Phys. Lett.*, 95, 3, 2009.
19. Radaelli, G., Petti, D., Plekhanov, E., Fina, I., Torelli, P., Salles, B.R., Cantoni, M., Rinaldi, C., Gutiérrez, D., Panaccione, G., Varela, M., Picozzi, S., Fontcuberta, J., Bertacco, R., Electric control of magnetism at the Fe/BaTiO3 interface. *Nat. Commun.*, 5, 5, 2014.
20. Yamauchi, K., Sanyal, B., Picozzi, S., Interface effects at a half-metal/ferroelectric junction. *Appl. Phys. Lett.*, 91, 2, 2007.
21. Tsujikawa, M. and Oda, T., Finite electric field effects in the large perpendicular magnetic anisotropy surface Pt/Fe/Pt(001): A first-principles study. *Phys. Rev. Lett.*, 102, 2, 2009.
22. Niranjan, M.K., Duan, C.G., Jaswal, S.S., Tsymbal, E.Y., Electric field effect on magnetization at the Fe/MgO(001) interface. *Appl. Phys. Lett.*, 96, 2, 2010.
23. He, K.H., Chen, J.S., Feng, Y.P., First principles study of the electric field effect on magnetization and magnetic anisotropy of FeCo/MgO(001) thin film. *Appl. Phys. Lett.*, 99, 2, 2011.
24. Weisheit, M., Fähler, S., Marty, A., Souche, Y., Poinsignon, C., Givord, D., Electric field-induced modification of magnetism in thin-film ferromagnets. *Sci. (80-.)*, 315, 349, 2007.
25. Maruyama, T., Shiota, Y., Nozaki, T., Ohta, K., Toda, N., Mizuguchi, M., Tulapurkar, A.A., Shinjo, T., Shiraishi, M., Mizukami, S., Ando, Y., Suzuki, Y., Large voltage-induced magnetic anisotropy change in a few atomic layers of iron. *Nat. Nanotechnol.*, 4, 158, 2009.
26. Chang, C.C., Lin, C.F., Chiou, J.M., Ho, T.H., Tai, Y., Lee, J.H., Chen, Y.F., Wang, J.K., Chen, L.C., Chen, K.H., Erratum: Effects of cathode buffer layers on the efficiency of bulk-heterojunction solar cells (Applied Physics Letters (2010) 96 (263506)). *Appl. Phys. Lett.*, 97, 49901, 2010.
27. Huang, X. and Dong, S., Ferroelectric control of magnetism and transport in oxide heterostructures. *Mod. Phys. Lett. B*, 28, 1430010, 2014.
28. Kanai, S., Yamanouchi, M., Ikeda, S., Nakatani, Y., Matsukura, F., Ohno, H., Electric field-induced magnetization reversal in a perpendicular-anisotropy CoFeB-MgO magnetic tunnel junction. *Appl. Phys. Lett.*, 101, 3, 2012.

29. Duan, C.G., Velev, J.P., Sabirianov, R.F., Mei, W.N., Jaswal, S.S., Tsymbal, E.Y., Tailoring magnetic anisotropy at the ferromagnetic/ferroelectric interface. *Appl. Phys. Lett.*, 92, 2, 2008.
30. Goncalves-Ferreira, L., Redfern, S.A.T., Atacho, E., Salje, E.K.H., The intrinsic elasticity of twin walls: Ferrielectric twin walls in ferroelastic CaTiO3. *Appl. Phys. Lett.*, 94, 2, 2009.
31. Goncalves-Ferreira, L., Redfern, S.A.T., Artacho, E., Salje, E., Lee, W.T., Trapping of oxygen vacancies in the twin walls of perovskite. *Phys. Rev. B - Condens. Matter Mater. Phys.*, 81, 1, 2010.
32. Zubko, P., Catalan, G., Buckley, A., Welche, P.R.L., Scott, J.F., Erratum: Strain-gradient-induced polarization in srTiO3 single crystals (Physical Review Letters (2007) 99 (167601)). *Phys. Rev. Lett.*, 100, 1, 2008.
33. Guo, H., Li, Z.W., Zhao, L., Hu, Z., Chang, C.F., Kuo, C.Y., Schmidt, W., Piovano, A., Pi, T.W., Sobolev, O., Khomskii, D.I., Tjeng, L.H., Komarek, A.C., Antiferromagnetic correlations in the metallic strongly correlated transition metal oxide LaNiO3. *Nat. Commun.*, 9, 4, 2018.
34. Bruno, F.Y., Garcia-Barriocanal, J., Varela, M., Nemes, N.M., Thakur, P., Cezar, J.C., Brookes, N.B., Rivera-Calzada, A., Garcia-Hernandez, M., Leon, C., Okamoto, S., Pennycook, S.J., Santamaria, J., Electronic and magnetic reconstructions in La07Sr0.3MnO3/SrTiO3 heterostructures: A case of enhanced interlayer coupling controlled by the interface. *Phys. Rev. Lett.*, 106, 1, 2011.
35. Liu, Y., Cuellar, F.A., Sefrioui, Z., Freeland, J.W., Fitzsimmons, M.R., Leon, C., Santamaria, J., Te Velthuis, S.G.E., Emergent spin filter at the interface between ferromagnetic and insulating layered oxides. *Phys. Rev. Lett.*, 111, 5, 2013.
36. Seo, J.W., Prellier, W., Padhan, P., Boullay, P., Kim, J.Y., Lee, H., Batista, C.D., Martin, I., Chia, E.E.M., Wu, T., Cho, B.G., Panagopoulos, C., Tunable magnetic interaction at the atomic scale in oxide heterostructures. *Phys. Rev. Lett.*, 105, 4, 2010.
37. Wang, B.X., Rosenkranz, S., Rui, X., Zhang, J., Ye, F., Zheng, H., Klie, R.F., Mitchell, J.F., Phelan, D., Antiferromagnetic defect structure in LaNi O3-δ single crystals. *Phys. Rev. Mater.*, 2, 2018.
38. Driza, N., Blanco-Canosa, S., Bakr, M., Soltan, S., Khalid, M., Mustafa, L., Kawashima, K., Christiani, G., Habermeier, H.U., Khaliullin, G., Ulrich, C., Le Tacon, M., Keimer, B., Long-range transfer of electron-phonon coupling in oxide superlattices. *Nat. Mater.*, 11, 675, 2012.
39. Salje, E.K.H., Mild and wild ferroelectrics and their potential role in neuromorphic computation. *APL Mater*, 9, 9, 2021.
40. Weiss, J. and Marsan, D., Three-dimensional mapping of dislocation avalanches: Clustering and space/time coupling. *Sci. (80-.)*, 299, 89, 2003.
41. Miguel, M.C., Vespignani, A., Zapperi, S., Weiss, J., Grasso, J.R., Intermittent dislocation flow in viscoplastic deformation. *Nature*, 410, 667, 2001.

42. Sawe, E., Phase transitions in ferroelastic and co-elastic crystals. *Ferroelectrics*, 104, 111, 1990.
43. Lu, G., Li, S., Ding, X., Salje, E.K.H., Piezoelectricity and electrostriction in ferroelastic materials with polar twin boundaries and domain junctions. *Appl. Phys. Lett.*, 114, 1, 2019.
44. Salje, E.K.H., Li, S., Stengel, M., Gumbsch, P., Ding, X., Flexoelectricity and the polarity of complex ferroelastic twin patterns. *Phys. Rev. B*, 94, 6, 2016.
45. Li, S., Ding, X., Ren, J., Moya, X., Li, J., Sun, J., Salje, E.K.H., Strain-controlled thermal conductivity in ferroic twinned films. *Sci. Rep.*, 4, 2, 2014.
46. Vives, E., Ortín, J., Mañosa, L., Ràfols, I., Pérez-Magrané, R., Planes, A., Distributions of avalanches in martensitic transformations. *Phys. Rev. Lett.*, 72, 1694, 1994.
47. Bonnot, E., Vives, E., Mañosa, L., Planes, A., Romero, R., Acoustic emission and energy dissipation during front propagation in a stress-driven martensitic transition. *Phys. Rev. B - Condens. Matter Mater. Phys.*, 78, 1, 2008.
48. Gallardo, M.C., Manchado, J., Romero, F.J., Del Cerro, J., Salje, E.K.H., Planes, A., Vives, E., Romero, R., Stipcich, M., Avalanche criticality in the martensitic transition of Cu67.64 Zn16.71 Al15.65 shape-memory alloy: A calorimetric and acoustic emission study. *Phys. Rev. B - Condens. Matter Mater. Phys.*, 81, 5, 2010.
49. Kuntz, M.C., Perković, O., Dahmen, K.A., Roberts, B.W., Sethna, J.P., Hysteresis avalanches, and noise. *Comput. Sci. Eng.*, 1, 73, 1999.
50. Proville, L., Depinning of a discrete elastic string from a random array of weak pinning points with finite dimensions. *J. Stat. Phys.*, 137, 717, 2009.
51. Salje, E.K.H. and Ishibashi, Y., Mesoscopic structures in ferroelastic crystals: Needle twins and right-angled domains. *J. Phys. Condens. Matter*, 8, 8477, 1996.

12

Magnetoelectric, Dielectric, and Optical Characteristics of Ferroelectric and Antiferroelectric Materials

M. S. Hasan[1], Sabahat Urossha[2], M. Zulqarnain[1] and S. S. Ali[2*]

[1]Department of Physics, The University of Lahore, Lahore, Pakistan
[2]School of Physical Sciences, University of the Punjab, Lahore, Pakistan

Abstract

Ferroelectric and antiferroelectric materials have attracted the attention of researchers in the last couple of decades due to their amazing applications such as magnetoelectric storage devices, positioners and generators, as well as magnetic field instruments. Ferroelectric and antiferroelectric materials display numerous kinds of physical attributes that include magnetoelectric, dielectric, and optical characteristics. The magnetoelectric impact describes interaction among electromagnetic fields in a material. In ferroelectric and antiferroelectric materials, this coupling may cause emergence of novel magnetic domains or alteration of operating domains as a result of an electric field being applied. These materials may additionally exhibit high dielectric constants, which make them useful in capacitive applications such as energy storage devices and high-frequency filters. Additionally, their optical properties, such as birefringence and nonlinear optical response, make them attractive for applications in electro-optic modulators and photonic devices. The behavior of ferroelectric and antiferroelectric materials is strongly influenced by the nature of their domain walls and defects, which can affect their magnetoelectric, dielectric, and optical properties. Understanding the mechanisms that govern these properties is crucial for the development of novel materials utilizing tailored characteristics for particular uses. In this chapter, we focused on the use of advanced characterization techniques, such as magnetoelectric, dielectric, and optical properties of ferroelectric and antiferroelectric materials. Continued research within this area is essential for the development of

**Corresponding author*: shahbaz.sps@pu.edu.pk

Inamuddin, Tariq Altalhi and Mohammad Abu Jafar Mazumder (eds.) Ferroic Materials-Based Technologies, (279–304) © 2024 Scrivener Publishing LLC

novel materials accompanying optimized characteristics for certain purposes in electronics, energy storage, and photonics applications.

***Keywords*:** Magnetoelectric, polarization, real dielectric constant, imaginary dielectric constant, optical band-gap energy

12.1 Introduction

Ferroelectric and antiferroelectric materials have attracted the attention of researchers in recent years in solid-state physics and materials science due to their wonderful magnetic, electric, dielectric, and optical properties. The inter- and intra-molecular relationships have exposed ferroelectric and antiferroelectric materials for numerous applications in the fields of data storage, microwave absorption, soft magnetism, and many other technical fields. The ancient magnetic materials were known as magnetite and were called magnetic ceramics or ferrites. Two similar pieces of these materials exert a force of attraction or repulsion when they were brought closer. Ferromagnetic materials are related to the group of materials with spontaneous net magnetization, even after removal of an exterior magnetic field. A ferromagnetic material becomes magnetized in the direction of applied field [1]. Antiferromagnetic materials consisted of metals, alloys, and other ionic solids; the magnetic atoms or ions showing magnetism in a single direction are lost by usual magnetic atoms or ions aligned in the opposite direction [2]. Magnetoelectrics (ME) are the multiferroics that keep double order parameters, i.e., magnetic and electrical polarization at a similar temperature [3]. With the same materials for ferroelectric (FE) and ferromagnetic (FM) orders, the following situations are required [4]:

i. The availability of enough structural building blocks enables ionic migrations resembling those seen in FE, namely, off-centered displacements associated to spontaneous polarization in FE.
ii. The magnetic interaction mechanisms for magnetic arrangement are frequently of extreme alternative type.
iii. Symmetrical condition.

The FE and AFE have reliable usages. The one-phase ME multiferroism is relatively exceptional due to its limited applications because of the deceptively opposing necessities concerning the d-orbit FE and magnetic ordering [5]. Currently, various processes have been adopted for the explanation of contradiction associated to the possible solutions of dual ordering [6, 7]. $BiFeO_3$ is a one-phase magnetoelectric material consisting of FE (T_C = 1100 K) and AFM (T_N = 633 K) temperatures; it also demonstrates feeble ferro/ferrimagnetic properties at certain ranges of temperatures [8]. $BiFeO_3$, in bulk form at room temperature, deformed rhombohedral perovskite structures R3c (a = b = c = 5.63 Å, α = β = γ = 59.4°). Even though they have a multiferroic nature, only weak dielectric and ferroelectric characteristics are observed in bulk ceramics form at room temperature. Such behavior is attributed to the semi-conducting nature with higher dielectric losses due to the absence of electrical mechanism [3]. Although the enhancement in structural complication provides a higher degree of freedom and is difficult to understand, FE and AFE materials are still attractive sources for multifunctional devices. The multiferroic properties such as ferroelectricity, ferromagnetism, and ferroelasticity happen simultaneously within identical phases [9]. They are highly applicable in magnetic field sensors, magnetoelectric memory chips for sensors, actuators, generators, etc. [10, 11]. Generally, single-phase magnetic FE materials have low magnetic ordering temperatures; this feature makes them a potential candidate for multifunctional devices [12, 13]. $BiFeO_3$ is a significant multiferroic simple perovskite material with numerous properties due to high FE (T_C = 1100 K) and AFM (T_N = 643 K) temperatures [14]. This material is wonderful for multifunctional devices, but it has also disadvantages like current leakage, greater dielectric constant and tangent loss, appearance of multiphases, and unstable structures [15]. Such materials are not suitable for electronic gadgets [16]. In order to overcome this issue, various steps are taken like doping of ions with other FE compounds [17].

The choice of FE and AFE compounds or materials depends on various factors such as synthesis process, doping of elements, and overwhelming properties like magnetoelectric, dielectric, and optical. Depending on the properties, the fabricated materials can be applied for various applications. Furthermore, the advancement in characterization technique for magnetic, electric, dielectric, and optical trends leads these materials to determine further in-depth properties [18]. The magnetoelectric characterization techniques can be used to find saturation magnetization, remanence, coercivity, polarization, etc. [19]. Dielectric properties can be used to find different factors like dielectric constant, dielectric loss, ac conductivity, impedance, etc. [20]. Optical properties are utilized to determine optical band gap, dielectric constants, refractive indices, optical conductivity, skin depth, extinction coefficient, etc. In this chapter, we report the ferroelectric and antiferroelectric materials for their various applications depending on the magnetoelectric, dielectric, and optical properties.

12.2 Magnetoelectric Properties of FE and AFE Materials

The magnetoelectric phenomenon that describes connection among magnetic and electric fields inside a material has drawn a lot of attention over the past few years because of its potential for applications in magnetic storage and sensing devices. Ferroelectric and antiferroelectric materials have been shown to possess robust magnetoelectric interaction, which makes them intriguing alternatives for development of next-generation devices with enhanced performance. The multiferroic materials reveal FE or AFE characteristics with the combinations of ferromagnetic or antiferromagnetic characteristics in a similar phase. As a result, electric polarization is induced due to applied MF and change is induced because of electric field. Being inside the MF, electric polarization is induced, or when there is an electric field, the induced magnetization of materials needs instantaneous existence for long-term ordering of magnetic moments and electric dipoles. This phenomenon leads to synchronicity ordering of FM and FE; such materials may be used for various purposes and can be applied as FE and FM materials in electronic gadgets. The conversion of FE and FM materials into a single material is responsible for the coupling characteristics because of the coupling effect among magnetization and electric polarization [21]. Yet, very few materials are present in nature that have the ability to expose dual spontaneous magnetization and electric polarization [22]. Around normal temperature, materials presenting higher ME coupling are practically optimum candidates, and also, at room temperature, ME coupling is generally smaller and it is difficult to trace materials at room temperature with larger ME coupling [21].

12.2.1 Magnetoelectric Characteristics of $BaTiO_3$-$BiFeO_3$ Materials

The multiferroic materials $BaTiO_3$-$BiFeO_3$ in the form of solid solution have attracted the attention of researchers due to innovative applications in electronic devices. Naturally, a very small number of materials showing both FE and FM properties simultaneously are present [23, 24]. Perhaps at room temperature, $BiFeO_3$ is the single material manifesting both FE and FM behavior with $T_C = 1100$ K and $T_N = 643$ K [8, 25]. Yet, there are two features that impede them for technical usage: (i) $BiFeO_3$ has lower electrical resistivity owing to a decrease in the trivalent and divalent iron ions and Bi evaporation in the annealing process. This causes the creation of oxygen spaces for charge compensation as a result of dielectric losses and increase and decrease in saturation magnetization [26, 27]. (ii) The preparation of single-phase $BiFeO_3$ particles is difficult because of the secondary phase's stability in the same range of temperatures, e.g., $Bi_2Fe_4O_9$, $Bi_{25}FeO_{39}$, etc. [27]. Figure 12.1 shows polarization and electric field (P–E) loop for 0.96BaTi–0.04BaFO sintered at different temperatures [24]. As depicted in Figure 12.1, it is clear that polarization P_r and E_c both increased with the increase of sintering temperature from 11.72 μC/cm² and 21.42 kV/cm to 16.74 μC/cm² and 30.28 kV/cm, respectively. Generally, the ceramic materials having greater crystallite size and density expose the ferroelectric properties [28]. From Figure 12.2, the values of P_r increased in the range of 4.34–6.74 μC/cm² for x = 0.01 to x = 0.04 accordingly, and then decline quickly to 6.04 μC/cm² until the value of x approaches 0.07. In the same way, the values of E_c rise up to 30.10 kV/cm for x = 0.04 and decline to 15.21 kV/cm for x = 0.07. For higher values of Ti, BaTi exhibits better FE characteristics where P_r = 11.38 μC/cm². With the insertion of BaFO, Ti-O bonds were disturbed due to existence of various valence states and dissimilarity in ionic radii of Fe and Ti. This fact causes the reduction in P_r in the ranges of x = 0.0–0.02 [29]. For the samples x = 0.03–0.04, better FE properties are demonstrated; such behavior is attributed to higher polarization [30].

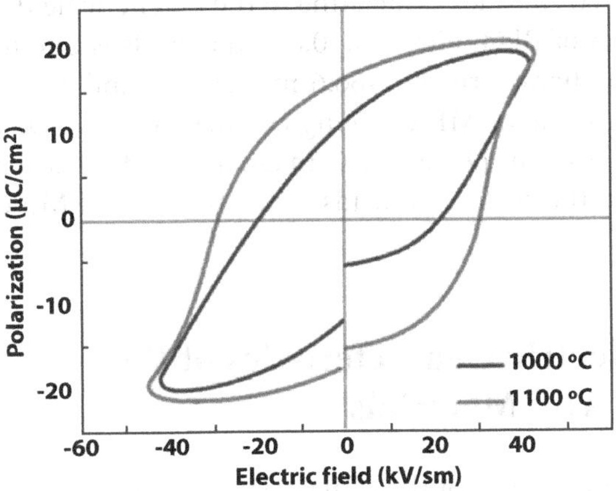

Figure 12.1 Different temperatures were employed to sinter ceramic 0.96BaTi–0.04BaFO to get modified P–E loops. Reused from [24].

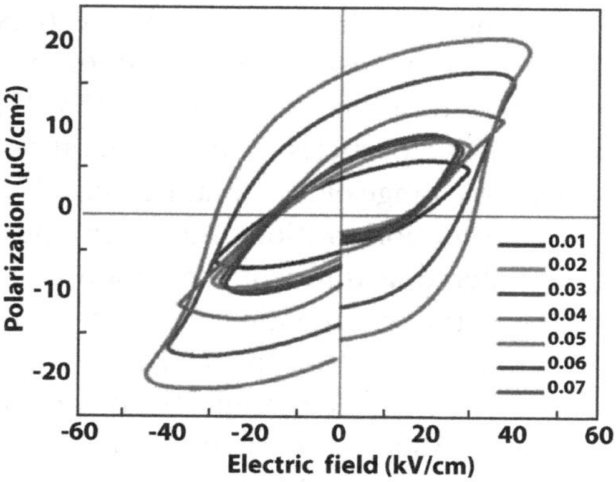

Figure 12.2 P–E loops for BaTi–xBaFO (x = 0–0.07) ceramic sintered at 1373 K. Reused from [24].

12.2.2 Magnetoelectric Characteristics of $BiFeO_3$–$BaTiO_3$–$LaFeO_3$ Materials

The ceramics materials with formula $0.7BiFeO_3$–$0.3BaTiO_3$ have stronger FE properties and weaker ME properties. The magnetoelectric characteristics of such material were enhanced by successful doping of different elements. In order to investigate the ME dipole contact, the coefficient of ME coupling is observed by the application of the process known as dynamic procedure. Figure 12.3a–d [31] shows the ME coupling coefficient against the frequency. The values for ME coupling coefficient increased in the range of 0–10^6 Hz, besides associating to the magnetic field. The increase in ME connection of BFO–BTO–La–0.025 materials is due to enhancement in FE and FM characteristics; 466.6 mV/cm.Oe and 53.9 mV/cm.Oe are the maximum value of ME coupling coefficient at 10^3 Oe and 1000 kHz. According to usage of DC and AC magnetic field, greater electric field is induced due to the strain and leads to the increase of ME effects in materials [32, 33].

12.3 Dielectric Characteristics of FE and AFE Materials

The dielectric properties of these materials, including their high dielectric constants, have also attracted significant attention in the field of capacitive energy storage, where they have been used in the development of high-performance capacitors and supercapacitors. Additionally, their

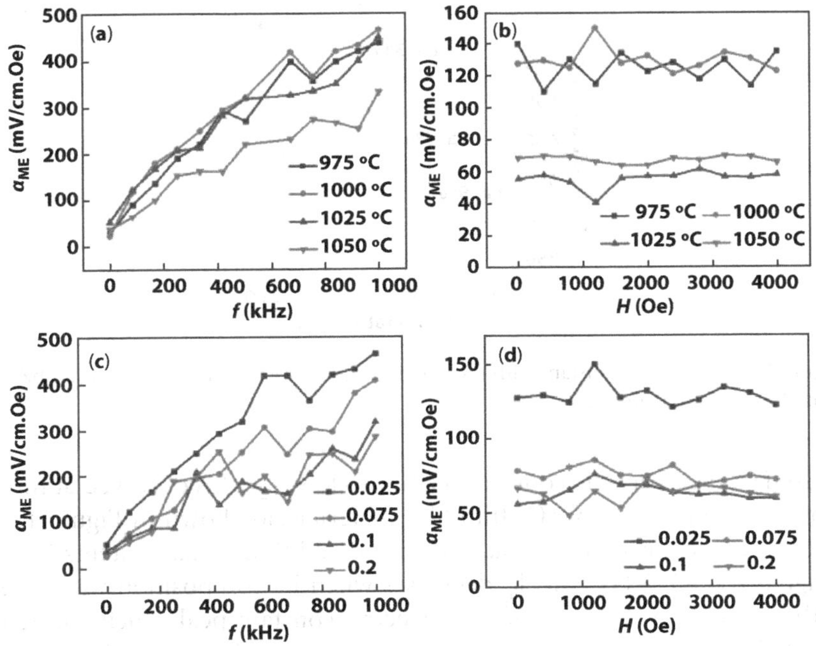

Figure 12.3 (a–d) ME coefficient based on (a) frequency ranges and (b) BFO–BTO–La–0.025 material sintering at 1248 K to1323 K being subjected to DC magnetic field, and (c) frequencies and (d) DC magnetic field of BFO–BTO–La–x (x = 0.0 to 0.2) materials. Reused from [31].

optical properties, such as birefringence and nonlinear optical response, make them attractive for applications in electro-optic modulators and photonic devices. Dielectric properties have a significant role in microwave frequency range and have appropriate usage for microwave devices. The stabilization of frequency in microwave devices might have greater dielectric constant, smaller dielectric losses, and lower temperature coefficient at resonance frequency [34].

12.3.1 Dielectric Characteristics of La-Modified AFE $PbZrO_3$ Thin Films

Figure 12.4 demonstrates how dielectric constant changes as temperature rises for La-modified AFE $PbZrO_3$ thin films [35]. At room temperature, $PbZrO_3$ is an AFE material and shows the dual polarization hysteresis properties in manifestation of applied MF. The dielectric properties of $PbZrO_3$ with the doping of La show AFE to have a paraelectric nature. In the range

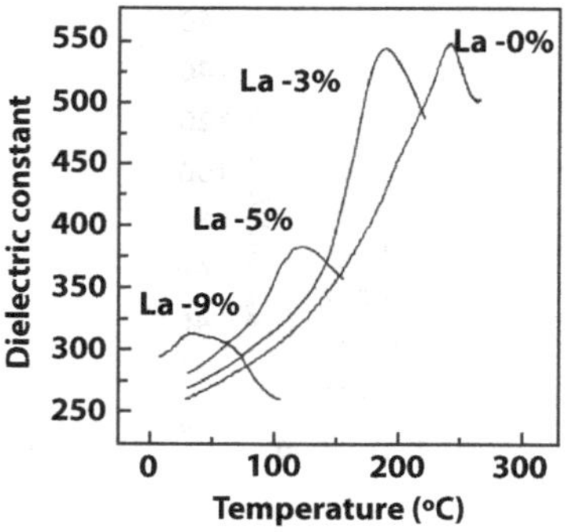

Figure 12.4 Dielectric constant with the rise of temperature for La-modified AFE $PbZrO_3$ thin films. Reused from [35].

of 10^3 Hz, the dielectric constant relative to low signal is observed at room temperature for the $PbZrO_3$ thin film materials, as shown in Figure 12.4. The dielectric constant increases with the addition of La contents in the range of 220–325 Hz. The T_C decreases when La composition was added with a reduction in height of the dielectric constant peak. Such nature of the dielectric constant is generally seen in relaxor FE materials at maximum concentration of lanthanum [36]. At various concentrations of La in the current composition, from the spontaneous values of polarization, the dielectric constant values were supposed to be greater as compared to the calculated values. Such behavior might be ascribed to the domain mechanism of reversible and irreversible nature. In the calculation of dynamic hysteresis, the entire macroscopic characteristics are a combination of reversible and irreversible components, while in lower frequency ac signal dielectric calculations, solitary reversible constituent contributions are observed.

12.3.2 Dielectric Characteristics of BNTC22 Materials

The range of $Bi_{0.487}Na_{0.427}K_{0.06}Ba_{0.026}TiO_{3-x}CeO_2$ (BNTC22), where x = ¼ 1.8, 2.2, 2.6, 3.0 wt% compounds, is presented as a source energy storage materials. Figure 12.5a–c shows the curves for dielectric constant and dielectric losses [37]. Figure 12.5a depicts graphs for dielectric constant at different temperatures, and the compositional percentages are illustrated. All samples have variance peaks related to optimum values of the dielectric constant at maximum temperature, T_m. The phase transition in peak is observed from AFE to paraelectric [38]. By the addition of CeO_2 concentration, both dielectric constant and T_m decline. On the other hand,

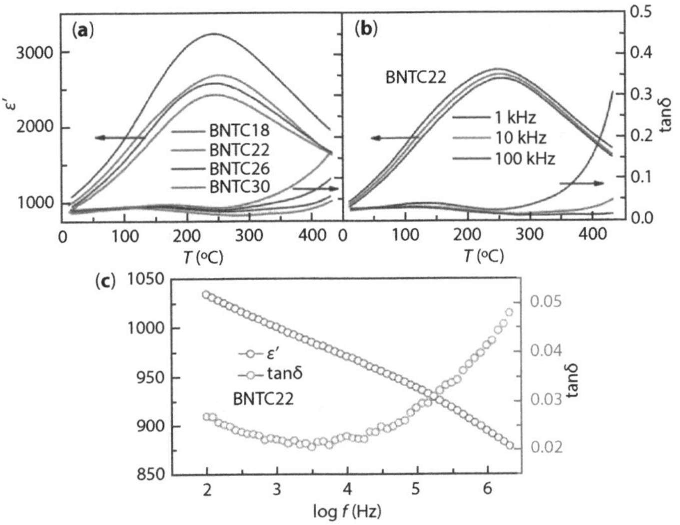

Figure 12.5 (a–c) Values of dielectric constant and losses as a consequence of (a) temperature recorded at 10^3 Hz for $Bi_{0.487}Na_{0.427}K_{0.06}Ba_{0.026}TiO_{3\text{-}x}CeO_2$ material, (b) calculated temperature at different range of frequency, and (c) frequency observed at normal temperature. Reused from [37].

no particular change is observed in dielectric loss as given in Figure 12.5a. The obtained lowest value of dielectric loss for the BNTC22 is suitable for the increase of energy storage. Hence, the outcomes reveal that the materials are AFE in nature. The dielectric loss and dielectric constant at various frequency ranges are provided in Figure 12.5b. Such nature of the graph indicates that the dielectric relaxation character in current materials existed. Such factors are exposed in Figure 12.5c. Due to the net polarization, with a rise in frequency, the value of dielectric constant dropped [25, 38]. Also, with the change of frequency, the dielectric loss shows a variable trend.

12.3.3 Dielectric Properties of PLZST-AFE Material

PLZST-AFE is an antiferroelectric material with single-phase crystals and is a combination of different metals like lead, lanthanum, zirconium, tin and titanium. They were prepared by the application of the flux process. The dielectric constant depending on the temperature is illustrated in Figure 12.6a. The range of temperature was adjusted between room temperature and 250°C, as can be seen from Figure 12.6a. At 125°C and 200°C, two anomalies were detected. The higher elevated temperature is known as

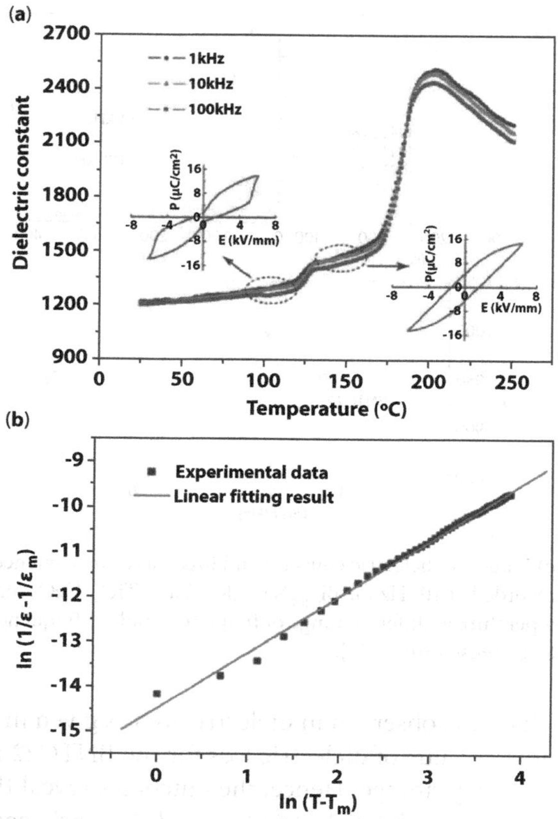

Figure 12.6 (a) Dielectric constant as a function of temperature with the inset showing PE at 125°C, (b) ln(1/ε-1/ε_m) associating to the ln(T-T_m) for PLZST nanomaterials. Reused from [41].

T_c, beyond which a typical non-polar PE phase emerged. The phase shift occurring at small ranges of temperature anomaly of approximately 125°C is identified through P–E loops. Below 125°C, a double loop is observed, indicating the presence of an AFE state, depicted as an inset of Figure 12.6a. Conversely, the greater phase of 125°C is likely to be ferroelectric (FE), as evidenced by the formation of a distinct loop. Throughout the entire range of testing temperatures, frequency distribution is present, which could be due to the presence of FE micro-domains within the AFE matrix [39]. The peak value shown by the dielectric constants is associated to the frequency, while the T_C remains nearly constant across the three measured frequencies. Beyond the temperature T_m, the dielectric constants start to decrease in accordance with an improved Curie-Weiss law [40].

$$\frac{1}{\varepsilon}-\frac{1}{\varepsilon_m}=\frac{(T-T_m)^{\alpha}}{C} \tag{12.1}$$

wherein the parameters "α" and "C" are constants specific to the specimen alone. The variable "α" provides valuable details for distinguishing the specimen: a value of 1 indicates a normal AFE state, whereas a value of 2 suggests a whole relaxor phase. Figure 12.6b illustrates plotting of ln(1/ε-1/ε_m) associated with the ln(T-T_m) at a frequency of 10^5 Hz. The amount "α" is discovered via the linear regression slope of the curve based on the aforementioned equation. In this case, it is calculated to be 1.36, indicating that the PLZST crystal is in an intermediary phase somewhere between normal and relaxed AFE [41].

12.3.4 Dielectric Characteristics of $BiMgFeCeO_6$ Materials

Figure 12.7 demonstrates the frequency-dependent behavior of the relative dielectric constants or permittivity (ε_r) and tangent loss (tanδ) of material up to 500°C. As the frequency increases, both ε_r (Figure 12.7a) and tangent loss (Figure 12.7b) slowly decrease and eventually reach a fixed number beyond a particular frequency range, which constitutes a polar dielectric's distinctive feature. At lower frequencies, higher values of ε_r can be attributed to various factors, such as grain boundaries, interface dislocation effects, and oxygen vacancies within the material. The amount of ε_r decreases steadily with increasing frequency, indicating the emergence of polarization as an outcome of domain transition [42]. The four main kinds of polarization—interfacial, dipolar, atomic in nature, and

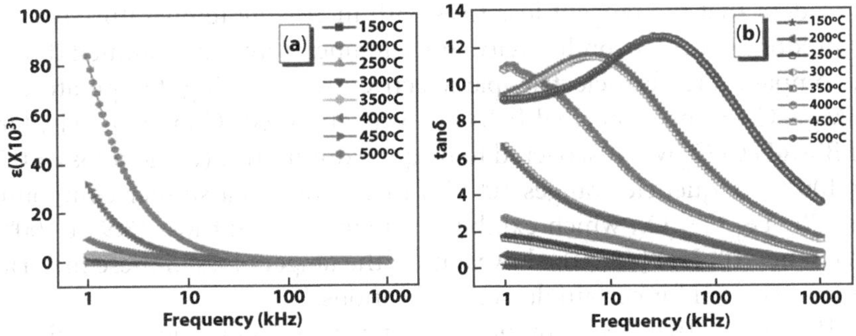

Figure 12.7 (a, b) Dielectric constants and losses against applied frequency range. Reused from [46].

electronic—distinctively take place over a low-frequency range, increasing the material's permittivity. However, as frequency exceeds a certain limit, the different contributions from these polarizations cease, and only electronic polarization significantly contributes to the dielectric constant, leading to a subsequent decrease in ε_r. The reduction in permittivity and tangent loss with increasing frequency can be understood through a Maxwell–Wagner type of dielectric structure, which is based on Koop's phenomenological theory [43, 44]. This theory suggests a two-layer model consisting of conducting grains and grain interfaces with poor conductivity in ceramics material. Whereas electrons remain particularly energetic at grain faces throughout the low-frequency zone, they are much active at individual grains that are responsible for conduction of electricity in the higher-frequency zone. In the lower-frequency section, higher energy is required in order to transport charge carriers due to high resistance at grain boundaries, leading to a higher tanδ. By contrast, in the high-frequency region, the low resistance requires less energy for movement of charge carriers, leading to lower dielectric loss. Comparatively, the relative dielectric constant of $BiMgFeCeO_6$ ($\varepsilon_r = 12 \times 10^3$ at 400°C and frequency 10^3 Hz) is smaller than that of a similar compound, $Bi_{0.5}Pb_{0.5}Fe_{0.5}Ce_{0.5}O_3$ ($\varepsilon_r = 22 \times 10^3$ at a similar temperature and frequency [45]).

Figure 12.8a depicts the relationship of temperature with relative dielectric constant (ε_r) and tanδ. The value of ε_r remains relatively stable below 250°C, but progressively surges up to 500°C. This temperature-dependent behavior is also observed in tangent loss. Initially, tangent loss is low over low temperatures, but it exhibits a sharper increase at greater temperatures. The significant rise in the tangent loss upon higher temperatures is based on several factors, including presence of unidentified impurities (such as an oxygen vacancy), dispersion of electrically triggered charged carrier, and contamination stages inside samples [47]. In Figure 12.8b, it is evident that the tangent loss rises with increasing temperature because of predominance of conductivity. Since temperature range studied did not encompass the ferroelectric–paraelectric transforming temperature of $BiMgFeCeO_6$, no abnormal behavior was observed. Comparatively, tanδ of $BiMgFeCeO_6$ was discovered to be quite diminutive (tangent loss = 0.76 at 400°C, frequencies ranges 10^6 Hz) in contrast to a similar compound, $Bi_{0.5}Pb_{0.5}Fe_{0.5}Ce_{0.5}O_3$, which exhibited a higher tangent loss (3.9 at 400°C, frequency 10^6 Hz) [48]. The low value of the tangent loss in these materials holds potential for certain device applications.

The destructive-free complicated impedance technique is an effective approach for studying electrical characteristics of ionic conductors and electro-ceramics, including grain impedance, boundaries of grains, and

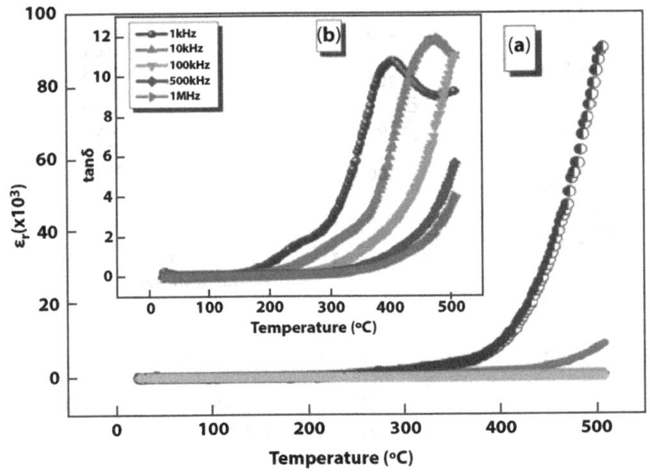

Figure 12.8 (a, b) Trends of dielectric constant and tangent loss within increase of temperature for $BiMgFeCeO_6$ nanomaterials. Reused from [46].

electrodes. This method provides valuable details regarding resistive and reactive impedances of material. By analyzing impedance calculations and the system's AC reaction to a sinusoidal disturbance at different frequencies and temperatures, impedance analysis can be used to characterize these materials [47, 49–52]. Figure 12.9 illustrates the frequency-dependent behavior of Z' and Z" at selected temperatures. The dielectric characteristics of material are frequently explained through complex dielectric constants (ε* = ε' - jε"), complex impedance (Z* = Z - jZ"), complex electric modulus (M* = M' - jM"), and dielectric loss that is interrelated. The real (Z') and imaginary (Z") parts of complex impedance might be described using the equations below [53, 54],

$$Z' = \frac{R}{1+(\omega\tau)^2} \tag{12.2}$$

$$Z'' = \frac{R\omega\tau}{1+(\omega t)^2} \tag{12.3}$$

Here, R= resistance, ω = angular frequency of electric field, and τ = RC = relaxation time.

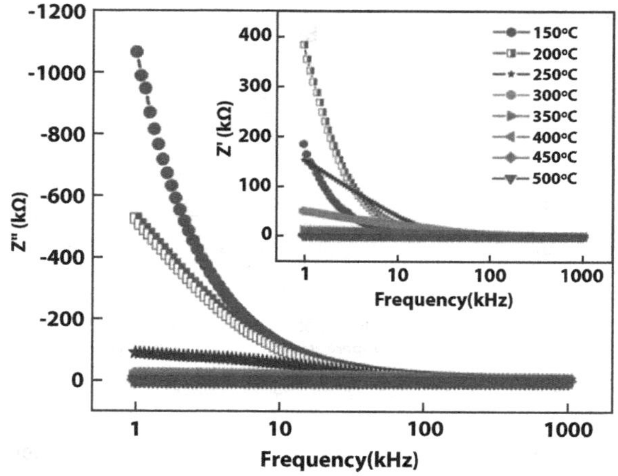

Figure 12.9 Real and imaginary impedances versus frequency. Reused from [46].

τ may be calculated via utilizing the equation $\tau = R_gC_g$, where R_g and C_g represent the resistance and capacitance of grains/bulk, respectively. At lower-frequency ranges, the amount of real impedance decreases whenever temperature increases, suggesting the NTCR reaction of material [55]. Conversely, at the high-frequency zone, NTCR behavior undergoes significant changes. Over a specific frequency, real impedance plot merges fully. Given various temperatures, real impedance values do not match, implying relief of space charge [56]. With increasing temperature, real impedance value falls, demonstrating the sample's relaxation characteristic [57]. The transition from 200°C to 250°C exhibits similar characteristics to that of $Bi_{0.8}Tb_{0.1}Pb_{0.1}Fe_{0.9}Ti_{0.1}O_3$ [58]. The relaxation process is attributed to the existence of stationary substances and electrons at reduced temperature, and deficiencies and voids at greater temperature [59].

12.4 Optical Characteristics of FE and AFE Materials

Optical properties of materials play a significant role in order to find their potential applications in optical devices [60]. The optical band-gap energy can be determined by application of Tauc's equation [61, 62] as

$$\alpha h\nu = B((h\nu - E_g)^m \tag{12.4}$$

Here, B, hv, and m indicate the constant value, energy of photons, and constant associated to the transition, respectively. The value of "m" might be 2 or 0.5 depending on the direct and indirect transition [63]. Absorption coefficient (α) [64] is determined applying the subsequent relationship as

$$\alpha = 2.303\frac{log(A)}{t} \tag{12.5}$$

where "A" is absorbance and "t" is the path distance of light where the absorbance occurs [65].

12.4.1 Optical Characteristics of KNaX Materials

The optical characteristics of FE and AFE materials occur from the interaction with electromagnetic waves, which can be studied by dielectric functions $\varepsilon(\omega)$ as

$$\varepsilon(\omega) = \varepsilon_1(\omega) - i\varepsilon_2(\omega) \tag{12.6}$$

The other optical factors like optical conductivity $\sigma(\omega)$, refractive index $n(\omega)$, extinction coefficient $k(\omega)$, optical reflectivity $R(\omega)$, and energy loss spectrum $L(\omega)$ are determined using the imaginary portion of Kramers Kronig transformation [66]. The material KNaX with X = S, Se, and Te consisted of an orthorhombic crystal structure having the space group *Pnma*; the three tensor modules ($\varepsilon^{xx}(\omega)$, $\varepsilon^{yy}(\omega)$, and $\varepsilon^{zz}(\omega)$) are obligatory to execute the optical parameters [67]. Figure 12.10 [67] shows ε' and ε'' for the material KNaX with X = S, Se, and Te against the photon energy for three crystal directions up to 30 eV. The dielectric function moves towards the small energy portions when shifting from S to Te. Such movement is homogeneous with the band structure and density of states (DOS). In Figure 12.10a, the anisotropy among $\varepsilon_2^{zz}(\omega)$ and dual $\varepsilon_2^{xx}(\omega)$ and $\varepsilon_2^{yy}(\omega)$ parameters is given in small energy regions. It means $\varepsilon_2^{zz}(\omega)$ is more leading in the given materials. The imaginary part of the dielectric constant is a combination of noticeable peaks with few extra peaks on the right positioned in energy portions 4.09 eV–7.8 eV for KNaS, 3.76 eV–7.63 eV for KNaSe, and 4.06 eV to 7.28 eV for KNaTe. The values for most prominent peaks are illustrated in Table 12.1 [67]. These are related to direct inter-band transition valance band (occupied state) to conduction band (unoccupied state). Assuming the partial density of states (PDOS),

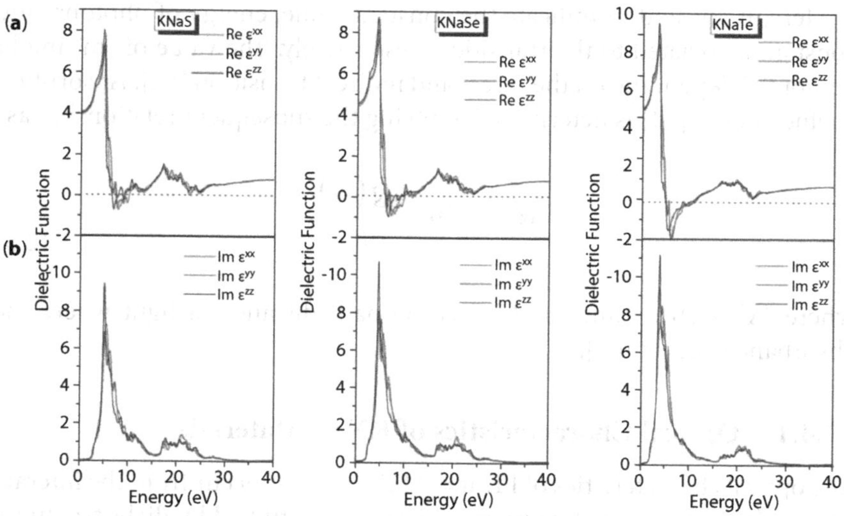

Figure 12.10 The ε′ (a) and ε″ (b) for KNaX (X = S, Se and Te) materials. Reused from [67].

the leading peaks of $\varepsilon_2^{zz}(\omega)$ positioned at 5.12 eV, 5.47 eV, and 3.99 eV are because of optical transition from occupied valance positions S-3p/Se-4p/Te-5p to unoccupied conduction states K-3d. The absorption peak is shown by KNaTe materials at 3.99 eV in the ultraviolet spectrum; this suggests that such material is applicable for optoelectronic gadgets. The threshold energy as a function of dielectric appears at 2.64 eV for KNaS, 2.29 eV for KNaSe, and 1.9 eV for KNaTe materials. The figure demonstrates the Drude nature for the compounds, as the real portion of dielectric function $\varepsilon_1(\omega)$ signs zero in certain ranges of energies and is more reliable for KNaTe material. At zero frequency, the real portion of dielectric function $\varepsilon_1(0)$ shows the dielectric reaction of the compound to static electric field. The Penn model [68] states that

$$\varepsilon_1(0) = 1 + \left(\frac{\hbar\omega_p}{E_g}\right)^2 \tag{12.7}$$

The obtained outcomes exhibit that band-gap energy is inversely proportional to $\varepsilon_1(0)$ and values of $\varepsilon_1^{xx}(0), \varepsilon_1^{yy}(0)$ and $\varepsilon_1^{zz}(0)$ as illustrated in Table 12.1.

Table 12.1 The utmost values of peaks for $\varepsilon_2^{xx}(\omega), \varepsilon_2^{yy}(\omega)$, and $\varepsilon_2^{zz}(\omega)$, the static dielectric constant $\varepsilon_1^{xx}(0), \varepsilon_1^{yy}(0)$, and $\varepsilon_1^{zz}(0)$, the dominant values of peaks for $L^{xx}(\omega)$, $L^{yy}(\omega)$, and $L^{zz}(\omega)$, and zero frequency limit reflectivity $R^{xx}(0)$, $R^{yy}(0)$, and $R^{zz}(0)$ for the materials KNaX with X = S, Se, and Te. Reused from [67].

	KNaS	**KNaSe**	**KNaTe**		**KNaS**		**KNaSe**		**KaTe**	
$\varepsilon_2^{xx}(\omega)$	5.37	4.77	4.52	$L^{xx}(\omega)$	13.56	24.47	13.16	24.61	3.87	7.72
$\varepsilon_2^{yy}(\omega)$	5.07	4.74	3.99	$L^{yy}(\omega)$	13.33	24.44	12.77	24.21	3.66	7.66
$\varepsilon_2^{zz}(\omega)$	5.12	4.74	3.99	$L^{zz}(\omega)$	14.04	24.92	13.42	24.64	3.86	7.66
$\varepsilon_1^{xx}(0)$	4.00	4.41	4.90	$R^{xx}(0)$	0.110		0.126		0.142	
$\varepsilon_1^{yy}(0)$	3.97	4.41	4.96	$R^{yy}(0)$	0.111		0.126		0.144	
$\varepsilon_1^{zz}(0)$	4.08	4.53	5.00	$R^{zz}(0)$	0.114		0.130		0.146	

12.4.2 Optical Characteristics of PLZST Materials

Optical transmission spectra versus wavelength in the range of 0.2–11.0 μm for the material (Pb, La) (Zr, Sn, Ti)O_3 or (PLZST) are illustrated in Figure 12.11 [41]. The PLZST materials with 001 coordination show the optical transmittance with the value 65% between 0.9 μm and 0.7 μm. The transmission deadline appears nearly at 0.4 μm or accurately at 360 nm in UV area with the lower energy gap having a value of 3.3 eV, comparable to the outcomes for PIN-PT particles with a wavelength of 0.4 μm [69]. In the infrared region at 10 μm, optical transmission disappears entirely. The cutoff wavelength in UV is substantially greater in comparison to the cutoff in IR, which has been mostly seen in oxygen-octahedral perovskites [70]. In the PLZT materials, the optical transparency region is greater than the $KTiOPO_4$ materials [71]. The results show that the materials are excellent candidates for the optical crystals in a wider wavelength state.

12.4.3 Optical Characteristics of $BiMgFeCeO_6$ Nanomaterials

Figure 12.12 depicts UV-visible optical absorption spectrum of the $BiMgFeCeO_6$ nanomaterial. The determination of band gap is commonly carried out using the Tauc plot method. In this study, the sample is considered to have a direct allowed transition with a value of n = 1/2, as stated in Ref. [72]. However, according to the relevant literature [73, 74], ferrites

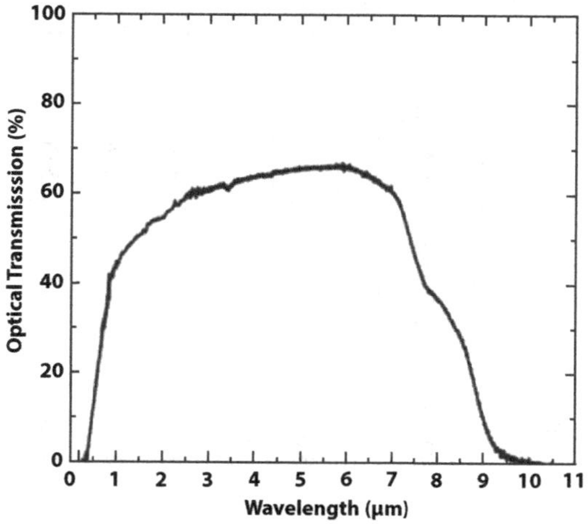

Figure 12.11 Wavelength-dependent optical transmission spectrum. Reused from [41].

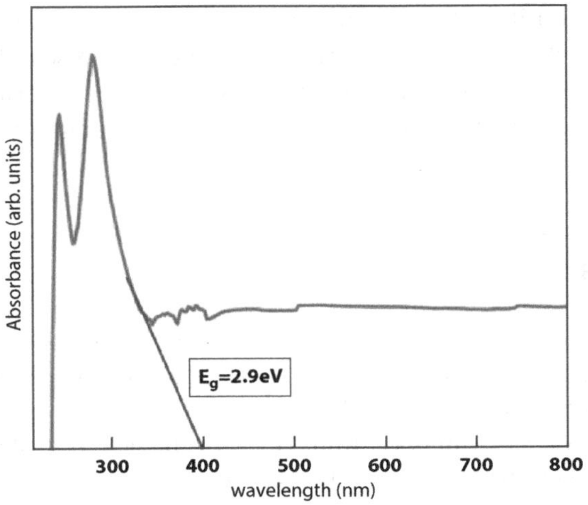

Figure 12.12 Optical band-gap energy of $BiMgFeCeO_6$ nanomaterial. Reused from [46].

typically exhibit an indirect allowed electronic transition. In the case of $BiMgFeCeO_6$, which contains bismuth ferrite, absorption is observed around the wavelength of 400 nm. By applying the equation $(h\upsilon\alpha)^2 - h\upsilon = E_g$, the energy band gap is estimated to be approximately 2.9 eV, matching related materials reported in previous studies [75].

12.5 Conclusion

Ferroelectric and antiferroelectric materials exhibit a rich variety of magnetoelectric, dielectric, and optical characteristics that distinguish them as highly attractive for a variety of technological purposes. The coupling between magnetic and electric fields in these materials can lead to new magnetic domains or the modification of existing domains in response to an applied electric field, while their high dielectric constants make them useful for capacitive applications such as energy storage devices and high-frequency filters. Additionally, their optical properties, such as birefringence and nonlinear optical response, make them promising candidates for applications in electro-optic modulators and photonic devices. The behavior of these materials is highly affected by the nature of their domain walls and defects, which must be understood in order to optimize their properties for specific applications. Overall, the study of magnetoelectric, dielectric, and optical characteristics of ferroelectric and antiferroelectric materials has significant implications upon creation of novel substances with characteristics tailored to certain uses. Further research in this area is needed to uncover new phenomena and to design materials with optimized properties, thereby expanding the potential range of technological applications for these materials.

References

1. Goldman, A., Applications and functions of ferromagnetic materials, in: *Handbook of Modern Ferromagnetic Materials*, A. Goldman, (Ed.), pp. 1–15, Springer US, Boston, MA, 1999.
2. Ryu, J., Priya, S., Uchino, K., Kim, H.E., Magnetoelectric effect in composites of magnetostrictive and piezoelectric materials. *J. Electroceramics*, 8, 107, 2002.
3. Buscaglia, M.T., Mitoseriu, L., Buscaglia, V., Pallecchi, I., Viviani, M., Nanni, P., Siri, A.S., Preparation and characterisation of the magneto-electric xBiFeO3-(1 - x)BaTiO3 ceramics. *J. Eur. Ceram. Soc.*, 26, 3027, 2006.
4. Kumar, A.S., Lekha, C.S.C., Vivek, S., Nandakumar, K., Anantharaman, M.R., Nair, S.S., Effect of CoFe2O4 weight fraction on multiferroic and magnetoelectric properties of (1 – x)Ba0.85Ca0.15Zr0.1Ti0.9O3–xCoFe2O4 particulate composites. *J. Mater. Sci. Mater. Electron.*, 30, 8239, 2019.
5. Hill, N.A. and Filippetti, A., Why are there any magnetic ferroelectrics? *J. Magn. Magn. Mater.*, 242–245, 976, 2002.
6. Ederer, C. and Spaldin, N.A., A new route to magnetic ferroelectrics. *Nat. Mater.*, 3, 849, 2004.
7. Efremov, D.V., Van Den Brink, J., Khomskii, D.I., Bond- versus site-centred ordering and possible ferroelectricity in manganites. *Nat. Mater.*, 3, 853, 2004.
8. Fischer, P., Polomska, M., Sosnowska, I., Szymanski, M., Temperature dependence of the crystal and magnetic structures of BiFeO3. *J. Phys. C Solid State Phys.*, 13, 1931, 1980.
9. Schmid, H., Multi-ferroic magnetoelectrics. *Ferroelectrics*, 162, 317, 1994.
10. Liu, Z., Feng, Z., Yan, H., Wang, X., Zhou, X., Qin, P., Guo, H., Yu, R., Jiang, C., Antiferromagnetic piezospintronics. *Adv. Electron. Mater.*, 5, 1900176, 2019.
11. Rehspringer, J.L., Bursik, J., Niznansky, D., Klarikova, A., Characterisation of bismuth-doped yttrium iron garnet layers prepared by sol-gel process. *J. Magn. Magn. Mater.*, 211, 291, 2000.
12. Fiebig, M., Fröhlich, D., Kohn, K., Leute, S., Lottermoser, T., Pavlov, V.V., Pisarev, R.V., Determination of the magnetic symmetry of hexagonal manganites by second harmonic generation. *Phys. Rev. Lett.*, 84, 5620, 2000.

13. Kimel, A.V., Pisarev, R.V., Bentivegna, F., Rasing, T., Ultrafast optical spectroscopy of hexagonal manganites RMnO3 (R = Y, Er, Sc). *Ferroelectrics*, 279, 135, 2002.
14. Karimi, S., Reaney, I.M., Han, Y., Pokorny, J., Sterianou, I., Crystal chemistry and domain structure of rare-earth doped BiFeO3 ceramics. *J. Mater. Sci.*, 44, 5102, 2009.
15. Lan, C., Jiang, Y., Yang, S., Magnetic properties of la and (La, Zr) doped BiFeO3 ceramics. *J. Mater. Sci.*, 46, 734, 2011.
16. Das, R., Gopal Khan, G., Mandal, K., Enhanced ferroelectric, magnetoelectric, and magnetic properties in Pr and Cr co-doped BiFeO 3 nanotubes fabricated by template assisted route. *J. Appl. Phys.*, 111, 104115, 2012.
17. Xu, J.M., Wang, G.M., Wang, H.X., Ding, D.F., He, Y., Synthesis and weak ferromagnetism of Dy-doped BiFeO3 powders. *Mater. Lett.*, 63, 855, 2009.
18. Akhtar, M., Hasan, M.S., Amin, N., Morley, N.A., Arshad, M.I., Tuning structural, electrical, dielectric and magnetic properties of Mg–Cu–Co ferrites via dysprosium (Dy3+) doping. *J. Rare Earths*, 42, 137–146, 2024.
19. Hasan, M.S., Ali, S.S., Khan, M.I., Rizwan, M., Zulqarnain, M., Hussain, A., Structural, optical, electrical and magnetic tuning based on Zn substitution at A site in yttrium doped spinel ferrites. *Mater. Chem. Phys.*, 301, 127538, 2023.
20. Zulqarnain, M., Ali, S.S., Hira, U., Feng, J.F., Khan, M.I., Rizwan, M., Javed, K., Farid, G., Hasan, M.S., Superparamagnetic contributions, optical band gap tuning and dominant interfacial resistive mechanisms in ferrites nanostructures. *J. Alloys Compd.*, 894, 162431, 2022.
21. Kumar, M. and Yadav, K.L., Study of room temperature magnetoelectric coupling in Ti substituted bismuth ferrite system. *J. Appl. Phys.*, 100, 074111, 2006.
22. Hill, N.A., Why are there so few magnetic ferroelectrics? *J. Phys. Chem. B*, 104, 6694, 2000.
23. Zhai, J., Cai, N., Shi, Z., Lin, Y., Nan, C.W., Magnetic-dielectric properties of NiFe2O4/PZT particulate composites. *J. Phys. D. Appl. Phys.*, 37, 823, 2004.
24. Zhang, M., Zhang, X., Qi, X., Li, Y., Bao, L., Gu, Y., Effects of sintering temperature and composition on dielectric, ferroelectric, and magnetoelectric properties of BaTiO3–BiFeO3 solid solutions. *Ceram. Int.*, 43, 16957, 2017.
25. Palai, R., Katiyar, R.S., Schmid, H., Tissot, P., Clark, S.J., Robertson, J., Redfern, S.A.T., Catalan, G., Scott, J.F., β phase and γ-β metal-insulator transition in multiferroic BiFeO3. *Phys. Rev. B - Condens. Matter Mater. Phys.*, 77, 014110, 2008.
26. Sagdeo, A., Mondal, P., Upadhyay, A., Sinha, A.K., Srivastava, A.K., Gupta, S.M., Chowdhury, P., Ganguli, T., Deb, S.K., Correlation of microstructural and physical properties in bulk BiFeO3 prepared by rapid liquid-phase sintering. *Solid State Sci.*, 18, 1, 2013.
27. Wang, Y.P., Zhou, L., Zhang, M.F., Chen, X.Y., Liu, J.M., Liu, Z.G., Room-temperature saturated ferroelectric polarization in BiFeO3 ceramics synthesized by rapid liquid phase sintering. *Appl. Phys. Lett.*, 84, 1731, 2004.

28. Lin, D., Zheng, Q., Li, Y., Wan, Y., Li, Q., Zhou, W., Microstructure, ferroelectric and piezoelectric properties of Bi0.5K0.5TiO3-modified BiFeO3-BaTiO3 lead-free ceramics with high Curie temperature. *J. Eur. Ceram. Soc.*, 33, 3023, 2013.
29. Schileo, G., Luisman, L., Feteira, A., Deluca, M., Reichmann, K., Structure-property relationships in BaTiO3-BiFeO3-BiYbO3 ceramics. *J. Eur. Ceram. Soc.*, 33, 1457, 2013.
30. Guo, Y., Xiao, P., Luo, L., Jiang, N., Lei, F., Zheng, Q., Lin, D., Structure, ferroelectric and piezoelectric properties of Bi 0.5(Na0.8K0.2)0.5TiO3 modified BiFeO3-BaTiO3 lead-free piezoelectric ceramics. *J. Mater. Sci. Mater. Electron.*, 25, 3753, 2014.
31. Zhang, M., Zhang, X., Qi, X., Zhu, H., Li, Y., Gu, Y., Enhanced ferroelectric, magnetic and magnetoelectric properties of multiferroic BiFeO3–BaTiO3–LaFeO3 ceramics. *Ceram. Int.*, 44, 21269, 2018.
32. Kowal, K., Guzdek, P., Kowalczyk, M., Jartych, E., Compositional dependence of hyperfine interactions and magnetoelectric coupling in (BiFeO3) x-(BaTiO3)1-x solid solutions. *Nukleonika*, 62, 117, 2017.
33. Mandal, S.K., Singh, S., Debnath, R., Nath, A., Dey, P., Magnetoelectric coupling, dielectric and electrical properties of xLa0.7Sr0.3MnO3 – (1–x) Pb (Zr0.58Ti0.42)O3 (x = 0.05 and 0.1) multiferroic nanocomposites. *J. Alloys Compd.*, 720, 550, 2017.
34. Mujtaba, A., Khan, M.I., Hasan, M.S., Ali, S.S., Shahid, W., Fatima, M., Abd-Rabboh, H.S.M., Sadaf, A., Alwadai, N., Tailoring the structural, optical, photoluminescence, dielectric and electrical properties of Zn0·6Ni0·2Mg0·2Fe2-xLaxO4 (x = 0.00, 0.0125, 0.0250, 0.0375). *J. Mater. Res. Technol.*, 23, 4538, 2023.
35. Bharadwaja, S.S.N., Saha, S., Bhattacharyya, S., Krupanidhi, S.B., Dielectric properties of La-modified antiferroelectric PbZrO3 thin films. *Mater. Sci. Eng. B Solid-State Mater. Adv. Technol.*, 88, 22, 2002.
36. Dai, X., Xu, Z., Viehland, D., Normal to relaxor ferroelectric transformations in lanthanum-modified tetragonal-structured lead zirconate titanate ceramics. *J. Appl. Phys.*, 79, 1021, 1996.
37. Liu, G., Fan, H., Dong, G., Shi, J., Chang, Q., Enhanced energy storage and dielectric properties of Bi0.487Na0.427K0.06Ba0.026TiO3-xCeO2 antiferroelectric ceramics. *J. Alloys Compd.*, 664, 632, 2016.
38. Liu, L., Fan, H., Ke, S., Chen, X., Effect of sintering temperature on the structure and properties of cerium-doped 0.94(Bi0.5Na0.5)TiO3-0.06BaTiO3 piezoelectric ceramics. *J. Alloys Compd.*, 458, 504, 2008.
39. Ujma, Z., Handerek, J., Pawelczyk, M., Hassan, H., Kugel, G.E., Carabatos-Nedelec, C., The antiferroelectric-ferroelectric-paraelectric phase sequence in lead-lanthanum zirconate-titanate ceramics with 8% Ti content. *J. Phys. Condens. Matter*, 6, 6843, 1994.
40. Uchino, K. and Nomura, S., Critical exponents of the dielectric constants in diffused- phase-trans ition crystals. *Ferroelectrics*, 44, 55, 1982.

41. Li, Y., Li, Q., Wang, L., Yang, Z., Gao, J., Chu, X., Domain, dielectric and optical studies in antiferroelectric (Pb, La) (Zr, Sn, Ti)O3 single crystals. *Solid State Commun.*, 152, 1791, 2012.
42. Pradhan, D.K., Misra, P., Puli, V.S., Sahoo, S., Pradhan, D.K., Katiyar, R.S., Studies on structural, dielectric, and transport properties of Ni 0.65Zn0.35Fe2O4. *J. Appl. Phys.*, 115, 243904, 2014.
43. Koops, C.G., On the dispersion of resistivity and dielectric constant of some semiconductors at audiofrequencies. *Phys. Rev.*, 83, 121, 1951.
44. Pawar, R.P. and Puri, V., Structural, electrical and dielectric properties of (Sr 1-xCax) MnO3 ($0 \leq x \leq 1.0$) ceramics. *Ceram. Int.*, 40, 10423, 2014.
45. Li, X., Deng, Z., Yin, Y., Zhu, L., Xu, D., Wang, Y., Teng, F., Efficiency enhancement of polymer solar cells with Ag nanoparticles incorporated into PEDOT:PSS layer. *J. Mater. Sci. Mater. Electron.*, 25, 140, 2014.
46. Parida, K., Dehury, S.K., Choudhary, R.N.P., Structural, electrical and magneto-electric characteristics of BiMgFeCeO6 ceramics. *Phys. Lett. Sect. A Gen. At. Solid State Phys.*, 380, 4083, 2016.
47. Sen, S. and Choudhary, R.N.P., Impedance studies of Sr modified BaZr0.5Ti0.95O 3 ceramics. *Mater. Chem. Phys.*, 87, 256, 2004.
48. Sahoo, S., Mahapatra, P.K., Choudhary, R.N.P., The structural, electrical and magnetoelectric properties of soft-chemically-synthesized SmFeO3 ceramics. *J. Phys. D. Appl. Phys.*, 49, 35302, 2015.
49. Brahma, S., Choudhary, R.N.P., Thakur, A.K., AC impedance analysis of LaLiMo 2O 8 electroceramics. *Phys. B Condens. Matter*, 355, 188, 2005.
50. Fang, L., Zhang, H., Huang, T.H., Yuan, R.Z., Liu, H.X., Preparation, structural, and dielectric properties of Ba 5YZnM9O30(M = Nb, Ta) ceramics. *J. Mater. Sci.*, 40, 533, 2005.
51. Macdonald, J.R., Impedance spectroscopy. *Ann. Biomed. Eng.*, 20, 289, 1992.
52. Suchanicz, J., The low-frequency dielectric relaxation Na0.5Bi0.5TiO3 ceramics. *Mater. Sci. Eng. B*, B55, 114, 1998.
53. Idrees, M., Nadeem, M., Hassan, M.M., Investigation of conduction and relaxation phenomena in LaFe 0.9Ni0.1O3 by impedance spectroscopy. *J. Phys. D. Appl. Phys.*, 43, 155401, 2010.
54. Rouahi, A., Kahouli, A., Challali, F., Besland, M.P., Vallée, C., Yangui, B., Salimy, S., Goullet, A., Sylvestre, A., Impedance and electric modulus study of amorphous TiTaO thin films: Highlight of the interphase effect. *J. Phys. D. Appl. Phys.*, 46, 65308, 2013.
55. James, A.R., Prakash, C., Prasad, G., Structural properties and impedance spectroscopy of excimer laser ablated Zr substituted BaTiO3 thin films. *J. Phys. D. Appl. Phys.*, 39, 1635, 2006.
56. Suman, C.K., Prasad, K., Choudhary, R.N.P., Complex impedance studies on tungsten-bronze electroceramic: Pb 2Bi3LaTi5O18. *J. Mater. Sci.*, 41, 369, 2006.

57. Chatterjee, S., Mahapatra, P.K., Choudhary, R.N.P., Thakur, A.K., Complex impedance studies of sodium pyrotungstate - Na2W 2O7. *Phys. Status Solidi Appl. Res.*, 201, 588, 2004.
58. Thakur, S., Rai, R., Bdikin, I., Valente, M.A., Impedance and modulus spectroscopy characterization of Tb modified Bi0.8A0.1Pb0.1Fe0.9Ti0.1O3 ceramics. *Mater. Res.*, 19, 1, 2016.
59. Jonscher, A.K., The "universal" dielectric response. *Nature*, 267, 673, 1977.
60. Hasan, M.S., Arshad, M.I., Ali, A., Mahmood, K., Amin, N., Ali, S.S., Khan, M.I., Mustafa, G., Khan, M.J., Saleem, M., Mg and la co-doped ZnNi spinel ferrites for low resistive applications. *Mater. Res. Express*, 6, 16302, 2019.
61. Amin, N., Ul Hasan, M.S., Majeed, Z., Latif, Z., Ajaz un Nabi, M., Mahmood, K., Ali, A., Mehmood, K., Fatima, M., Akhtar, M., Arshad, M.I., Bibi, A., Iqbal, M.Z., Jabeen, F., Bano, N., Structural, electrical, optical and dielectric properties of yttrium substituted cadmium ferrites prepared by co-precipitation method. *Ceram. Int.*, 46, 20798, 2020.
62. Hasan, M.S., Ali, S.S., Rizwan, M., Khan, M.I., Naeem Ullah, H.M., Irfan, M.I., Structural, optical, electrical and magnetic properties of Cu0.2Zn0.2Ni0.6-x MgxFe2O4 (x = 0.00, 0.15, 0.30, 0.45, 0.60) soft ferrites. *J. Alloys Compd.*, 956, 170392, 2023.
63. Farooq, W.A., Sajjad Ul Hasan, M., Khan, M.I., Ashraf, A.R., Abdul Qayyum, M., Yaqub, N., Almutairi, M.A., Atif, M., Hani, A., Structural, optical and electrical properties of Cu0.6CoxZn0.4-xFe2O4 (x = 0.0, 0.1, 0.2, 0.3, 0.4) Soft Ferrites. *Molecules*, 26, 1399, 2021.
64. Nabi, M.A.U., Moin, M., Hasan, M.S., Arshad, M.I., Bibi, A., Amin, N., Mahmood, K., Ali, S.S., Study of electrical transport properties of cadmium-doped Zn–Mn soft ferrites by co-precipitation method. *J. Supercond. Nov. Magn.*, 34, 1813, 2021.
65. Arshad, M.I., Hasan, M.S., Rehman, A.U., Akhtar, M., Tung, L.D., Amin, N., Mahmood, K., Ali, A., Trakoolwilaiwan, T., Thanh, N.T.K., Structural, optical, electrical, dielectric, molecular vibrational and magnetic properties of La3+ doped Mg–Cd–Cu ferrites prepared by Co-precipitation technique. *Ceram. Int.*, 48, 14246, 2022.
66. Hameau, S., Guldner, Y., Verzelen, O., Ferreira, R., Bastard, G., Zeman, J., Lemaître, A., Gérard, J.M., Strong electron-phonon coupling regime in quantum dots: Evidence for everlasting resonant polarons. *Phys. Rev. Lett.*, 83, 4152, 1999.
67. Belghit, R., Belkhir, H., Kadri, M.T., Heciri, D., Bououdina, M., Ahuja, R., Structural, elastic, electronic and optical properties of novel antiferroelectric KNaX (X = S, Se, and Te) compounds: First principles study. *Phys. B Condens. Matter*, 545, 18, 2018.
68. Penn, D.R., Wave-number-dependent dielectric function of semiconductors. *Phys. Rev.*, 128, 2093, 1962.

69. Tu, C.S., Hung, C.M., Wang, F.T., Chien, R.R., Yang, S.W., Dielectric and optical behaviors in relaxor ferroelectric Pb(In1/2Nb1/2)1-xTixO3 crystal. *Solid State Commun.*, 138, 190, 2006.
70. Wan, X., Luo, H., Wang, J., Chan, H.L.W., Choy, C.L., Investigation on optical transmission spectra of (1 - x)Pb(Mg 1/3Nb2/3)O3-xPbTiO3 single crystals. *Solid State Commun.*, 129, 401, 2004.
71. Guo, R. and Bhalla, A.S., Hypersonic anomalies and optical properties of (formula presented) and (formula presented) single crystals. *Phys. Rev. B - Condens. Matter Mater. Phys.*, 59, 251, 1999.
72. Marques, V.S., Cavalcante, L.S., Sczancoski, J.C., Paris, E.C., Teixeira, J.M.C., Varela, J.A., De Vicente, F.S., Joya, M.R., Pizani, P.S., Siu Li, M., Santos, M.R.M.C., Longo, E., Synthesis of (Ca,Nd)TiO3 powders by complex polymerization, rietveld refinement and optical properties. *Spectrochim. Acta - Part A Mol. Biomol. Spectrosc.*, 74, 1050, 2009.
73. Gao, F., Yuan, Y., Wang, K.F., Chen, X.Y., Chen, F., Liu, J.M., Ren, Z.F., Preparation and photoabsorption characterization BiFeO 3 nanowires. *Appl. Phys. Lett.*, 89, 102506, 2006.
74. Radojković, A., Savić, S.M., Jović, N., Ćirković, J., Despotović, Ribić, A., Branković, Z., Branković, G., Structural and electrical properties of BaCe0.9Ee0.1O2.95 electrolyte for IT-SOFCs. *Electrochim. Acta*, 161, 153, 2015.
75. Srivastav, S.K. and Gajbhiye, N.S., Low temperature synthesis, structural, optical and magnetic properties of bismuth ferrite nanoparticles. *J. Am. Ceram. Soc.*, 95, 3678, 2012.

13

Ferroic Characteristics of Metal-Halide Perovskites

Nawishta Jabeen[1*], Ahmad Hussain[2†], Najam ul Hassan[3] and Jazib Ali[4]

[1]*Department of Physics, Fatima Jinnah Women University Rawalpindi, Rawalpindi, Pakistan*
[2]*Department of Physics, The University of Lahore, Sargodha Campus, Sargodha, Pakistan*
[3]*Department of Physics, Division of Science and Technology, University of Education, Lahore, Pakistan*
[4]*Center for Hybrid and Organic Solar Energy (CHOSE), University of Rome Tor Vergata, Rome, Italy*

Abstract
With the rapid growth in smart electronic, photovoltaic, and optoelectronic devices, the need for materials to demonstrate suitable electronic and optical properties has increased in recent years. Metal-halide perovskites (MHPs) have emerged as auspicious and innovative materials with a semiconducting nature to show the strong characteristics of photovoltaic and optoelectronic devices/applications. MHPs have proven to be pioneer materials to open up a novel era of advanced light-dependent characteristics like absorption, transmission, emission, and modulation for future photovoltaic and optoelectronic technology. Recently, various solution-based MHPs have been discovered and reported by researchers, which demonstrate intriguing characteristics such as ferroelasticity, ferroelectricity, strong light absorption, and high charge mobility. MHPs have gained tremendous popularity recently due to their debatable phenomenon of existence of ferroelectricity. In theoretical studies, due to the perovskite structure of MHPs, the existence of ferroelectricity is highly predicted. In this chapter, the theoretical and experimental evidence of ferroic nature of MHPs will be discussed, and ferroelectricity within the dimensions dependent on MHPs and their applications in devices will be examined. Dimensionality of the fabricated MHPs can influence

**Corresponding author*: nawishta.jabeen@fjwu.edu.pk
†*Corresponding author*: ahmed_00277@yahoo.com

Inamuddin, Tariq Altalhi and Mohammad Abu Jafar Mazumder (eds.) Ferroic Materials-Based Technologies, (305–326) © 2024 Scrivener Publishing LLC

the ferroelectric performance, and different dimensional MHPs are discussed with their reported ferroelectricity values to demonstrate a comprehensive analysis. From the detailed analysis, it can be concluded that such rapid progress in MHPs with ferroelectric characteristics has opened up many new horizons to design novel electronic, photovoltaic, and photoelectric devices.

Keywords: Metal-halide perovskite, ABX_3, ferroelectricity, Curie temperature, domain switching

13.1 Introduction

In the era of smart optoelectronic devices and photovoltaics, metal-halide perovskites (MHPs) have gained significant attention to provide reliable and efficient energy sources due to their large surface area, flexibility, and substantial optical characterizations. Power conversion efficiency (PCE) is considered as the fundamental parameter for the optoelectronic devices; among MHP solar cells, this value has surpassed 25% [1]. MHPs are presented with the general formula ABX_3, where A is a positively charged organic cation like CS^+, $NH_2CH_3NH_2^+$, or $CH_3NH_3^+$, B is a divalent metallic cation like Pb or Sn, while X is a monovalent halide anion like Br^-, Cl^-, or I^-; it is reported that monovalent halide anions contribute larger charge mobility and light-absorbing ability as compared to other metallic cations or polymer-based counterparts [2]. Optoelectronic and photovoltaic characterizations of the MHPs are dependent on specific polymorphs of materials; all MHPs possess different forms of polymorphs. Controlling the parameters to relate polymorphism is important for the assured reliable performance of MHPs along with their long-lasting stability. Scientists have reported that usually at room temperature, MHPs like $CsSnI_3$, $FAPbI_3$, and $CsPbI_3$ transform to insulating non-perovskite phases, which is not suitable to be utilized in optoelectronic/photovoltaic devices as active layers [3]. At elevated temperatures, MHPs are usually found in cubic structure symmetry. Transformation of MHPs towards low-symmetry polymorphs under decreasing temperature causes BX_6 octahedral tilting. The various investigations regarding pressurization and crystal size reduction are conducted to understand the concepts of thermodynamic stability and temperature-dependent polymorph transition in MHPs.

Lev Perovski was the first Russian scientist who discovered the perovskite $CaTiO_3$, where X-ions make the corner sharing octahedra with B-ions at the center, while A-ions inhabit the planetary between octahedra

(Figure 13.1) [4]. Tolerance factor (t) calculation is considered to be the best presentation of the stability of the ABX_3 perovskites with the following equation [5]:

$$t = \frac{R_A + R_X}{\sqrt{2}\left(R_B + R_X\right)}$$

where R_A, R_B, and R_X, are the ionic radii of A, B, and X ions, respectively. Moreover, for the perfect formation of perovskite materials, the selection of dependence of anion on cation radii is referred to conjunction of octahedral factor (μ) as follows [5]:

$$\mu = \frac{R_M}{R_X}$$

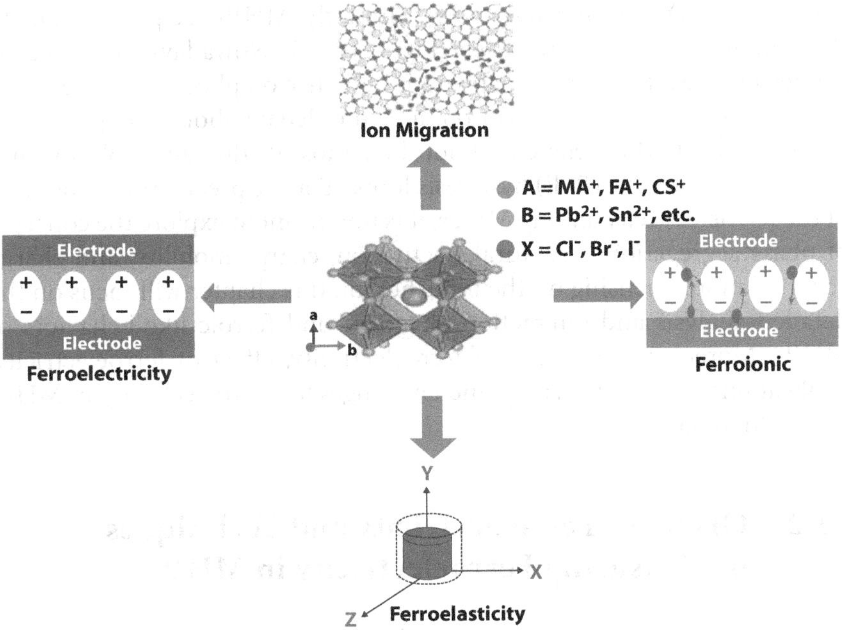

Figure 13.1 Schematic illustration of the ferroelectricity and ionic properties in MHPs. Ion migration image is used after permission [9], Copyright 2016, ACS Publication.

Travis *et al.* reported the joint reviewed octahedral factors and tolerance factors focused on precise estimate perovskite structures at ambient conditions and defined the threshold values for perovskites ($t \leq 1.06$ and $\mu > 0.41$) [5]. The electronic and optical property-related applications are dependent on the dimensions of MHPs. Similarly, the electronic and optoelectronic utilizations of hybrid MHPs are also the most explored topic these days.

Ferroelectrics are materials that show spontaneous polarization, which can be reversed under the influence of the external electric field. In other words, ferroelectricity is the fundamental approach of the materials (ferroelectric) that is implemented for binary information storage applications and is the auspicious approach for adoption in non-volatile memories [6]. Perovskite oxides (ABO_3) are renowned materials for their high ferroelectric performances. Lead-based perovskites (i.e., $PbZrTiO_3$) are frequently studied due to their ferroelectric (remnant polarization, P_r ~40μC/cm^2) and piezoelectric (d_{33} ~630 pC N^{-1}) properties [7], but due to their toxic nature, they have restricted applications. Lead-free ferroelectric perovskites have gained popularity with the passage of time. Hussain *et al.* have reported the maximum polarization (P_{max}) of 48 μC cm^{-2} in the $0.6Bi_{1.05}FeO_3$–$0.40Bi_{0.5}K_{0.5}TiO_3$ ceramic system [8]. Similarly, MHPs are popular due to their multifunctional properties (Figure 13.1). Nowadays, ferroelectric properties of MHPs and their apparent influence on photovoltaic behavior have received significant attention, and the debate about this procedure is unsolved yet. This debate is about the nanoscale domain evolution and polarization–electric (P–E) hysteresis loops. For the precise understanding of the ferroelectricity among MHPs, it is important to explore the concepts of phase segregation, structural fluctuation, charge mobility, and charge accumulation. Focusing on the topic herein, this chapter will focus on the detailed analysis and interaction of ferroic and ferroic-like behaviors of MHPs. Moreover, the origin of ferroelectricity, effect of ferroelectricity, applications of ferroelectricity, and challenges for ferroelectricity in MHPs will be highlighted.

13.2 Origin of Ferroelectricity and Techniques of Measuring Ferroelectricity in MHPs

Ferroelectrics are a subclass of pyroelectrics, followed by piezoelectrics, and they all are from dielectric materials (Figure 13.2a). Moreover, ferroelectric materials belong to 10 symmetry breaking point groups. Although

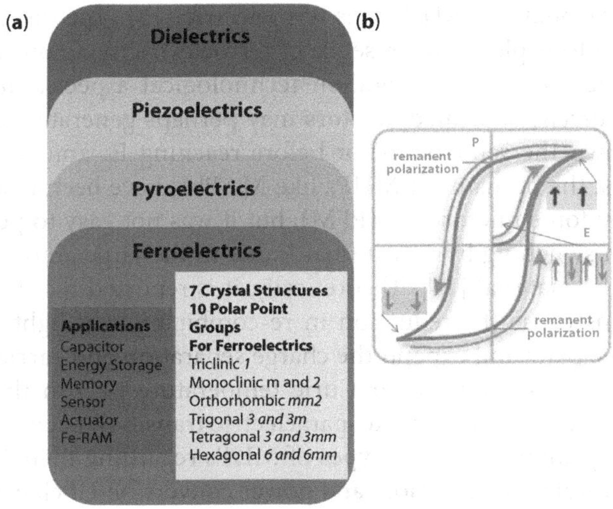

Figure 13.2 (a) Standard relationship between dielectric, piezoelectric, pyroelectric, and ferroelectrics. (b) Obvious P–E hysteresis loop presentation for ferroelectrics.

it is believed that ferroelectricity usually occurs in crystalline solids, some monoclinic symmetry liquid crystals consisting of C and H of chiral molecules also demonstrate ferroelectricity [10]. Ferroelectric materials possess lots of tiny microscopic regions (electric domains) and the polarization (P) becomes zero when domains are randomly oriented (Figure 13.2b). These domains tend to align themselves under the influence of external applied electric field (E), and achieve unidirectional orientation when the value of E crosses the coercive electric field (E_C); such polarization is called saturated/spontaneous polarization (P_S), and vice versa, for the E in the opposite direction [11].

Numerous techniques are employed to analyze the ferroelectricity among the MHPs including polarization versus electric field (P–E) loops, piezoresponse force microscopy, conductive atomic force microscopy, and Kelvin probe force microscopy. Some researchers have reported the presence of ferroelectricity among $CH_3NH_3PbI_3$, while others relate this effect to the leakage current, ion migration, or charge trapping. Scientists have achieved several milestones in the research of ferroelectricity among MHPs. Coll *et al.* highlighted that $CH_3NH_3PbI_3$ is not ferroelectric at room temperature because polarization retention time is very short [12]. Rakita *et al.* clarified that the imaginary part of polar behavior in dielectrics is dominant in semiconductors and the experimental P–E loop in $CH_3NH_3PbI_3$ single crystal is due to the imaginary part of permittivity, concluding that the tetragonal $CH_3NH_3PbI_3$ is ferroelectric [13]. The discussion and investigation to explore the presence of ferroelectricity among MHPs are important due to future generation technological aspects. In addition, the inside current of semiconductors may perhaps generate enough heat to break down the semiconductor before reaching E_C among MHPs. In recent times, the domains of MHPs like $MAPbI_3$ have been detected by a piezoelectric force microscope (PFM), but it was not easy to perform the polarization switching by an external E due to the large amount of inner current of the material [14]. Theoretically, it is reported that ferroelectric nanodomains cause the reduction in re-combination of light influential electron/hole pairs and expand the charge separation and carrier lifetime. A series of MHPs exhibit their Curie temperature (T_C) in the range of 370–380 K where ferroelectric-to-paraelectric transition occurs. The comparison study among different types of MHPs regarding their Curie temperature, saturated polarization, and power conversion efficiency (PCE) is presented in Table 13.1.

Table 13.1 Some basic physical properties of MHPs ferroelectrics.

Material	T_C (K)	P_S (μC cm^{-2})	PCE or J_{SC}	Ref.
$MAPbI_3$ film	327		PCE ~ 20%	[13]
$MAPbI_{3-x}Cl_x$ film			PCE ~ 19%	[15]
$FAPbBr_3$			PCE ~ 6.1%	[16]
$CsPbBr_3$ (under 15 GPa)		23	PCE ~ 7.78%	[17]
$(PEA)_2(MA)_4Pb_5I_{16}$	44	~0.25		[18]
$(EA)_2(MA)_2Pb_3Br_{10}$ crystal	75	3.7	J_{SC} ~ 4.1 ìA cm^{-2}	[19]
TC-$MnCl_3$ crystal	06	4		[20]
R_2PbI_4 crystal	83	13.96		[21]
$MAPb_{1-x}FexI_3$ (x = 0–0.07)		~2		[22]
$(EA)_4Pb_3Br_{10}$ crystal	374	3	I_{SC} ~ 15 nA	[23]
MD-NH_4I_3 crystal	48	19	-	[24]
$(TF)_{0.26}(TC)_{0.74}CdCl_3$ crystal	370	~8	-	[25]
S_2PbI_4 crystal	73.2	13.96	-	[21]

13.3 Ferroelectricity in MHPs and Large Polaron Phenomenon

Lead-based MHPs fabricated at room temperature by the aqueous solution method have shown a significant characteristic to be utilized in light-emitting diodes (LEDs), photovoltaics, and solar cells. In spite of tremendous defect density, lead-based MHPs like $CH(NH_2)_2PbI_3$ (i.e., $FAPbI_3$), $CsPbBr_3$, and $MAPbI_3$ have demonstrated outstanding charge mobility similar to conventional Si and GaAs semiconductors [26].

Recently, Miyata *et al.* have reported the influence of large polarons on the properties of the materials; the authors also reported that such large polarons hinder the Coulomb potential among the crystals and decrease carrier scattering due to charge defects [27]. Two types of polarons exist among ferroelectrics: (1) ferroelectric large polaron and (2) Fröhlich large polaron. For the ferroelectric large polaron model, a greater electric field of the range 0.1 to 1 MV/cm in the vicinity of a charge is employed to induce collective polarization and polar nanodomains in the ferroelectrics, while for the Fröhlich large polaron, polarization is produced due to ionic lattice or charged defects, which causes the linear response to an external applied electric field, resulting in the dipole moment in a unit cell; it decreases with the increase in space due to inserted charges [27]. Here, it can be concluded that the ferroelectric large polaron model offers a novel opportunity for defect-tolerant hybrid/composite MHPs to be utilized in semiconductor devices.

Hermes *et al.* were the first to demonstrate the twin domains for lateral and vertical PFM that contribute to ferroelasticity and ferroelectricity (Figure 13.3a), adding to the wide-ranging concerns to discuss the ferroelectric nature of MHPs. Preferably, lateral and vertical PFM detect in-plane and out-of-plane ferroelectric polarization; nonetheless, it is mostly expected that non-ferroelectric phenomena also contribute to PFM signals. However, in $CH_3NH_3PbI_3$, no switching among twin-domain contrast was detected either by out-of-plane electric field or by in-plane electric field. Twin-domain switching can possibly occur by implementing electric poling, which can cause the production of ferroelectric phenomenon, electrostrictive effect, and ion migration during poling. Such twin-domain phenomenon has a strong influence on the optoelectronic properties of the MHPs [28]. Afterwards, Huang *et al.* exhibited that adjacent twin domains affect the surface potential and photo current (Figure 13.3b); heterogeneity in $CH_3NH_3PbI_3$ can influence the photovoltaic behavior of $CH_3NH_3PbI_3$ solar cells. Such variation in photocurrent and surface potential has been related to the intrinsic properties of adjacent twin domains; their polar property show negative shift of

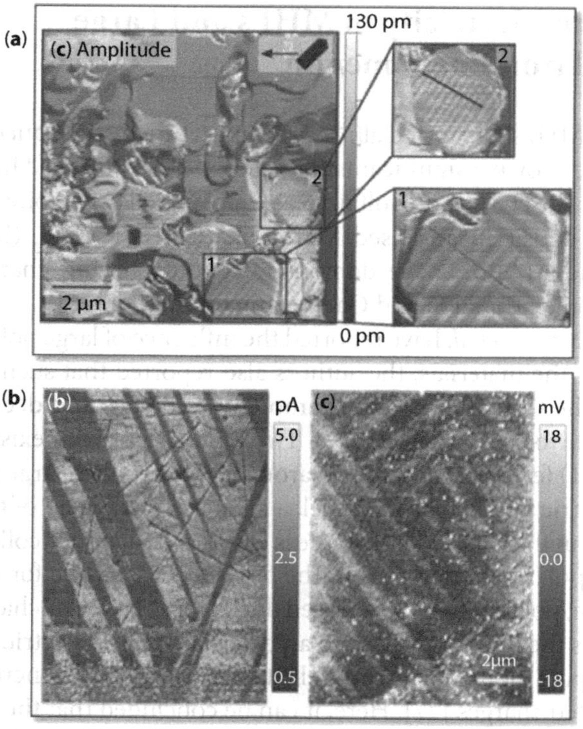

Figure 13.3 Nanoscale investigation of ionic and ferroic nature of MHPs. (a) Magnified PFM amplitude analysis, used after permission [28], Copyright 2016, ACS Publication. (b) Distribution of photo-current without DC-bias patterns and distribution of surface potential under light of ferroic domain, used after permission [29], Copyright 2018, Nature.

surface potential and minute photocurrent. Conversely, the other existing nonpolar domains demonstrate a large photocurrent and a positive shift in surface potential [29]. Röhm *et al.* proposed that variation of photocurrent can be due to ferroelectric polarization-induced electric field. Furthermore, the authors describe that in $CH_3NH_3PbI_3$ (Cl), the polarization component in vertical PFM can cause the increment of charge carrier extraction up to 25% [30]. Strelcov *et al.* studied the poly-crystals and single crystal of $CH_3NH_3PbI_3$ and showed the behavior of twin domains in the material under external stress [31]. Liu *et al.* investigated the twin-domain structure of $CH_3NH_3PbI_3$ using different innovative PFM methods like interferometric displacement sensor PFM, dual AC Resonance Tracking and PFM, and band excitation PFM. These measurements revealed that elastic deviations among ferroelastic twin domains occur within twin-domain contrast in PFM images by influencing cantilever dynamics in the course of calculation [32].

13.4 Theoretical Analysis of Ferroic Domain Evolution

It is reported that ferroelectric nanodomains in MHPs are the reason to restrict the hole distribution and charge distribution, resulting in the formation of individual paths for holes and electrons; it is best to optimize photovoltaic properties after the reduction of possible charge recombination given such phenomenon [32]. Theoretically, such kind of confinement is supposed to be due to the accumulation of polarization-driven charges near domain walls. Fan *et al.* reported the inconsistency in the ferroelectricity of $CH_3NH_3PbI_3$ among experimental analysis and theoretical simulation, referring to the orientation disorder of polar $CH_3NH_3^+$ in a real system [33]. In theoretical calculations, it is observed that organic $CH_3NH_3^+$ counterparts in MHPs are about ~6 Å from each other. Alongside the significant permanent electric dipoles, dipole–dipole interaction energy was also calculated (25 meV). From the analysis, it has been observed that such low dipole–dipole interaction energy is compatible with the thermal energy of the material at ambient conditions; hence, it can be concluded that a very complex kind of ferroelectric property is expected due to the rotation of domains [34]. Frost *et al.* implemented Monte Carlo simulations to analyze the anti-ferroelectric alignment of dipoles at low-temperature transition to the paraelectric alignment at high temperature, and the authors proposed that such transition results in the ferroelectric domain structure achieving equilibrium and such polarizability is produced due to the molecular rotation of CH_3NH_3 in the MHPs. Additionally, domains in ferroelectrics have been estimated to fluctuate under the influence of applied electric field and temperature because of the rotation of CH_3NH_3 . In conclusion, it was theoretically expected that MHPs may demonstrate intrinsic ferroelectric characteristics [35].

13.5 Experimental Evidences of Ferroelectricity on MHPs

From a theoretical study of the ferroic nature in MHPs, it can be concluded that lots of experimental evidences are required to fill the gap and to estimate the influence of ferroelectricity reported during computational simulations. Theoretically, it has been suggested that polarization can occur in ferroelastic domain walls regardless of whether the ferroelastic domains are polar or non-polar, and assist the charge separation in $CH_3NH_3PbI_3$ applications [36].

Liu *et al.* suggested that some other approaches in ferroelastic twin domain can also influence the optoelectronic performances of MHPs. Firstly, electronic properties of MHPs can be influenced by implementing the strain condition, just like adjacent ferroelastic twin domains with different strain conditions can disturb and modify the local electronic performances of MHPs. Secondly, ionic dispersal along with twin domains can disturb the optoelectronic performances of the MHPs, as local electronic band structure and charge dynamics of the materials may change the local chemical composition [32].

Stoumpos *et al.* reported the nonohmic hysteresis analysis under dark conditions of $CH_3NH_3PbI_3$ by altering current polarity. Nonohmic hysteretic properties of the material were initiated by the re-orientation of permanent dipoles of CH_3NH_3 by employing an external electric field. As a result, the permanent dipoles are aligned in CH_3NH_3 with the influence of an external electric field, leading to the formation of ferroelectric domains, and finally, the ferroelectricity is produced in the material [37]. Kutes *et al.* investigated the direct confirmation of the existence of ferroelectric domains in β-$CH_3NH_3PbI_3$ by employing PFM. The ferroelectric domains reported were about ~100 nm, which were reversely switched by implementing DC bias with β-$CH_3NH_3PbI_3$ [38]. Moreover, Xiao *et al.* measured the switchable photocurrent by implementing low electric field in MHP photovoltaic devices. Such low-electric-field-generated switchable photocurrent may be attributed to the ferroelectric behavior in the perovskite solar cells [39].

Liu *et al.* successfully fabricated the biaxial 2D trilayered hybrid perovskite ferroelectric $\{(EA)_2(MA)_2Pb_3Br_{10}\}$, a material that has shown the following ferroelectric properties: a T_C of ~375 K, a P_S of ~3.7 μC cm^{-2}, and a high photocurrent density of ~4.1 μA cm^{-2} [40]. Wang *et al.* fabricated the 2D trilayered hybrid perovskite ferroelectric ($EA_4Pb_3Cl_{10}$, EA = ethylammonium), a material that presented a T_C of ~415 K, a P_S of ~4.5 μC cm^{-2}, and a wide bandgap (3.39 eV). The investigated material is suitable for use in self-powered photodetectors [41].

13.6 Ferroic Behavior in MHPs Depending on Dimensions

Moreover, focusing on the dimensions of the MHPs, there are four categories, namely, zero-dimensional (0D), one-dimensional (1D), two-dimensional (2D), and three-dimensional (3D), contingent on spatial arrangements of BX_6 octahedral tilting.

13.6.1 Ferroelectricity in Zero-Dimensional (0D) MHPs

Halide perovskites with 0D consist of an organic or inorganic cation matrix, isolating the metal-halide clusters or metal-halide octahedral anions. Owing to such isolation of metallic halide octahedra, these 0D halide perovskites demonstrate several fascinating characteristics, including large strong exciton binding energy, Stokes shifts, and significant stability under ambient environmental conditions. The discovery of ferroic nature in such materials may result in the development of novel properties within 0D halide perovskites. Cd-based ferroelectrics are regarded as the most explored 0D MHP ferroelectrics. Commonly, Cd-based materials exhibit weaker spontaneous polarizations (<1 μC cm^{-2}). Wang *et al.* explored the 0D lead-iodide-based perovskite DMA_4PbI_6, which showed a P_S of ~0.3 μC cm^{-2} at ambient environment, which is even less than the P_S value for Cd-based ferroelectrics [42]. Huang *et al.* reported the organic–inorganic hybrid molecular $[cyclopentylammonium]_2CdBr_4$(**1**) ferroelectric, with a T_C of ~340.3 K and a P_S of ~0.57 μC cm^{-2} at room temperature [43]. Li *et al.* reported the organic–inorganic lead halide $[BrCH_2CH_2N(CH_3)_3][PbBr_3]$ perovskite ferroelectric, attained by mixing the quasi-spherical (2-bromoethyl) trimethylammonium cation to the infinite 1D zigzag $[PbBr_3]^-$ anion, which exhibited a switchable ferroelectric phase transition (0.36 μC cm^{-2}) at 286–302 K [44].

Luo *et al.* [31] established a 0D lead-free hybrid compound family like $(NMP)_3Sb_2Cl_{9-9x}Br_{9x}$, with x from 0 to 1. Three typical compounds were presented by researchers, $(NMP)_3Sb_2Br_9$ (NMPB, x = 1), $(NMP)_3Sb_2Cl_6Br_3$ (NMPCB, x = 1/3), and $(NMP)_3Sb_2Cl_9$ (NMPC, x = 0), possessing the same crystal structures. For example, for NMPC, the isolated Sb_2C19_3-dioctahedron is the basic unit of the crystal structure of NMPC, and three protonated organic cations are diligently linked through N–H···Cl hydrogen bonds. The T_C of three as-synthesized samples was nearly ~320 K, with a P_S of ~5.2–7.6 μC cm^{-2} [45]. To date, not enough research work is performed on 0D MHPs, presenting that there is still so much opportunity left for the development of 0D halide perovskite ferroelectrics with the exploration of their splendid characterizations.

13.6.2 Ferroelectricity in One-Dimensional (1D) MHPs

Recently, for the investigation of ferroelectric analysis in the 1D MHP nanostructure, ferroelectric behaviors are significantly explored. 1D ferroelectric halide perovskite materials with low dimensions enable the creation of single-domain nanostructures; as the size of the crystal reduces, ferroelectric behavior improves. Zhang *et al.* have studied the series of Mn-based 1D halide perovskites. Firstly, they designed the 1D ferroelectric (Pyrrolidinium)$MnBr_3$ and reported a P_S of ~6 μC cm^{-2} [46]. Later, Zhang *et al.* completely replaced the Br with Cl ions for (Pyrrolidinium)$MnBr_3$ to analyze the influence on the T_C. As expected, (Pyrrolidinium)$MnCl_3$ demonstrated a high T_C of ~295 K, but P_S was reduced to ~5.5 μC cm^{-2} [47]. It is well known that T_C has a strong influence to restrict the ferroelectric applications of the materials, because T_C controls the P_S. For the improvement in the value of T_C, Ye *et al.* have utilized the 3-pyrrolinium organic cations to replace pyrrolidinium in (Pyrrolidinium)$MnCl_3$. The as-fabricated (3-pyrrolinium)$MnCl_3$ demonstrated a high T_C of ~376 K and an improved P_S of ~6.2 μC cm^{-2} [48]. Li *et al.* reported the ferroelectric (2-CE)$PbBr_3$1D halide perovskite material, which has presented a ferroelectric phase transition above room temperature with a P_S of ~6.20 μC cm^{-2} [49]. Li *et al.* designed the organic–inorganic (3-pyrrolinium)$CdBr_3$ hybrid perovskite, which has shown excellent ferroelectric behavior (with an excellent P_S of ~7.0 μC cm^{-2}) [50]. The presented results, ideas, and strategies reveal the excellent tunability of low-dimensional (1D) halide perovskite ferroelectrics; this discussion also leads to new ideas for the development of suitable multifunctional chiral ferroelectrics.

13.6.3 Ferroelectricity in Two-Dimensional (2D) MHPs

MHPs with 2D structures have significant liberty for the selection of functional organic cations as these materials are slightly inhibited by Goldschmidt's tolerance factor (t); such phenomenon is beneficial to improve ferroelectric properties. Dion-Jacobson (D-J) and Ruddlesden-Popper (R–P) layered structures are the members included in 2D MHPs. In layered 2D halide perovskites, the organic spacer cations control the choice of arrangements, flexibility, and separation of inorganic stacking. 2D MHPs have demonstrated significance to be utilized in ferroelectric applications and devices. Liao *et al.* reported the fabrication of monolayered $(BEA)_2PbCl_4$ perovskite (with noticeable ferroelectric behavior), which has demonstrated a high P_S of ~13 μC cm^{-2} and also a good T_C of ~438 K [51]. Ye *et al.* designed the $(CHA)_2PbBr_{4-4x}I_{4x}$ (CHA = cyclohexylammonium, x = 0–1) material and reported bandgap regulations in the range of 3.05–2.74 eV; this material has shown a considerable P_S of ~8.5 μC cm^{-2} at the x = 0.1125 composition, with a T_C of ~378 K [52]. Shi *et al.* synthesized the new 2D $(2\text{-FBA})_2PbCl_4$ (2-FBA = 2-fluorobenzylammonium) ferroelectric halide perovskite; the sample has shown a high T_C of ~448 K and a P_S of ~5.35 μC cm^{-2} [53]. Wu *et al.* have fabricated the

Cs-based bilayer 2D halide perovskite by alloying n-butylamine (BA) into the $CsPbBr_3$. The as-fabricated $BA_2CsPb_2Br_7$ demonstrated a significant P_S of ~4.2 μC cm^{-2} and a T_C of ~412 K [54]. Li *et al.* fabricated the $(BA)_2(FA)Pb_2Br_7$, a new bilayered halide perovskite ferroelectric; later, the authors substituted the formamidinium (FA) in place of Cs as $BA_2CsPb_2Br_7$, and the modified sample demonstrated a T_C of ~322 K and a P_S of ~3.8 μC cm^{-2} [55].

Other than bilayer 2D halide perovskite materials, scientists have also explored the trilayered 2D halide perovskites with the general formula $R_2A_{n-1}B_nX_{3n+1}$, with n = 3 for the ferroelectric analysis. Ji *et al.* reported the fabrication of $(BA)_2(EA)_2Pb_3Br_{10}$ ferroelectric materials, with a P_S of ~5 μC cm^{-2} and a T_C of ~380 K [56]. Wang *et al.* reported the fabrication of the trilayered 2D $(IA)_2(MA)_2Pb_3Cl_{10}$ (IA = isoamylammonium) halide perovskite ferroelectric sample, exhibiting a P_S of 5 μC cm^{-2} by 3D to 2D tailoring dimensions [57]. Li *et al.* reported the fabrication of a lead-free $BA_2(MA)_2Sn_3Br_{10}$ halide perovskite semiconductor, with a high P_S of ~11.76 μC cm^{-2} and a T_C of ~318 K [58].

Moreover, scientists have also designed the multiaxial 2D $BA_2CsAgBiBr_7$ halide double perovskite ferroelectric possessing the four polarization directions, exhibiting a P_S of 5.2 μC cm^{-2} and a T_C of ~273 K [59]. Zhang *et al.* designed the $(DFPIP)_4AgBiI_8$ (DFPIP = 4,4-difluoropiperidinium) double perovskite ferroelectric, which demonstrated a P_S of 10.5 μC cm^{-2} and a T_C of 422 K [60]. Recently, Jabeen *et al.* have fabricated the single and multilayered $CH_3NJH_3PbIBr_2$ mix halide perovskite thin films [61]. The appearance of a pseudo-cubic phase is the main reason for the ferroelectric behavior of the thin films. It is reported that the multilayer $CH_3NH_3PbIBr_2$ thin film has demonstrated a leaky ferroelectric property with a leakage current of ~5.06×10^{-6} A and a resistivity of ~1.60×10^{6} Ω.cm along with a P_S of ~0.5 μC cm^{-2} at 50 kV cm^{-1} (Figure 13.4).

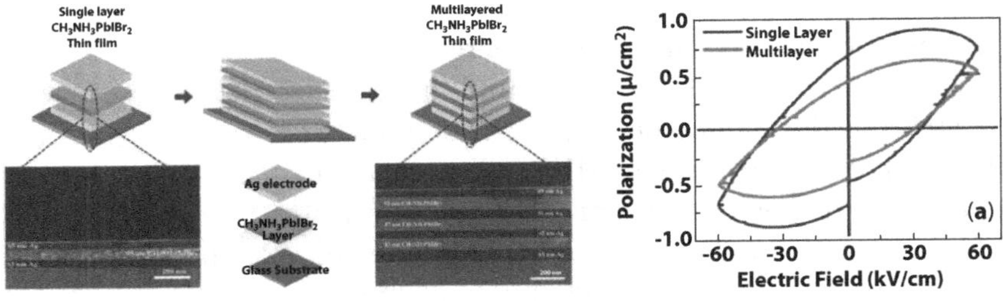

Figure 13.4 Schematic and fabricated micrographs of single and multilayered $CH_3NH_3PbIBr_2$ thin films, and their reported ferroelectric performance [61].

As described, ferroelectric results and schemes are helpful for the readers to investigate an alternative similar kind of MHPs, which can demonstrate better ferroelectric performances associated to their synergistic effects for ordered organic cations and distortion of inorganic octahedral to be utilized in multifunctional applications and devices.

13.6.4 Ferroelectricity in Three-Dimensional (3D) MHPs

Description of the Goldschmidt's tolerance factor (t) for 3D halide perovskites demonstrates much lesser structural flexibility, which restricts the development of 3D MHP ferroelectrics. The first-principles density function theory has been implemented to calculate the theoretical ferroelectricity in 3D halide perovskites. Frost *et al.* estimated the high value of polarization (38 μC cm^{-2}) for $MAPbI_3$ [62]. Later on, Fan *et al.* estimated the polarization of ~8 μC m^{-2} by first-principles calculations conducted on 3D periodic boundary conditions for $MAPbI_3$ [33]. The authors also discussed the discrepancies in their results as compared to Frost's calculations.

Piezoresponse force microscopy (PFM) is an important, updated, and efficient tool used to observe ferroelectric behavior experimentally in materials. Kutes *et al.* observed the ~100-nm ferroelectric domains for $MAPbI_3$ PFM (Figure 13.5a–e) [63].

Li *et al.* suggested that in orthorhombic $CsPbBr_3$, the ferroelectric properties can be derived by distorting $[PbBr_6]^{4-}$ octahedron and off-centering the Cs^+ ions, resulting in a P_S of 0.25 μC cm^{-2} and a T_C of 298 K [64]. He *et al.* measured the polarization of 5.73 μC cm^{-2} and T_C of 432 K in (3-hydroxypyrrolidinium)$_2$RbBiBr$_6$ [65]. Such discussion and summarized results are important to understand the concepts to utilize such schemes in ferroelectric applications.

13.7 Potential Applications

Ferroelectric photovoltaic is the emerging potential application of the MHPs nowadays [61]. Keeping in mind the ferroelectric performances of the MHPs by their spontaneous polarization, there are several applications where MHPs can be utilized. MHPs are potential materials for self-generated photodetectors because of their significant properties, such as high light–polarization interactions, excellent photovoltage, smooth photovoltaic current, and rapid photoactivity. Implementation of MHP ferroelectrics for high-performance self-generated photodetectors is explored by so many researchers. Ji *et al.* synthesized the 2D ferroelectric $(BPA)_2PbBr_4$

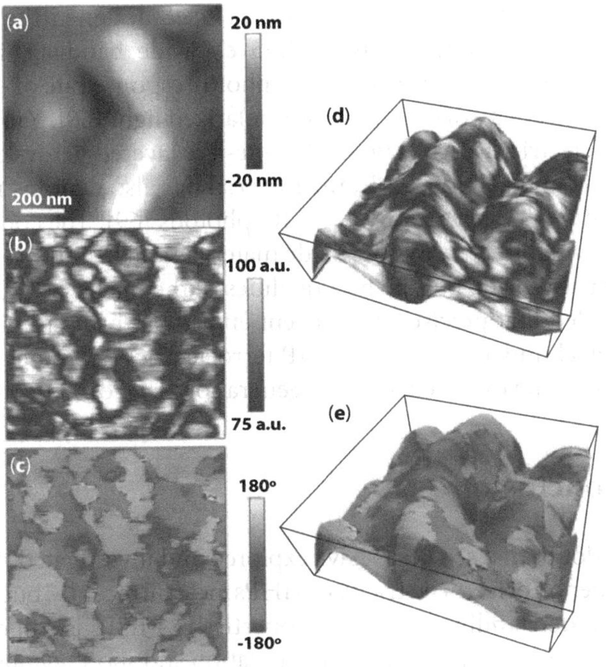

Figure 13.5 PFM and AFM analysis of $MAPbI_3$ thin film. (a) Topology by AFM. (b) Amplitude by PFM. (c) Phase analysis by PFM. (d) Superimposed-topography-amplitude. (e) Superimposed-topography-phase, used with permission [63], Copyright 2014, American Chemical Society.

(BPA=3-bromopropylammonium), which is capable of working as a ferroelectricity-driven self-powered photodetector [66].

X-ray detection is as basic requirement in the field of medical sciences; MHP ferroelectrics have been used as new-generation X-ray detecting materials due to their two specific qualities: (1) large attenuation coefficients have been observed in MHPs due to their cations (e.g., Bi and Pb) and large atomic number (Z value) values, and (2) MHPs are capable of generating a large amount of free charges and hence possess better charge transportation ability, which is promoted by internal electric fields of spontaneous polarization.

Zhang *et al.* fabricated the $(EATMP)PbBr_4$ phosphonium-based ferroelectric, with a high electrostrictive coefficient (~3.96 $m^4\ C^{-2}$); this is considered as the ferroelectric that can be utilized in future-generation efficient devices, like transducers, actuators, and sensors [67]. Xiong and co-workers fabricated the $(TMCM)MnCl_3$ ferroelectric with a P_S of 4 $\mu C\ m^{-2}$ and a strong piezoelectric coefficient (d_{33}) of 185 $pC\ N^{-1}$; such materials can be utilized in piezoelectric sensors and actuators [48]. Li *et al.* designed the $(BA)_2(MA)Pb_2Br_7$ ferroelectric halide with a ~2.55 eV wide bandgap. In addition, $(BA)_2(MA)Pb_2Br_7$ demonstrated a rapid photo-response rate of ~20 μs and a high photo-detectivity of ~10^9 Jones for polarized light [68]. Yao *et al.* developed a ferroelectric $(BEA)_2CsPb_2Br_7$ fatigue-free layered perovskite with a P_S of 6.5 μC cm^{-2} and a T_C of 425 K; the material has also demonstrated a high switching ratio of current for over 3×10^5 photons [69]. The combination of such MHPs with other perovskite oxide materials and bismuth layered ferroelectrics can also bring out the best qualities of the materials [70–72]. It has been concluded that precise and efficient efforts can bring improvement in the mechanical understanding of MHP ferroelectrics; the range of practical applications can be explored for next-generation devices.

13.8 Conclusion

There is no doubt that scientists have explored and investigated the concepts and phenomenon of ferroelectricity in MHPs in detail recently, but an inclusive systematic understanding of ferroic properties in MHPs is still challenging. Mostly, it is found that spontaneous polarization and spontaneous strain play a vital role for the production of ferroelectricity in MHPs, which is a suitable property for optoelectronic applications, sensors, and actuators. Moreover, it is found that whenever ferroelectricity couples with ion migration, electric charge, structural dynamics, and strain in MHPs, the concept of developing ferroelectricity becomes more complex. Nonetheless, owing to the rigorous structural necessities of MHP ferroelectrics, phase transition techniques, and difficult-to-predict molecular packing, a molecular design scheme has faced significant restrictions for the selection of ferroelectric counterparts. In this chapter, comprehensive literature reviews of the lead-based and lead-free MHPs, the origin of ferroelectricity production in MHPs, the controversy of ferroelectricity in MHPs, theoretical and experimental analysis of ferroelectricity in MHPs, and ferroelectric concepts in MHPs depending on the dimensions of the materials are discussed. From the detailed analysis, it can be concluded that ionic transfer and structural dynamics have a strong influence on the ferroelectricity of MHPs. The coupling of ionic transfer with charge accumulation, strain gradient, and lattice distortion has the ability to suppress or vary the ferroelectric polarization in MHPs; similarly, the coupling of ionic ferroelectricity with ferroelasticity controls the strain condition of MHPs. Although certain MHP ferroelectrics with a reliable performance have been developed and utilized in different optoelectronic and photovoltaic applications, for large-scale ferroelectric applications, there are still a few restrictions. This chapter presents more potential ways of controlling ferroic–ionic interaction in MHPs and can provide opportunities to be utilized in next-generation devices.

References

1. Jiao, Y., Yi, S., Wang, H., Li, B., Hao, W., Pan, L., Shi, Y., Li, X., Liu, P., Zhang, H., Gao, C., Zhao, J., Lu, J., Strain engineering of metal halide perovskites on coupling anisotropic behaviors. *Adv. Funct. Mater.*, 31, 2006243, 2021.
2. Sun, M., Zhao, C., Shu, J., Yin, X., Synthesis, properties, and applications of metal halide perovskite-based nanomaterials, in: *Functional Nanomaterials*, pp. 225–273, 2022.
3. Lai, M., Lei, T., Zhang, Y., Jin, J., Steele, J.A., Yang, P., Phase transition dynamics in one-dimensional halide perovskite crystals. *MRS Bull.*, 46, 310, 2021.
4. Katz, E.A., Perovskite: Name puzzle and German-Russian odyssey of discovery. *Helv. Chim. Acta*, 103, 307, 2020.
5. Travis, W., Glover, E.N.K., Bronstein, H., Scanlon, D.O., Palgrave, R.G., On the application of the tolerance factor to inorganic and hybrid halide perovskites: A revised system. *Chem. Sci.*, 7, 4548, 2016.
6. Han, S.T., Zhou, Y., Roy, V.A.L., Towards the development of flexible nonvolatile memories. *Adv. Mater.*, 25, 5425, 2013.
7. Qaiser, M.A., Hussain, A., Xu, Y., Wang, Y., Wang, Y., Yang, Y., Yuan, G., CuO added Pb0.92Sr0.06Ba0.02(Mg1/3Nb2/3)0.25(Ti0.53Zr0.47)0.75O3 ceramics sintered with Ag electrodes at 900 °c for multilayer piezoelectric actuator. *Chin. Phys. B*, 26, 37702, 2017.
8. Hussain, A., Zheng, P., Xu, X., Chen, X., Li, T., Wang, Y., Yuan, G., Yin, J., Liu, Z., Structural and electrical properties of multiferroic (1-x)BiFeO3-x Bi0.5K0.5TiO3 ceramics. *J. Alloys Compd.*, 678, 228, 2016.
9. Yuan, Y. and Huang, J., Ion migration in organometal trihalide perovskite and its impact on photovoltaic efficiency and stability. *Acc. Chem. Res.*, 49, 286, 2016.
10. Meyer, R.B., Liebert, L., Strzelecki, L., Keller, P., Ferroelectric liquid crystals. *J. Physique Lett.*, 36, 69, 1975.
11. Jabeen, N., Nawaz, S., Qaiser, M.A., Rana, M.A., Hassan, F., Abbas, Z., Ahmed, F., Hussain, A., Stable piezoelectric response of 0-3 type $CaBi_2Nb_2O_9$:*x*wt%$BiFeO_3$ composites for high-temperature piezoelectric applications. *J. Asian Ceram. Soc.*, 9, 289, 2021.
12. Coll, M., Gomez, A., Mas-Marza, E., Almora, O., Garcia-Belmonte, G., Campoy-Quiles, M., Bisquert, J., Polarization switching and light-enhanced piezoelectricity in lead halide perovskites. *J. Phys. Chem. Lett.*, 6, 1408, 2015.

13. Rakita, Y., Bar-Elli, O., Meirzadeh, E., Kaslasi, H., Peleg, Y., Hodes, G., Lubomirsky, I., Oron, D., Ehre, D., Cahen, D., Tetragonal CH3NH3Pbi3 is ferroelectric. *Proc. Natl. Acad. Sci. U. S. A.*, 114, E5504, 2017.
14. Zhang, C.C., Yuan, S., Lou, Y.H., Okada, H., Wang, Z.K., Physical fields manipulation for high-performance perovskite photovoltaics. *Small*, 18, 2107556, 2022.
15. Chen, Q., Zhou, H., Fang, Y., Stieg, A.Z., Song, T.B., Wang, H.H., Xu, X., Liu, Y., Lu, S., You, J., Sun, P., McKay, J., Goorsky, M.S., Yang, Y., The optoelectronic role of chlorine in $CH_3NH_3PbI_3$ (Cl)-based perovskite solar cells. *Nat. Commun.*, 6, 310, 2015.
16. Ding, R., Liu, H., Zhang, X., Xiao, J., Kishor, R., Sun, H., Zhu, B., Chen, G., Gao, F., Feng, X., Chen, J., Chen, X., Sun, X., Zheng, Y., Flexible piezoelectric nanocomposite generators based on formamidinium lead halide perovskite nanoparticles. *Adv. Funct. Mater.*, 26, 7708, 2016.
17. Chen, W., Zhang, J., Xu, G., Xue, R., Li, Y., Zhou, Y., Hou, J., Li, Y., A semi-transparent inorganic perovskite film for overcoming ultraviolet light instability of organic solar cells and achieving 14.03% efficiency. *Adv. Mater.*, 30, 1800855, 2018.
18. Zhang, X., Zhang, M., Lu, G., Charge stripe formation in molecular ferroelectric organohalide perovskites for efficient charge separation. *J. Phys. Chem. C*, 120, 23969, 2016.
19. Liu, X., Wang, S., Long, P., Li, L., Peng, Y., Xu, Z., Han, S., Sun, Z., Hong, M., Luo, J., Polarization-driven self-powered photodetection in a single-phase biaxial hybrid perovskite ferroelectric. *Angew. Chem.*, 131, 14646, 2019.
20. You, Y.M., Liao, W.Q., Zhao, D., Ye, H.Y., Zhang, Y., Zhou, Q., Niu, X., Wang, J., Li, P.F., Fu, D.W., Wang, Z., Gao, S., Yang, K., Liu, J.M., Li, J., Yan, Y., Xiong, R.G., An organic-inorganic perovskite ferroelectric with large piezoelectric response. *Sci. (80-.)*, 357, 306, 2017.
21. Yang, C.K., Chen, W.N., Ding, Y.T., Wang, J., Rao, Y., Liao, W.Q., Tang, Y.Y., Li, P.F., Wang, Z.X., Xiong, R.G., The first 2D homochiral lead iodide perovskite ferroelectrics: [*R*- and *S*-1-(4-chlorophenyl)ethylammonium]$_2$Pbl$_4$. *Adv. Mater.*, 31, 1808088, 2019.
22. Ippili, S., Jella, V., Kim, J., Hong, S., Yoon, S.G., Enhanced piezoelectric output performance via control of dielectrics in Fe^{2+}-incorporated $MAPbI_3$ perovskite thin films: Flexible piezoelectric generators. *Nano Energy*, 49, 247, 2018.
23. Wang, S., Liu, X., Li, L., Ji, C., Sun, Z., Wu, Z., Hong, M., Luo, J., An unprecedented biaxial trilayered hybrid perovskite ferroelectric with directionally tunable photovoltaic effects. *J. Am. Chem. Soc.*, 141, 7693, 2019.
24. Kong, L.D., Zhang, S., Chen, Y.P., Zhang, S.N., Ji, L., Doroshenko, V., Wang, P.J., Tao, L., Ge, M.Y., Liu, C.Z., Song, L.M., Lu, F.J., Qu, J.L., Li, T.P., Xu, Y.P., Cao, X.L., Chen, Y., Bu, Q.C., Cai, C., Chang, Z., Chen, G., Chen, L., Chen, T.X., Chen, Y.B., Cui, W., Cui, W.W., Deng, J.K., Dong, Y.W., Du, Y.Y., Fu, M.X., Gao, G.H., Gao, H., Gao, M., Gu, Y.D., Guan, J., Guo, C.C., Han,

D.W., Huang, Y., Huo, J., Jia, S.M., Jiang, L.H., Jiang, W.C., Jin, J., Jin, Y.J., Li, B., Li, C.K., Li, G., Li, M.S., Li, W., Li, X., Li, X.B., Li, X.F., Li, Y.G., Li, Z.W., Liang, X.H., Liao, J.Y., Liu, B.S., Liu, G.Q., Liu, H.W., Liu, X.J., Liu, Y.N., Lu, B., Lu, X.F., Luo, Q., Luo, T., Ma, X., Meng, B., Nang, Y., Nie, J.Y., Ou, G., Sai, N., Shang, R.C., Song, X.Y., Sun, L., Tan, Y., Tuo, Y.L., Wang, C., Wang, G.F., Wang, J., Wang, L.J., Wang, W.S., Wang, Y.S., Wen, X.Y., Wu, B.B., Wu, B.Y., Wu, M., Xiao, G.C., Xiao, S., Xiong, S.L., Xu, H., Yang, J.W., Yang, S., Yang, Y.J., Yang, Y.J., Yi, Q.B., Yin, Q.Q., You, Y., Zhang, A.M., Zhang, C.M., Zhang, F., Zhang, H.M., Zhang, J., Zhang, T., Zhang, W.C., Zhang, W., Zhang, W.Z., Zhang, Y., Zhang, Y.F., Zhang, Y.J., Zhang, Y., Zhang, Z., Zhang, Z., Zhang, Z.L., Zhao, H.S., Zhao, X.F., Zheng, S.J., Zheng, Y.G., Zhou, D.K., Zhou, J.F., Zhu, Y.X., Zhuang, R.L., Two complete spectral transitions of swift J0243.6+6124 observed by insight-HXMT. *Astrophys. J.*, 902, 18, 2020.

25. Liao, W.Q., Zhao, D., Tang, Y.Y., Zhang, Y., Li, P.F., Shi, P.P., Chen, X.G., You, Y.M., Xiong, R.G., A molecular perovskite solid solution with piezoelectricity stronger than lead zirconate titanate. *Sci. (80-.)*, 363, 1206, 2019.
26. Ambrosio, F., Meggiolaro, D., Mosconi, E., De Angelis, F., Charge localization, stabilization, and hopping in lead halide perovskites: Competition between polaron stabilization and cation disorder. *ACS Energy Lett.*, 4, 2013, 2019.
27. Miyata, K. and Zhu, X.Y., Ferroelectric large polarons. *Nat. Mater.*, 17, 379, 2018.
28. Hermes, I.M., Bretschneider, S.A., Bergmann, V.W., Li, D., Klasen, A., Mars, J., Tremel, W., Laquai, F., Butt, H.J., Mezger, M., Berger, R., Rodriguez, B.J., Weber, S.A.L., Ferroelastic fingerprints in methylammonium lead iodide perovskite. *J. Phys. Chem. C*, 120, 5724, 2016.
29. Huang, B., Kong, G., Esfahani, E.N., Chen, S., Li, Q., Yu, J., Xu, N., Zhang, Y., Xie, S., Wen, H., Gao, P., Zhao, J., Li, J., Ferroic domains regulate photocurrent in single-crystalline $CH_3NH_3PbI_3$ films self-grown on FTO/TiO_2 substrate. *NPJ Quantum Mater.*, 3, 312, 2018.
30. Röhm, H., Leonhard, T., Hoffmann, M.J., Colsmann, A., Ferroelectric domains in methylammonium lead iodide perovskite thin-films. *Energy Environ. Sci.*, 10, 950, 2017.
31. Strelcov, E., Dong, Q., Li, T., Chae, J., Shao, Y., Deng, Y., Gruverman, A., Huang, J., Centrone, A., $CH_3NH_3PbI_3$ perovskites: Ferroelasticity revealed. *Sci. Adv.*, 3, 312, 2017.
32. Liu, Y., Kim, D., Ievlev, A.V., Kalinin, S.V., Ahmadi, M., Ovchinnikova, O.S., Ferroic halide perovskite optoelectronics. *Adv. Funct. Mater.*, 31, 2102793, 2021.
33. Fan, Z., Xiao, J., Sun, K., Chen, L., Hu, Y., Ouyang, J., Ong, K.P., Zeng, K., Wang, J., Ferroelectricity of $CH_3NH_3PbI_3$ perovskite. *J. Phys. Chem. Lett.*, 6, 1155, 2015.
34. Liu, S., Guo, R., Xie, F., The effects of organic cation rotation in hybrid organic-inorganic perovskites: A critical review. *Mater. Des.*, 221, 110951, 2022.

35. Frost, J.M., Butler, K.T., Walsh, A., Molecular ferroelectric contributions to anomalous hysteresis in hybrid perovskite solar cells. *APL Mater.*, 2, 313, 2014.
36. Liu, Y., Collins, L., Proksch, R., Kim, S., Watson, B.R., Doughty, B., Calhoun, T.R., Ahmadi, M., Ievlev, A.V., Jesse, S., Retterer, S.T., Belianinov, A., Xiao, K., Huang, J., Sumpter, B.G., Kalinin, S.V., Hu, B., Ovchinnikova, O.S., Chemical nature of ferroelastic twin domains in $CH_3NH_3PbI_3$ perovskite. *Nat. Mater.*, 17, 1013, 2018.
37. Stoumpos, C.C., Malliakas, C.D., Kanatzidis, M.G., Semiconducting tin and lead iodide perovskites with organic cations: Phase transitions, high mobilities, and near-infrared photoluminescent properties. *Inorg. Chem.*, 52, 9019, 2013.
38. Kutes, Y., Ye, L., Zhou, Y., Pang, S., Huey, B.D., Padture, N.P., Direct observation of ferroelectric domains in solution-processed CH3NH3PbI3 perovskite thin films. *J. Phys. Chem. Lett.*, 5, 3335, 2014.
39. Xiao, Z., Yuan, Y., Shao, Y., Wang, Q., Dong, Q., Bi, C., Sharma, P., Gruverman, A., Huang, J., Giant switchable photovoltaic effect in organometal trihalide perovskite devices. *Nat. Mater.*, 14, 193, 2015.
40. Liu, X., Wang, S., Long, P., Li, L., Peng, Y., Xu, Z., Han, S., Sun, Z., Hong, M., Luo, J., Polarization-driven self-powered photodetection in a single-phase biaxial hybrid perovskite ferroelectric. *Angew. Chem.*, 131, 14646, 2019.
41. Wang, S., Li, L., Weng, W., Ji, C., Liu, X., Sun, Z., Lin, W., Hong, M., Luo, J., Trilayered lead chloride perovskite ferroelectric affording self-powered visible-blind ultraviolet photodetection with large zero-bias photocurrent. *J. Am. Chem. Soc.*, 142, 55, 2020.
42. Wang, Y., Tang, Z., Liu, C., Jiang, J., Liu, W., Zhang, B., Gao, K., Cai, H.L., Wu, X., Room temperature ferroelectricity and blue photoluminescence in zero dimensional organic lead iodine perovskites. *J. Mater. Chem. C*, 9, 223, 2021.
43. Huang, C.R., Luo, X., Liao, W.Q., Tang, Y.Y., Xiong, R.G., An above-room-temperature molecular ferroelectric: $[Cyclopentylammonium]_2CdBr_4$. *Inorg. Chem.*, 59, 829, 2020.
44. Li, D.F., Deng, X.H., Ma, Y.X., Liu, D.S., Huang, X.F., Zhao, X.F., Chen, W.T., Sui, Y., A near-room-temperature hybrid organic–inorganic lead halide perovskite ferroelectric $[BrCH_2CH_2N(CH_3)_3][PbBr_3]$ and its flexible composite film. *J. Phys. Chem. C*, 126, 728, 2022.
45. Ji, C., Sun, Z., Zeb, A., Liu, S., Zhang, J., Hong, M., Luo, J., Bandgap narrowing of lead-free perovskite-type hybrids for visible-light-absorbing ferroelectric semiconductors. *J. Phys. Chem. Lett.*, 8, 2012, 2017.
46. Zhang, Y., Liao, W.Q., Fu, D.W., Ye, H.Y., Liu, C.M., Chen, Z.N., Xiong, R.G., The first organic-inorganic hybrid luminescent multiferroic: (Pyrrolidinium) MnBr3. *Adv. Mater.*, 27, 3942, 2015.
47. Zhang, Y., Liao, W.Q., Fu, D.W., Ye, H.Y., Chen, Z.N., Xiong, R.G., Highly efficient red-light emission in an organic-inorganic hybrid ferroelectric: $(Pyrrolidinium)MnCl_3$. *J. Am. Chem. Soc.*, 137, 4928, 2015.

48. Ye, H.Y., Zhou, Q., Niu, X., Liao, W.Q., Fu, D.W., Zhang, Y., You, Y.M., Wang, J., Chen, Z.N., Xiong, R.G., High-temperature ferroelectricity and photoluminescence in a hybrid organic-inorganic compound: (3-pyrrolinium) $MnCl_3$. *J. Am. Chem. Soc.*, 137, 13148, 2015.
49. Li, D.F., Zhao, P.J., Deng, X.H., Wu, Y.Z., He, X.L., Liu, D.S., Li, Y.X., Sui, Y., A new organic-inorganic hybrid perovskite ferroelectric $[ClCH_2CH_2N(CH_3)_3][PbBr_3]$ and Its PVDF matrix-assisted highly-oriented flexible ferroelectric films. *New J. Chem.*, 46, 19391, 2022.
50. Li, P.F., Liao, W.Q., Tang, Y.Y., Ye, H.Y., Zhang, Y., Xiong, R.G., Unprecedented ferroelectric-antiferroelectric-paraelectric phase transitions discovered in an organic-inorganic hybrid perovskite. *J. Am. Chem. Soc.*, 139, 8752, 2017.
51. Liao, W.Q., Zhang, Y., Hu, C.L., Mao, J.G., Ye, H.Y., Li, P.F., Huang, S.D., Xiong, R.G., A lead-halide perovskite molecular ferroelectric semiconductor. *Nat. Commun.*, 6, 316, 2015.
52. Ye, H.Y., Liao, W.Q., Hu, C.L., Zhang, Y., You, Y.M., Mao, J.G., Li, P.F., Xiong, R.G., Bandgap engineering of lead-halide perovskite-type ferroelectrics. *Adv. Mater.*, 28, 2579, 2016.
53. Shi, P.P., Lu, S.Q., Song, X.J., Chen, X.G., Liao, W.Q., Li, P.F., Tang, Y.Y., Xiong, R.G., Two-dimensional organic-inorganic perovskite ferroelectric semiconductors with fluorinated aromatic spacers. *J. Am. Chem. Soc.*, 141, 18334, 2019.
54. Wu, Z., Ji, C., Li, L., Kong, J., Sun, Z., Zhao, S., Wang, S., Hong, M., Luo, J., Alloying n-Butylamine into CsPbBr3 to give a two-dimensional bilayered perovskite ferroelectric material. *Angew. Chem. - Int. Ed.*, 57, 8140, 2018.
55. Li, L., Shang, X., Wang, S., Dong, N., Ji, C., Chen, X., Zhao, S., Wang, J., Sun, Z., Hong, M., Luo, J., Bilayered hybrid perovskite ferroelectric with giant two-photon absorption. *J. Am. Chem. Soc.*, 140, 6806, 2018.
56. Ji, C., Wang, S., Wang, Y., Chen, H., Li, L., Sun, Z., Sui, Y., Wang, S., Luo, J., 2D hybrid perovskite ferroelectric enables highly sensitive X-ray detection with low driving voltage. *Adv. Funct. Mater.*, 30, 1905529, 2020.
57. Wang, J., Ma, Y., Wang, Z., Liu, X., Han, S., Liu, Y., Guo, W., Luo, J., Sun, Z., Unusual ferroelectric-dependent birefringence in 2D trilayered perovskite-type ferroelectric exploited by dimensional tailoring. *Matter*, 5, 194, 2022.
58. Li, L., Liu, X., He, C., Wang, S., Ji, C., Zhang, X., Sun, Z., Zhao, S., Hong, M., Luo, J., A potential Sn-based hybrid perovskite ferroelectric semiconductor. *J. Am. Chem. Soc.*, 142, 1159, 2020.
59. Zheng, W., Wang, X., Zhang, X., Chen, B., Suo, H., Xing, Z., Wang, Y., Wei, H.L., Chen, J., Guo, Y., Wang, F., Emerging halide perovskite ferroelectrics. *Adv. Mater.*, 35, 317, 2023.
60. Wang, C.F., Li, H., Li, M.G., Cui, Y., Song, X., Wang, Q.W., Jiang, J.Y., Hua, M.M., Xu, Q., Zhao, K., Ye, H.Y., Zhang, Y., Centimeter-sized single crystals of two-dimensional hybrid iodide double perovskite

(4,4-difluoropiperidinium)4AgBiI8 for high-temperature ferroelectricity and efficient X-ray detection. *Adv. Funct. Mater.*, 31, 2009457, 2021.
61. Jabeen, N., Zaidi, A., Hussain, A., Hassan, N.U., Ali, J., Ahmed, F., Khan, M.U., Iqbal, N., Elnasr, T.A.S., Helal, M.H., Single- and multilayered perovskite thin films for photovoltaic applications. *Nanomaterials*, 12, 3208, 2022.
62. Frost, J.M., Butler, K.T., Brivio, F., Hendon, C.H., Van Schilfgaarde, M., Walsh, A., Atomistic origins of high-performance in hybrid halide perovskite solar cells. *Nano Lett.*, 14, 2584, 2014.
63. Kutes, Y., Ye, L., Zhou, Y., Pang, S., Huey, B.D., Padture, N.P., Direct observation of ferroelectric domains in solution-processed $CH_3NH_3PbI_3$ perovskite thin films. *J. Phys. Chem. Lett.*, 5, 3335, 2014.
64. Li, X., Chen, S., Liu, P.F., Zhang, Y., Chen, Y., Wang, H.L., Yuan, H., Feng, S., Evidence for ferroelectricity of all-inorganic perovskite $CsPbBr_3$ quantum dots. *J. Am. Chem. Soc.*, 142, 3316, 2020.
65. He, L., Xu, K., Shi, P.P., Liu, Z.B., Zhang, W., Ye, Q., A rare 3D hybrid bimetal halide ferroelectric: (3-Hydroxypyrrolidinium)$_2$RbBiBr$_6$. *Sci. China Mater.*, 65, 2879, 2022.
66. Ji, C., Dey, D., Peng, Y., Liu, X., Li, L., Luo, J., Ferroelectricity-driven self-powered ultraviolet photodetection with strong polarization sensitivity in a two-dimensional halide hybrid perovskite. *Angew. Chem.*, 132, 19095, 2020.
67. Zhang, H.Y., Zhang, Z.X., Chen, X.G., Song, X.J., Zhang, Y., Xiong, R.G., Large electrostrictive coefficient in a two-dimensional hybrid perovskite ferroelectric. *J. Am. Chem. Soc.*, 143, 1664, 2021.
68. Li, L., Liu, X., Li, Y., Xu, Z., Wu, Z., Han, S., Tao, K., Hong, M., Luo, J., Sun, Z., Two-dimensional hybrid perovskite-type ferroelectric for highly polarization-sensitive shortwave photodetection. *J. Am. Chem. Soc.*, 141, 2623, 2019.
69. Yao, Y., Peng, Y., Li, L., Zhang, X., Liu, X., Hong, M., Luo, J., Exploring a fatigue-free layered hybrid perovskite ferroelectric for photovoltaic non-volatile memories. *Angew. Chem. - Int. Ed.*, 60, 10598, 2021.
70. Jabeen, N., Hussain, A., Nawaz, S., Qaiser, M.A., Zafar, R., Ishtiaq, S., Ahmed, F., Masud, U., Sultan, T., Muntaha, S.T., Thermally stable piezoelectric performance of MnO2 inserted pseudo-tetragonal phase existent CaBi2Nb2O9-based ceramics. *Mater. Technol.*, 37, 2702, 2022.
71. Hussain, A., Nawaz, S., Jabeen, N., Zafar, R., Qaiser, M.A., Abbas, Z., Ahmed, F., Khan, M.U., Waseem, M., Aslam, S., Enhanced ferroelectric and piezoelectric response by MnO_2 added $Bi_{0.5}(K_{0.2}Na_{0.8})_{0.5}TiO_3$ ceramics. *J. Solid State Chem.*, 306, 122716, 2022.
72. Jabeen, N., Rehman, A.U., Hassan, N.U., Qaiser, M.A., Zaidi, A., Khan, M.U., Khan, I.A., Nouman, M., Boosting of magnetic, ferroelectric, energy storage efficiency, and piezoelectric properties of Zn intercalated $SrBi_4Ti_4O_{15}$-based ceramics. *Mater. (Basel)*, 15, 5057, 2022.

Index

Printed in the USA/Agawam, MA
September 16, 2026

910661.095